CAMBRIDGE MONOGRAPHS ON
APPLIED AND COMPUTATIONAL
MATHEMATICS

Series Editors
P. G. CIARLET, A. ISERLES, R. V. KOHN, M. H. WRIGHT

11   Generalized Riemann Problems in Computational Fluid Dynamics

The *Cambridge Monographs on Applied and Computational Mathematics* reflects the crucial role of mathematical and computational techniques in contemporary science. The series publishes expositions on all aspects of applicable and numerical mathematics, with an emphasis on new developments in this fast-moving area of research.

State-of-the-art methods and algorithms as well as modern mathematical descriptions of physical and mechanical ideas are presented in a manner suited to graduate research students and professionals alike. Sound pedagogical presentation is a prerequisite. It is intended that books in the series will serve to inform a new generation of researchers.

Generalized Riemann Problems
in Computational Fluid Dynamics

Numerical simulation of compressible, inviscid, time-dependent flow is a major branch of computational fluid dynamics. Its primary goal is to obtain accurate representation of the time evolution of complex flow patterns, involving interactions of shocks, interfaces, and rarefaction waves. The generalized Riemann problem (GRP) algorithm, developed by the authors for this purpose, provides a unifying "shell" that comprises some of the most commonly used numerical schemes for such flows. This monograph gives a systematic presentation of the GRP methodology, starting from the underlying mathematical principles, through basic scheme analysis and scheme extensions (such as reacting flow or two-dimensional flows involving moving or stationary boundaries). An array of instructive examples illustrates the range of applications, extending from (simple) scalar equations to computational fluid dynamics. Background material from mathematical analysis and fluid dynamics is provided, making the book accessible to both researchers and graduate students of applied mathematics, science, and engineering.

Matania Ben-Artzi is a professor of mathematics at the Institute of Mathematics, The Hebrew University of Jerusalem. His area of interest is mathematical physics, where pure mathematical analysis, theory of partial differential equations, and numerical analysis are applied in the study of various fundamental differential equations of physics.

Joseph Falcovitz is a research Fellow at the Institute of Mathematics, The Hebrew University of Jerusalem. He retired in 1990 from the Rafael Ballistic Center, where he specialized in the development of computational methods of compressible flows and the simulation of terminal ballistics, blast wave phenomena, and fluid–structure interaction.

Generalized Riemann Problems in Computational Fluid Dynamics

MATANIA BEN-ARTZI and JOSEPH FALCOVITZ
The Hebrew University of Jerusalem

CAMBRIDGE
UNIVERSITY PRESS

CAMBRIDGE UNIVERSITY PRESS
Cambridge, New York, Melbourne, Madrid, Cape Town, Singapore,
São Paulo, Delhi, Dubai, Tokyo, Mexico City

Cambridge University Press
The Edinburgh Building, Cambridge CB2 8RU, UK

Published in the United States of America by Cambridge University Press, New York

www.cambridge.org
Information on this title: www.cambridge.org/9780521173278

First published 2003
First paperback edition 2010

A catalogue record for this publication is available from the British Library

Library of Congress Cataloging in Publication Data

Ben-Artzi, Matania, 1948–
Generalized Riemann problems in computational fluid dynamics / Matania Ben-Artzi,
Joseph Falcovitz.
p. cm. – (Cambridge monographs on applied and computational mathematics; 11)
Includes bibliographical references and index.
ISBN 0-521-77296-6
1. Fluid dynamics. 2. Riemann-Hilbert problems. I. Falcovitz, Joseph, 1937–.
II. Title. III. Series.
QA911 .B36   2003
532.05 – dc21   2002034811

ISBN 978-0-521-77296-9 Hardback
ISBN 978-0-521-17327-8 Paperback

Contents

List of Figures

xi

Preface

Computational fluid dynamics (CFD) is a relatively young branch of fluid dynamics, the other two being the experimental and the theoretical disciplines. Its rapid development was enabled by the spectacular progress in high power computers, as well as by a matching progress in numerical schemes.

The starting point for the formulation of CFD schemes is the governing equations. In fact, the term "fluid dynamical equations" is much too general and indeed ambivalent. In practice there exist numerous models of such equations. They reflect a variety of stipulations on the nature of the flow, such as compressibility, viscosity, or elasticity. They also involve various effects such as heat conduction or chemical reactions. A large portion of these models do not fall, mathematically speaking, under the category of "hyperbolic conservation laws," which is the subject matter of this monograph. We refer the reader to the book by Landau and Lifshitz [75] for a general survey of fluid dynamical models.

In this monograph we are concerned with time-dependent, inviscid, compressible flow, which is studied primarily in the "quasi-one-dimensional" geometric setting. This leads to a system of partial differential equations expressing the conservation of mass, momentum, and energy. There are various approaches to the numerical resolution of this system, such as the classical method of characteristics or the "artificial viscosity" scheme. Our focus here is on finite-difference approximations of the type referred to as "conservation law schemes." In the quasi-one-dimensional setting they are practically equivalent to "finite-volume schemes." At present, these schemes are the commonly preferred choice, since they are robust by virtue of their shock-capturing capability, and they may be readily extended to more than one space variable.

Rather than constructing a particular scheme, we try here to develop a methodology aimed at the derivation of high-resolution, second-order schemes, all of which extend the basic (first-order) Godunov scheme. It is based on the "heart of the matter," the analysis of the generalized Riemann problem (GRP). This problem is an extension of the classical Riemann problem (RP) of fluid dynamics, which is the initial value problem in which data consist of two *constant states* separated by a discontinuity at the origin. Loosely speaking, the corresponding GRP is defined by replacing those two states by *constant-gradient* states, also having a jump at the origin.

This monograph is devoted primarily to a mathematical introduction to conservation law schemes and to the development of the basic GRP methodology. However, we have also included (Part II) a number of applications to more representative cases of "scientific computing." Although these examples are not "algorithmically heavy," they serve to illustrate the kind of extensions (geometric or physical) that are invariably required for realistic CFD simulations.

Our collaboration on the GRP methodology began some twenty years ago. Over all these years we have benefited from numerous discussions with colleagues and students who helped us shape the method, its goals, and its presentation. Their ideas led oftentimes to joint work, as is witnessed by the bibliographic list. It is with real pleasure that we acknowledge their contribution to this monograph.

The preparation of this monograph was a demanding task to which we devoted considerable time and effort over the past few years. Throughout that time we have enjoyed the insights offered to us by A. Chorin, A. Birman, J.-P. Croisille, and O. Igra; their help is gratefully acknowledged. Special credit is due to J. Li whose participation was instrumental in shaping our treatment of (2-D) scalar conservation laws. Validation experiments are an important part of CFD research, and in that, as well as in other aspects, the cooperation of K. Takayama and other colleagues of the Shock Wave Research Center, Tohoku University, is gratefully acknowledged. Our thanks and appreciation are due to our colleague U. Feldman who set up the computer system used in the calculations and typesetting for this monograph.

Finally, it is with deep gratitude that we acknowledge the (silent) participation of our wives, Ofra and Linda, in this endeavor. Book writing invariably involves hard labor; their understanding and encouragement were instrumental in seeing it through to its successful conclusion.

1

Introduction

This monograph deals with the *generalized Riemann problem* (GRP) of mathematical fluid dynamics and its application to computational fluid dynamics. It shows how the solution to this problem serves as a basic tool in the construction of a robust numerical scheme that can be successfully implemented in a wide variety of fluid dynamical topics. The flows covered by this exposition may be quite different in nature, yet they share some common features; they all belong to the class of compressible, inviscid, time-dependent flows. Fluid dynamical phenomena of this type often contain a number of smooth flow regions separated by singularities such as shock fronts, detonation waves, interfaces, and centered rarefaction waves. One must then address various computational issues related to this class of fluid dynamical problems, notably the "capturing" of discontinuities such as shock fronts, detonation waves, or interfaces; resolution of centered rarefaction waves where flow gradients are unbounded; and evaluation of flow variables in irregular computational cells at the intersection of a moving boundary surface with an underlying mesh.

From the mathematical point of view, the various systems of equations governing compressible, inviscid, time-dependent flow phenomena may all be characterized as systems of "(nonlinear) hyperbolic conservation laws."

Hyperbolic conservation laws (in one space variable) are systems of time-dependent partial differential equations. The most common problem associated with such systems is the *initial value problem* (henceforth IVP), which is the following: Given the values of the unknown functions at time $t = 0$ (as functions of the space variable $x \in \mathbb{R}$), use the equations to determine the evolution in time of those functions. When the unknown functions are defined over the whole real line $\mathbb{R}$, one often refers to the IVP as the "Cauchy Problem." In contrast, when the unknown functions are defined only over a finite interval $\mathcal{D} \subseteq \mathbb{R}$, suitable "boundary conditions" must be imposed at the endpoints of $\mathcal{D}$. From

1

the physical point of view the latter is clearly the more realistic case. Thus, for example, $\mathcal{D}$ can represent a pipe of finite length, in which one studies the evolution (in time) of flow variables subject to the system of fluid dynamical equations. In this case, the boundary conditions consist of influx and outflux requirements imposed on the pressure, velocity, etc. at the edges of the pipe.

The solutions to the problems considered here possess one common fundamental property, that of "finite propagation speed"; that is, the waves travel at finite speeds. Mathematically speaking, when a change in the initial data is confined to the neighborhood of some point A, it is "felt" by the solution at any other point B only after a certain amount of time, an amount that depends on the distance between the points. It is precisely this feature that allows the construction of "conservation law schemes" for the (numerical) approximation of the solutions.

Although this monograph focuses on the resolution of compressible, inviscid flow problems, and the construction of suitable conservation law schemes, an effort is made to place the treatment in the broader (theoretical and numerical) perspective of hyperbolic conservation laws. However, the necessary background material from physics is also included. We refer the reader to the classical book by R. Courant and K. O. Friedrichs [30] for a thorough discussion of the mathematical aspects of compressible flow. This book also discusses in detail the derivation of the flow equations from the underlying physical conservation laws. For mathematical treatments of hyperbolic conservation laws, we refer to the books by Courant and Hilbert [31], Evans [36], Hörmander [63], Lax [75], and Smoller [103].

To simplify the discussion, we consider primarily the associated Cauchy problem, thus avoiding the further mathematical complications introduced by boundary conditions. Naturally, when dealing with real flow examples, boundary conditions will be needed, and the ways in which they are introduced into the numerical scheme will be explained.

The origin of the subject matter of this monograph can be traced back some forty years, to the early days of computational fluid dynamics. It can best be described by the opening sentence to Section 12.15 of the book by Richtmyer and Morton [96]: "In 1959, Godunov described an ingenious method for one-dimensional problems with shocks."

Godunov's method, as much as it was recognized for its novelty and robustness, suffered from some significant drawbacks. It was Bram van Leer, some twenty years later, who, in an important breakthrough [112], has shown how to modify Godunov's original construction and, indeed, has made it possible to implement the method as the most efficient tool (to date) in this area of computational fluid dynamics. In the simpler case of a scalar conservation law, these ideas will be explained in Chapter 3.

The monograph is divided into two parts, Part I (Basic Theory) and Part II (Numerical Implementation). Part I (Chapters 2–6) deals with the more basic aspects (theoretical and numerical) of systems of conservation laws and the development of the GRP method. Part II (Chapters 7–10) is devoted to several extensions (physical and geometric) of the GRP method for computational fluid dynamics. A more detailed discussion of the contents will follow. The reader will also find a brief summary at the beginning of each chapter.

In writing this monograph we have aimed at a wide readership, consisting not only of graduate students and researchers in applied mathematics but also of those working in various areas of physics and engineering. Yet, we have attempted to maintain a solid level of mathematical rigor. Notions such as "weak solutions" and "convergence of a scheme" are carefully introduced (Chapter 2) in suitable functional settings. We believe that, given the current mathematical level of modern numerical analysis, such concepts ought to be familiar to anyone working in this field. In particular, theorems related to the convergence of the Godunov scheme (in the scalar case) are proved in all mathematical detail (Section 2.2 and Appendix B). In this context we introduce (and compare numerically) some of the "classical" discrete schemes of hyperbolic equations, such as the Lax–Friedrichs and the Lax–Wendroff schemes. At the same time, our main objective in Chapters 2 and 3 is the introduction of the "high-resolution GRP scheme," by way of the Riemann and generalized Riemann problems. We refer to LeVeque [81] for introductory material on finite-difference schemes for conservation laws and to Richtmyer and Morton [96] for the general theory of finite-difference methods (primarily linear theory). A comprehensive survey of the convergence properties of finite-difference schemes to scalar conservation laws can be found in Godlewski and Raviart [54].

In Chapter 4 we introduce systems of conservation laws. The first section outlines the general mathematical background and can be skipped on first reading, as it is of a more mathematical nature. The physical systems of interest, those representing the basic conservation laws of compressible, inviscid flow in the "quasi-one-dimensional" setting, are introduced in the second section. This section is self-contained; the analysis of centered rarefaction waves, as well as the Rankine–Hugoniot shock conditions and the solution to the Riemann problem, is discussed in detail.

Chapter 5 is devoted to the analysis of the GRP in the context of the systems considered in Section 4.2. In Section 5.1 we study the solution to the linear GRP, which is the core of the GRP method. Given linear initial distributions of the flow variables on the two sides of a jump discontinuity, we determine their instantaneous time derivatives (at that singularity). Van Leer's idea to use this solution for the refinement of Godunov's scheme is implemented in the development of the GRP scheme in Section 5.2.

Chapter 6 is devoted to an investigation of the GRP scheme for fluid dynamics. Numerical results are compared to analytical or asymptotic solutions for a variety of wave interaction problems.

In Chapter 7 we introduce, in rather general terms, the operator-splitting method of Strang. It enables us to extend the GRP algorithm to two-dimensional (2-D) settings, while retaining its second-order accuracy. Chapter 8 deals with further geometric extensions, such as (one-dimensional) "tracking" of singularities and (2-D) moving boundaries.

In Chapter 9 we consider a reacting flow system. The basic set of conservation laws is augmented by a chemical reaction-rate equation, thus providing a simple model of combustion. The GRP algorithm is applied to this extended system.

As a concluding (numerical) example for this monograph, we consider in Chapter 10 a case of wave interaction with a segment of decreasing cross-sectional area in a two-dimensional duct. The major (GRP) numerical approaches developed in Chapters 5, 7, and 8, namely the quasi-1-D approximation and the fully 2-D scheme, are applied to this case. A comparative study of the two solutions sheds light on the nature of the fluid dynamical interaction, as well as on the nature of the quasi-1-D approximation.

Finally, a comment about the numbering system in this book. For the reader's convenience, all theorems, remarks, definitions, claims, etc. within each chapter are sequentially numbered. Thus, for example, Remark 2.22 comes after Definition 2.21 and is followed by Example 2.23.

Part I

Basic Theory

2

Scalar Conservation Laws

This chapter introduces the basic concepts of the present monograph. In Section 2.1 we review the general theory of (nonlinear) scalar conservation laws and introduce the fundamental notions of weak solutions and Rankine–Hugoniot jump conditions. In Section 2.2 we introduce the basic ideas of discrete approximations, such as accuracy and convergence.

2.1 Theoretical Background

In this chapter we overview the basic details concerning the simplified model of *scalar* conservation laws. This means that we are looking for the solution $u(x, t)$ of the Cauchy problem for a single partial differential equation of the type

$$\frac{\partial}{\partial t} u + \frac{\partial}{\partial x} f(u) = 0, \qquad x \in \mathbb{R}, \quad t > 0, \qquad (2.1)$$

$$u(x, 0) = u_0(x), \qquad x \in \mathbb{R}. \qquad (2.2)$$

The solution $u(x, t)$ as a function of the space variable $x \in \mathbb{R}$ is sought for all nonnegative time values. The function $f(u)$ is assumed to be smooth (namely, continuously differentiable at least as many times as needed in the analysis).

The term "conservation law" stems from the following argument. Integrating Equation (2.1) over a rectangle $0 \leq t \leq T$, $x_1 \leq x \leq x_2$, one gets

$$\int_{x_1}^{x_2} u(x, T) \, dx - \int_{x_1}^{x_2} u_0(x) \, dx = - \int_0^T f(u(x_2, t)) \, dt + \int_0^T f(u(x_1, t)) \, dt. \qquad (2.3)$$

Thinking of $u(x, t)$ as "mass density" (per unit length) we see that the integral $\int_{x_1}^{x_2} u(x, t) dx$ expresses the total mass in $[x_1, x_2]$ at time t, whereas

7

$\int_0^T f(u(x, t))dt$, for any fixed x, can be interpreted as the "mass flux" to the right at the point x over the time interval $[0, T]$. Thus, Equation (2.3) may be viewed as a "balance equation," stating that the gain in total mass in $[x_1, x_2]$ equals the net flux into the interval through its boundary points x_1 and x_2. Accordingly, we call $f(u)$ the "flux function." In particular, if we let $[x_1, x_2]$ expand to $\mathbb{R} = (-\infty, \infty)$, and we assume that the fluxes diminish to zero [e.g., if $f(0) = 0$ and $u(x_1, t)$, $u(x_2, t)$ vanish as $x_1 \to -\infty$, $x_2 \to +\infty$], Equation (2.3) reduces to

$$\int_{-\infty}^{\infty} u(x, t)\, dx = \int_{-\infty}^{\infty} u_0(x)\, dx, \qquad 0 \le t \le T. \tag{2.4}$$

Clearly, this equation expresses the conservation (in time) of the total mass over the real axis. We refer the reader to LeVeque [81], where Equation (2.1) serves as a model for traffic flow.

   A rigorous treatment of the problem (2.1), (2.2) should include a specification of the set of "admissible functions," i.e., the appropriate differentiability requirements needed to make sense out of the equation. In particular, it seems as if a "natural" requirement is that the partial derivatives u_t, u_x exist (and are continuous) at all points $(x, t) \in \mathbb{R} \times (0, \infty)$. However, this is not so, and indeed one of the basic features of conservation laws (both from the mathematical and the physical points of view) is the fundamental role played by discontinuous solutions. In the physical context, they manifest themselves as "shock waves" or "material interfaces." In this chapter we introduce the basic mathematical notions developed in the search for a systematic way in which such discontinuous functions can serve as solutions of (2.1), (2.2)–so-called weak solutions. The purpose here is to outline the main ideas and arguments of the theory, and the reader is referred to Evans [36] for more comprehensive presentations.

   We start by looking at the simple case $f(u) = au$, where $a \ne 0$ is a real constant. Equation (2.1) takes now the form of the (constant-speed) advection equation,

$$u_t + au_x = 0, \tag{2.5}$$

for which the solution is easily seen to be the "traveling wave"

$$u(x, t) = u_0(x - at), \qquad x \in \mathbb{R}, \quad t \ge 0. \tag{2.6}$$

In this case the "initial profile" $u_0(x)$ propagates unmodified at a speed a. This "constant-speed propagation" can be seen even more clearly if we note that

Equation (2.5) can be reformulated as

$$\frac{D}{Dt}u(x,t) = 0 \qquad \text{along} \qquad \frac{dx}{dt} = a. \tag{2.7}$$

The notation $\frac{D}{Dt}u(x,t)$ introduced in (2.7) designates the "total derivative" (also referred to as the "Lagrangian" or "convective" derivative), namely, the derivative $\frac{d}{dt}u\,(x(t),t)$, where $x = x(t)$ is any line in the family of straight lines satisfying the equation $\frac{d}{dt}x(t) = a$.

Definition 2.1 The lines satisfying

$$\frac{d}{dt}x(t) = a \tag{2.8}$$

are called the "characteristic lines" associated with Equation (2.5).

Thus we can rephrase the aforementioned observation by saying that the solution is constant along characteristic lines.

Remark 2.2 In the more general case where $a = a(x,t)$ in (2.5), one can still define the family of characteristic curves by (2.8), namely, $\frac{d}{dt}x(t) = a(x(t),t)$. Individual curves are uniquely determined by giving the initial point $x(0) = x_0$. As before, we verify the validity of Equation (2.7), where $\frac{D}{Dt}$ is now the derivative along the characteristic curve, implying that u is constant along such a curve. However, as is known from the theory of ordinary differential equations, the existence of the characteristic curves for all $t \geq 0$ is not guaranteed in this case. This is easily seen, for example, in the case $a(x,t) = x^2$ (try the curve passing through $x_0 = 1$, $t_0 = 0$).

Let us now go back to the nonlinear problem (2.1), (2.2). Assuming that u is a smooth (that is, in our case, continuously differentiable) solution, we can write the equation in the form

$$u_t + f'(u)u_x = 0. \tag{2.9}$$

We see that here $f'(u)$ plays the role of the coefficient a in (2.5), and as in (2.7) we get the invariance of the value of u along "characteristic curves," namely,

$$\frac{D}{Dt}u(x,t) = 0 \qquad \text{along} \qquad \frac{dx}{dt} = f'(u). \tag{2.10}$$

However, there is a fundamental difference between the characteristic curves

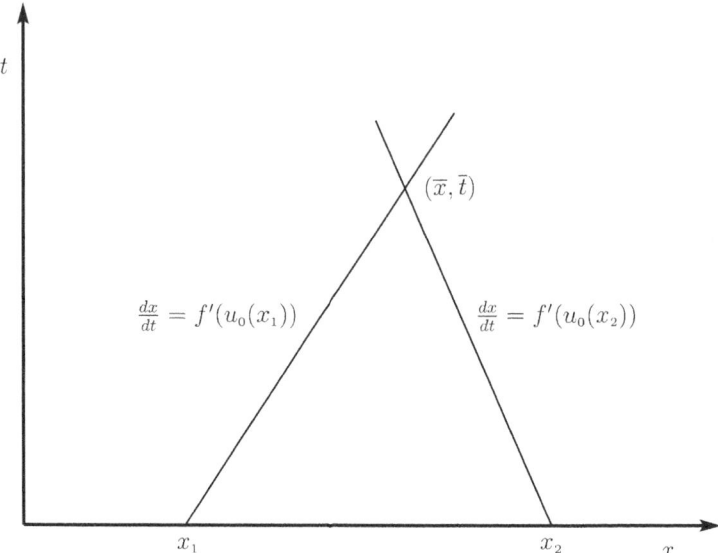

Figure 2.1. Characteristic curves in the nonlinear case.

in the linear case $[a = a(x, t)]$ and those of the nonlinear case at hand. Indeed, in the linear case these curves are determined uniquely by $a(x, t)$ and do not depend on the solution function u. However, in the situation given in (2.10) the slopes $f'(u)$ of these curves depend on the solution u itself! Thus, referring to the notation in Figure 2.1, the characteristic curve passing through $(x_1, 0)$ initially has a slope $f'(u_0(x_1))$. Its slope at later times is given by $f'(u(x(t), t))$. However, it follows from (2.10) that $u = constant = u_0(x_1)$ along this curve, so that the slope is constant and equal to $f'(u_0(x_1))$.

We conclude that in the nonlinear case the characteristic curves must be straight lines, at least as long as the solution exists and is smooth. Observe that in the linear case characteristics are straight lines only if $a = constant$.

Consider now another point x_2, where the initial value is $u_0(x_2)$. The characteristic (straight) line through this point has slope $f'(u_0(x_2))$ and it carries the constant value $u = u_0(x_2)$. However, as we see from Figure 2.1, if $x_2 > x_1$ and $f'(u_0(x_2)) < f'(u_0(x_1))$, the two straight lines will intersect at some $t = \bar{t} > 0$. At the point of intersection $(\bar{x}, \bar{t})$ we cannot expect the existence of a smooth (in fact, even continuous) solution, as the two constant values, $u_0(x_1)$ and $u_0(x_2)$, carried by the characteristic curves to that point, are in conflict. Note that this "breakdown in finite time" of the smooth solution does not depend on the smoothness of the initial function $u_0(x)$. From the preceding discussion, we can even assume that $u_0(x)$ is infinitely differentiable and compactly supported (that is, vanishes outside of a finite interval) and still face the same situation [by assigning to two points $x_1 < x_2$ values $u_0(x_1), u_0(x_2)$ such that $f'(u_0(x_1)) > f'(u_0(x_2))$].

Weak Solutions and Jump Conditions

We are thus led to one of the most fundamental aspects of the theory (and its practical application), namely, the inclusion of "discontinuous solutions" as members of the family of admissible solutions. Although this is forced on us in the nonlinear case, there is a natural need for such an extension even in the linear case. For example, we would like to refer to the traveling wave (2.6) as solving Equation (2.5) ($a = constant$) even when $u_0(x)$ is a "step function" (say $u_0(x) = 1$ for $x < 0$ and $u_0(x) = 0$ for $x \geq 0$).

The basic clue to the method that will allow us to activate such a generalization of the concept of a solution may be found in the derivation of the "balance equation" (2.3). As explained there, this equation is obtained by integrating (2.1) over the rectangle $Q^T_{x_1,x_2} = [x_1, x_2] \times [0, T]$. This can also be written as

$$\int_{\mathbb{R}} \int_0^\infty (u_t + f(u)_x) \, \chi^T_{x_1,x_2}(x, t) \, dx \, dt = 0,$$

(2.11)

$$\chi^T_{x_1,x_2}(x, t) = \begin{cases} 1 & \text{if } (x, t) \in Q^T_{x_1,x_2}, \\ 0 & \text{if } (x, t) \notin Q^T_{x_1,x_2}. \end{cases}$$

The modern theory of partial differential equations has taken this integrated version of the equation one step further, replacing the discontinuous function $\chi^T_{x_1,x_2}(x, t)$ by a smooth "test function" $\phi(x, t)$. This is the well-established procedure of defining "solutions in the sense of distributions" (see Evans [36]). In our case the family of test functions is C^1_0, that is, the class of functions ϕ that are continuously differentiable and vanish outside of some rectangle $Q^T_{-N,N}$ (where T and N depend on ϕ). Assuming that u is a smooth solution of (2.1), multiplying the equation by a test function $\phi(x, t)$, and integrating by parts over $\mathbb{R} \times [0, \infty]$, we obtain

$$\int_{\mathbb{R}} \int_0^\infty (u\phi_t + f(u)\phi_x) \, dx \, dt + \int_{\mathbb{R}} \phi(x, 0)u_0(x) \, dx = 0. \qquad (2.12)$$

The crucial idea in the introduction of discontinuous (henceforth called "weak") solutions is to reverse the procedure leading up to (2.12), by viewing the latter as defining the solution. The rigorous definition is the following:

Definition 2.3 The bounded function $u(x, t)$ is called a "weak solution" (sometimes also a "distribution solution") to the IVP (2.1), (2.2) if the equality (2.12) holds for every $\phi \in C_0^1$.

Example 2.4 The "moving step" $u(x, t) = u_0(x - at)$ [see (2.6)], where $u_0(x) = 1$ for $x < 0$ and $u_0(x) = 0$ for $x \geq 0$, is a weak solution to (2.5). The verification is left to the reader as an easy exercise. One should carry out the double integration (2.12) separately in the regions $x < at$ and $x > at$.

As already noted, Equation (2.12) is satisfied by any smooth solution to (2.1), (2.2). Conversely, in the following claim we show that if a weak solution u is a smooth function then it satisfies (2.1), (2.2). In other words, the set of smooth solutions of (2.1), (2.2) is identical to the set of smooth weak solutions. Moreover, any weak solution u that is continuous in a rectangle satisfies the balance equation (2.3) in that rectangle. Thus, the fundamental balance equation (2.3) is recovered from Equation (2.12) under the sole requirement of continuity of u.

In addition to the space C_0^1 of test functions we shall make frequent use of the space C^1 of continuously differentiable functions [in (x, t)]. In contrast to the concept of "weak solution" (where u is only assumed to be bounded), we refer to a C^1 solution of (2.1), (2.2) as a "classical solution."

Claim 2.5

(a) *If $u \in C^1$ is a weak solution to (2.1), (2.2) then it is a classical solution.*
(b) *If $u(x, t)$ is a weak solution [satisfying (2.12)] that is continuous in a rectangle Q_{x_1, x_2}^T then u satisfies the balance equation (2.3) in Q_{x_1, x_2}^T.*

Proof
(a) First, take in (2.12) any test function ϕ that vanishes identically near $t = 0$. Equation (2.12) then reads

$$\int_{\mathbb{R}} \int_0^\infty (u\phi_t + f(u)\phi_x) \, dx \, dt = 0 \qquad (2.13)$$

and since $u \in C^1$ (hence also $f(u) \in C^1$) we can integrate in (2.13) by parts. Since ϕ vanishes near $t = 0$, for sufficiently large x, t, the boundary terms vanish and (2.13) yields

$$\int_{\mathbb{R}} \int_0^\infty (u_t + f(u)_x) \phi \, dx \, dt = 0. \qquad (2.14)$$

Thus, the continuous function $v(x, t) = u_t + f(u)_x$ satisfies

$$\int_{\mathbb{R}} \int_0^\infty v(x, t)\phi(x, t) \, dx \, dt = 0,$$

for all test functions ϕ "supported" in the half-plane $t > 0$.[1] It is well known that this implies that $v \equiv 0$ in $\mathbb{R} \times [0, \infty)$, so that u satisfies (2.1). Now take any $\phi \in C_0^1$ and integrate by parts in (2.12). Since $u_t + f(u)_x = 0$ by the foregoing argument, we obtain

$$\int_{\mathbb{R}} (u_0(x) - u(x, 0))\,\phi(x, 0) \, dx = 0. \tag{2.15}$$

Since $\phi(x, 0)$ can be chosen as an arbitrary smooth and compactly supported function on $\mathbb{R}$, we conclude that $u(x, 0) - u_0(x) \equiv 0$, so that (2.2) is satisfied.

(b) Note that if $u \in C^1$ then it satisfies (2.1) by (a); hence (2.3) is obtained by integration. The point here is that we want to establish the validity of (2.3) for weak solutions that are only continuous, in which case Equation (2.1) has no classical meaning. The reader may skip the following proof on first reading.

Let k_0 be a large integer so that $1/k_0 < \min\left[\frac{1}{2}T, \frac{1}{3}(x_2 - x_1)\right]$. Let $\{\phi_k\}_{k=k_0}^\infty \subseteq C_0^1$ be a sequence of test functions approximating χ_{x_1,x_2}^T in the following sense:

(i) $\phi_k(x, t) \equiv 1$ for $(x, t) \in Q_{x_1+1/k, x_2-1/k}^{T-1/k}$, $k = k_0, k_0 + 1, \ldots$.

(ii) $\phi_k(x, t) \equiv 0$ outside the rectangle Q_{x_1, x_2}^T, $k = k_0, k_0 + 1, \ldots$.

(iii) $\frac{\partial}{\partial t}\phi_k(x, t) \leq 0$ in Q_{x_1, x_2}^T, $k = k_0, k_0 + 1, \ldots$, $\frac{\partial}{\partial x}\phi_k(x, t) \geq 0$ for $x \in \left[x_1, x_1 + \frac{1}{k}\right]$, and $\frac{\partial}{\partial x}\phi_k(x, t) \leq 0$ for $x \in \left[x_2 - \frac{1}{k}, x_2\right]$ (see Appendix B in Evans [36] for the construction of such functions).

Now, write (2.12) with $\phi = \phi_k$. Since $\frac{\partial}{\partial x}\phi_k, \frac{\partial}{\partial t}\phi_k$ vanish in $Q_{x_1+1/k, x_2-1/k}^{T-1/k}$, we get

$$\iint_{\Omega_k} \left(u\frac{\partial}{\partial t}\phi_k + f(u)\frac{\partial}{\partial x}\phi_k \right) dx \, dt + \int_{x_1+1/k}^{x_2-1/k} u_0(x) \, dx$$

$$+ \int_{x_1}^{x_1+1/k} \phi_k(x, 0)u_0(x) \, dx + \int_{x_2-1/k}^{x_2} \phi_k(x, 0)u_0(x) \, dx = 0, \tag{2.16}$$

where $\Omega_k = Q_{x_1, x_2}^T \setminus Q_{x_1+1/k, x_2-1/k}^{T-1/k}$.

---

[1] Recall that supp $g(x, t)$, the "support" of a C^1 function g defined in $R \times [0, \infty)$, is by definition the closure of the set $\{(x, t), g(x, t) \neq 0\}$. Thus, $g \in C_0^1$ if and only if supp g is bounded. If $g \in C_0^1$, supp $g \subseteq \mathbb{R} \times (0, \infty)$, then g vanishes near $t = 0$.

Let $M = \max \left\{ |u(x,t)|, (x,t) \in Q_{x_1,x_2}^T \right\}$. Clearly,

$$\left| \int_{x_1}^{x_1+1/k} \int_0^T u \frac{\partial}{\partial t} \phi_k \, dx \, dt + \int_{x_2-1/k}^{x_2} \int_0^T u \frac{\partial}{\partial t} \phi_k \, dx \, dt \right|$$

$$\leq M \left[\int_{x_1}^{x_1+1/k} \left(\int_0^T \left| \frac{\partial \phi_k}{\partial t} \right| dt \right) dx + \int_{x_2-1/k}^{x_2} \left(\int_0^T \left| \frac{\partial \phi_k}{\partial t} \right| dt \right) dx \right] \leq \frac{2M}{k},$$

which tends to 0 as $k \to \infty$. However, by the same reasoning,

$$\left| \int_{x_1+1/k}^{x_2-1/k} \int_{T-1/k}^T (u(x,t) - u(x,T)) \frac{\partial}{\partial t} \phi_k(x,t) \, dx \, dt \right|$$

$$\leq |x_2 - x_1| \times \max \left\{ |u(x,t) - u(x,T)|, x_1 \leq x \leq x_2, T - 1/k \leq t \leq T \right\},$$

which tends to 0 as $k \to \infty$, owing to the uniform continuity of u in Q_{x_1,x_2}^T. We obtain therefore

$$\lim_{k \to \infty} \int \int_{\Omega_k} u \frac{\partial}{\partial t} \phi_k \, dx \, dt = \lim_{k \to \infty} \int_{x_1}^{x_2} u(x,T) \int_{T-1/k}^T \frac{\partial}{\partial t} \phi_k \, dt \, dx$$

$$= - \int_{x_1}^{x_2} u(x,T) \, dx.$$

Similarly,

$$\lim_{k \to \infty} \int_0^T \int_{x_1}^{x_1+1/k} f(u) \frac{\partial}{\partial x} \phi_k \, dx \, dt$$

$$= \lim_{k \to \infty} \int_0^T f(u(x_1,t)) \int_{x_1}^{x_1+1/k} \frac{\partial}{\partial x} \phi_k \, dx \, dt = \int_0^T f(u(x_1,t)) \, dt,$$

$$\lim_{k \to \infty} \int_0^T \int_{x_2-1/k}^{x_2} f(u) \frac{\partial}{\partial x} \phi_k \, dx \, dt = \lim_{k \to \infty} \int_0^T f(u(x_2,t)) \int_{x_2-1/k}^{x_2} \frac{\partial}{\partial x} \phi_k \, dx \, dt$$

$$= - \int_0^T f(u(x_2,t)) \, dt,$$

$$\lim_{k \to \infty} \int_{x_1}^{x_1+1/k} \phi_k(x,0) u_o(x) \, dx = \lim_{k \to \infty} \int_{x_2-1/k}^{x_2} \phi_k(x,0) u_0(x) \, dx = 0.$$

Inserting all these limits in (2.16) we obtain the balance equation (2.3).   □

We have thus shown that the notion of a weak solution leads to the balance equation (2.3). However, the real interest (and applicability) of this notion lies

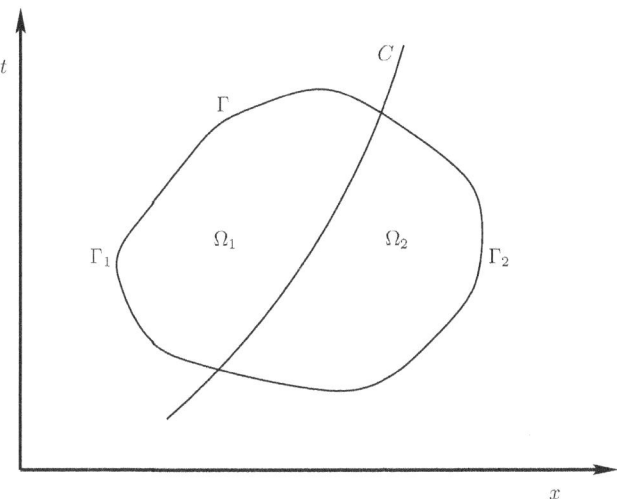

Figure 2.2. A jump discontinuity of a weak solution.

in the fact that it leads to considerably more general formulations of the balance equation. In particular, it extends the validity of the equations to domains where the solution has jump discontinuities. We shall not attempt at utmost generality but rather content ourselves with the following situation (which is illustrated in Figure 2.2).

Let $\Omega \subseteq \mathbb{R} \times [0, \infty)$ be a bounded domain with a piecewise smooth boundary Γ (i.e., Γ consists of finitely many continuously differentiable segments). Assume further that u is a weak solution that has the following structure: There exists a smooth curve C, which intersects Ω, dividing it into two subdomains Ω_1, Ω_2. We assume that u is continuous in each of the two subdomains, up to the boundary. This means that u attains continuous boundary values on C, when the latter is approached from either side. However, the boundary values obtained that way need not be equal. In other words, u may experience a *jump discontinuity* across C. A suitable modification of the proof of Claim 2.5(b) (which is not difficult but will not be discussed here in detail) yields the following claim.

Claim 2.6 *Let Ω, u be as described in the foregoing, and let u be a weak solution [satisfying (2.12)]. Then*

$$\oint_{\Gamma} - u(x, t) \, dx + f(u(x, t)) \, dt = 0. \tag{2.17}$$

Remark 2.7

(a) Note that if $u \in C^1$ in Ω then (2.17) is obtained simply by integrating (2.1) over Ω and using Green's theorem.

(b) When $\Omega = Q^T_{x_1, x_2}$, Equation (2.17) reduces to the balance equation (2.3) [with $u(x, t)$ replaced by $u_0(x)$ along the portion of Γ that lies on the x axis]. Thus, Equation (2.3) is valid even when u has a jump discontinuity across a curve C that intersects the rectangle.

(c) Clearly, Claim 2.6 can be generalized further by assuming that Ω is intersected by finitely many smooth curves C_1, C_2, ..., across which u has jump discontinuities. However, we shall not need this more general claim.

Claim 2.6 entails important consequences concerning the possible slopes of a curve C carrying a jump discontinuity of a weak solution u. To illustrate the situation, consider first the "moving step" function as in Example 2.4. The curve C is then given by the straight line $x = at$. Suppose that we try another weak solution $u(x, t) = u_0(x - bt)$ [to (2.5)], so that u has a jump discontinuity across $x = bt$. Invoking a rectangle as in Remark 2.7(b) we obtain $b = a$. (The simple verification is left to the reader.) Thus, in the case of the linear equation (2.5), the jump discontinuities of a weak solution are forced to move at the characteristic speed of $\frac{dx}{dt} = a$.

We now turn to the study of jump discontinuities of weak solutions to the general nonlinear equation (2.1). If $C: x = x(t)$ is a curve, across which u is discontinuous, it can actually be viewed as the trajectory of a *moving discontinuity*. Its slope $x'(t)$ is then the speed of propagation of the discontinuity. As we have seen earlier, the discontinuity propagates at characteristic speed in the linear case. We shall see now that the situation is very different in the nonlinear case. So, consider again the situation in Figure 2.2, where u is a weak solution satisfying (2.12), continuous in Ω_1, Ω_2, with a discontinuity along the smooth curve $C: x = x(t)$. Let Γ_1, Γ_2 be the two parts of the boundary of Ω, separated by C. Writing Equation (2.17) separately for the two subdomains Ω_1, Ω_2, we obtain

$$\int_{\Gamma_1} -u(x, t)\, dx + f(u(x, t))\, dt + \int_C -u_1(x, t)\, dx + f(u_1(x, t))\, dt = 0,$$

(2.18)

$$\int_{\Gamma_2} -u(x, t)\, dx + f(u(x, t))\, dt + \int_{-C} -u_2(x, t)\, dx + f(u_2(x, t))\, dt = 0.$$

In (2.18) we have denoted by u_1, u_2 the boundary values of u, as attained by approaching the discontinuity curve C from Ω_1, Ω_2, respectively. Observe that the closed curves $\Gamma_1 \cup C$ and $\Gamma_2 \cup (-C)$ are traversed counterclockwise.

Now, as noted earlier (Claim 2.6) the balance equation (2.17) is valid when applied to the boundary $\Gamma = \Gamma_1 \cup \Gamma_2$ of Ω. Thus, we have also

$$\oint_\Gamma -u(x, t)\, dx + f(u(x, t))\, dt = 0. \tag{2.19}$$

Comparing (2.18) and (2.19) we conclude

$$\int_C [u_2(x, t) - u_1(x, t)]\, dx - [f(u_2(x, t)) - f(u_1(x, t))]\, dt = 0. \tag{2.20}$$

However, Equation (2.20) can be applied to any arbitrarily small part of C by the same argument. Hence the integrand in (2.20) must vanish identically.

These considerations lead us to one of the most fundamental facts concerning the speed of propagation of discontinuities.

Corollary 2.8 (Rankine–Hugoniot jump condition) *Let* C: $x = x(t)$ *be a smooth trajectory traced out by a jump discontinuity of the weak solution* $u(x, t)$. *Then, for every* t, *the speed* $S = \frac{dx}{dt}$ *is given by*

$$S = \frac{[f(u)](t)}{[u](t)}, \tag{2.21}$$

where $[u](t) = u_2(x(t), t) - u_1(x(t), t), [f(u)](t) = f(u_2(x(t), t)) - f(u_1(x(t), t))$ *are the jumps of* u, $f(u)$, *respectively, across the discontinuity at time* t.

Proof When expressed as a function of t, the integrand in (2.20) is $[u](t) \cdot x'(t) - [f(u)](t)$, which must vanish identically, as noted above. $\square$

Remark 2.9 We emphasize the fact that the jump condition (2.21) had been obtained in the fluid dynamical context (Courant and Friedrichs [30]) much before the concept of a weak solution was introduced. Indeed, it is based on "conservation considerations," which are also at the root of our treatment, which derives from the balance equations (2.3) or (2.17).

Definition 2.10 A moving discontinuity satisfying the Rankine–Hugoniot condition is called a "shock wave" [for the weak solution $u(x, t)$].

Shocks, Rarefaction Waves, and Entropy

Remark 2.11 Observe that in the linear case $f(u) = au$ it was noted in the foregoing that $S = a$. Thus, shock waves must move at characteristic speeds.

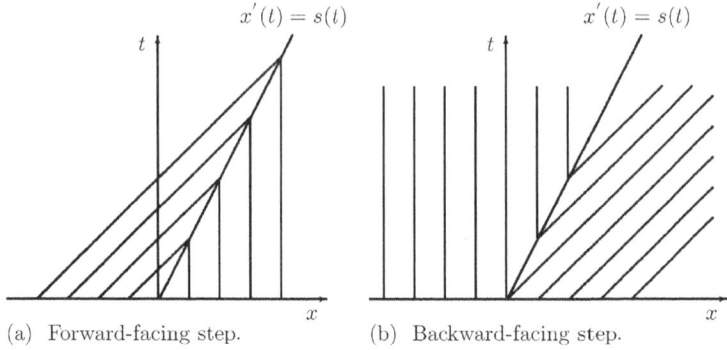

(a) Forward-facing step.   (b) Backward-facing step.

Figure 2.3. Characteristics for initial step function in Burgers' equation.

However, as we shall see in the next example, the situation is generally very different in the nonlinear case.

Example 2.12 (Burgers' equation) The simplest example for a nonlinear equation is given by the flux $f(u) = \frac{1}{2}u^2$. The resulting equation

$$u_t + \left(\tfrac{1}{2}u^2\right)_x = 0 \tag{2.22}$$

is known as the Burgers' equation [22].

Suppose we take as initial condition the step function $u_0(x)$ as in Example 2.4. Then, as in the linear case, we can verify that a corresponding weak solution is given by a moving step $u(x, t) = u_0(x - St)$ [see Figure 2.3(a)]. However, the Rankine–Hugoniot condition (2.21) is easily seen to enforce a speed given by,

$$S = \frac{[f(u)]\,(t)}{[u]\,(t)} = \frac{\frac{1}{2}(1 - 0)}{1 - 0} = \frac{1}{2}. \tag{2.23}$$

Thus, as is seen in Figure 2.3(a), the weak solution $u(x, t)$ involves a shock wave moving at a speed $S = \frac{1}{2}$, while the characteristic straight lines (see (2.10) and Figure 2.1) are "running into it" as time increases, from the left and from the right.

Consider now the same problem (2.22), but with the initial values of $u_0(x)$ interchanged, so that $u_0 = 1$ (resp. 0) for $x > 0$ (resp. $x < 0$). Clearly, the moving step $u_0(x - St)$, $S = \frac{1}{2}$ is still a weak solution of the equation, but the layout of the characteristic lines is very different, as is seen in Figure 2.3(b).

In particular, in the case shown in Figure 2.3(b) there is a whole sector, namely, $0 < x < t$, that is unattained by any characteristic line emanating from the initial line $t = 0$ (x axis). This situation is remedied when we consider the following alternative solution.

Example 2.13 (Burgers' equation–rarefaction wave) As in the preceding discussion, consider Equation (2.22) with initial data as in Figure 2.3(b). Let the function $u(x, t)$ be defined by

$$u(x, t) = \begin{cases} 0 & \text{if } x \leq 0, \\ \frac{x}{t} & \text{if } 0 < x \leq t, \\ 1 & \text{if } x > t. \end{cases} \tag{2.24}$$

It is a straightforward verification (left to the reader) that $u(x, t)$ of (2.24) is indeed a weak solution satisfying (2.12) with $f(u) = \frac{1}{2}u^2$ [simply carry out the space–time integration in (2.12) separately over the three regions indicated in (2.24)]. The field of characteristic lines (of slopes $f'(u) = u$) is now as displayed in Figure 2.4, and in particular there is no void region as in Figure 2.3(b). Note also the important fact that the jump discontinuity of the initial function $u_0(x)$ at $x = 0$ is diffused instantaneously and for $t > 0$ the solution $u(x, t)$ is continuous.

Definition 2.14 The solution (2.24) is called a "centered rarefaction wave" to the equation.

It is now easy to see how centered rarefaction waves are encountered also as weak solutions of the more general conservation law (2.1), (2.2). In fact, assume that $f(u)$ is *strictly convex* [meaning that $f''(u) \geq \mu > 0$ or that $f'(u)$ is strictly increasing with u]. Consider again the initial value problem where $u_0(x)$ is a

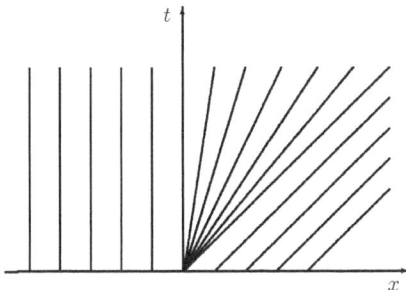

Figure 2.4. Rarefaction wave solution for Burgers' equation.

step function such that

$$u_0(x) = \begin{cases} u_{\text{L}}, & x < 0, \\ u_{\text{R}}, & x > 0, \end{cases} \tag{2.25}$$

where $u_{\text{L}} < u_{\text{R}}$ are two constants. The foregoing discussion in the case of Burgers' equation $[f(u) = \frac{1}{2}u^2]$ can now be applied directly to the case at hand. Thus, in addition to a single possible jump ("rarefaction shock") we have the centered rarefaction wave solution given by

$$u(x, t) = \begin{cases} u_{\text{L}}, & x \le f'(u_{\text{L}})t, \\ \lambda, & \text{along } \frac{dx}{dt} = f'(\lambda), u_{\text{L}} \le \lambda \le u_{\text{R}}, \\ u_{\text{R}}, & x \ge f'(u_{\text{R}})t. \end{cases} \tag{2.26}$$

Recall (see the discussion following (2.10)) that in this case all characteristic curves are straight lines carrying constant values of u, with $f'(u)$ as the slope.

The verification that (2.26) satisfies (2.12) is left to the reader. Observe that although $u(x, t)$ is continuous except at the singular point $(0, 0)$, it is not a C^1 function; it is readily checked that along the "acoustic lines" $x = f'(u_{\text{R}})t$ and $x = f'(u_{\text{L}})t$ (the "head" and "tail" lines of the rarefaction wave, respectively), the derivatives u_x, u_t undergo a jump discontinuity. Thus, u is not a "classical" solution of (2.1).

The term "rarefaction," inspired by the analogous situation in compressible fluid flow (see Chapter 4), comes from the fact that the "density u" at each fixed time level $t > 0$ "rarefies" from the high value $u = u_{\text{R}}$ for $x > f'(u_{\text{R}})t$ continuously to the low value $u = u_{\text{L}}$ at $x < f'(u_{\text{L}})t$. In contrast, the discontinuous solution represented in Figure 2.3(b) by the moving step $u_0(x - St)$ can be labeled as a "rarefaction shock." Thus, the introduction of the concept of a "weak solution," which was meant to overcome the difficulty of nonexistence of a smooth solution to the conservation law (while retaining the property of the balance equation), has now led us to the problem of *nonuniqueness*, namely, the existence of several possible weak solutions for the same initial data. We therefore need some "selection rule" to pick out a unique weak solution, considered to be the most appropriate one. From the physical viewpoint, based on the analogous fluid dynamical situation, the rarefaction shock solution is a nonphysical one, whereas the "physically correct" solution is the centered rarefaction wave. In these physical considerations, the nonphysical solution is shown to violate the basic entropy law of thermodynamics. It is a remarkable fact that, in the context of the scalar conservation law, we have a fully mathematical analog of the "entropy selection rule," which guarantees uniqueness of the weak solution and in particular forces the centered rarefaction wave as the unique solution in the situation discussed here. Unfortunately, these considerations do not easily generalize to the case of systems of nonlinear hyperbolic equations, and we will not make much use of them in this monograph. So, for the sake of completeness, we defer a more detailed discussion of this topic to Appendix A. However, one aspect of the entropy rule (indeed, one possible formulation of it), the *Lax entropy condition*, is of importance in our treatment and is described next.

Recall [see Figure 2.1 and the discussion following (2.10)] that shock waves (namely, jump discontinuities) are forced by characteristic lines of different slopes "running into each other" as time increases. In such a case a discontinuity of the solution is inevitable. However, the case of the centered rarefaction wave shows that if the characteristic lines emanating from the initial line ($t = 0$) are "spreading out," then a rarefaction shock can be replaced by a "rarefaction fan," thus avoiding a discontinuity. These considerations lead us to the following definition.

Definition 2.15 Let $C: x = x(t)$, $a \le t \le b$, be a smooth trajectory traced out by a jump discontinuity of the weak solution $u(x, t)$ [to Equation (2.1)]. We say that the (Lax) entropy condition is satisfied along C, if for any $t \in [a, b]$,

$$f'(u_+(x(t), t)) < S(t) = \frac{dx}{dt} < f'(u_-(x(t), t)). \qquad (2.27)$$

Observe that instead of the notation employed in Corollary 2.8, we have used here the notation

$$u_\pm(x(t), t) = \lim_{x \to x(t)\pm} u(x, t)$$

to indicate the values attained by u on the two sides of C.

As is seen from (2.9), (2.10) the two values of f' in (2.27) give the slopes of the characteristic lines on either side of the discontinuity. The two inequalities in (2.27) can therefore be summarized by saying that as time evolves, the discontinuity moves faster than the characteristic lines ahead of it and is slower than those trailing it (see Figure 2.5). Note in particular that whereas the Rankine–Hugoniot condition (2.21) is symmetric (the indices "1" and "2" can be interchanged) this is not true in the case of (2.27) where the signs "+" (ahead of the wave) and "−" (behind the wave) have definite meanings and cannot be interchanged.

We have remarked earlier that the slopes $f'(u)$ may be viewed as "sonic speeds," corresponding to the speed of a "signal of magnitude u." (This terminology is inspired by the fluid dynamical case, as we shall see later.) Thus, the entropy condition can be rephrased to say that the "discontinuity is supersonic

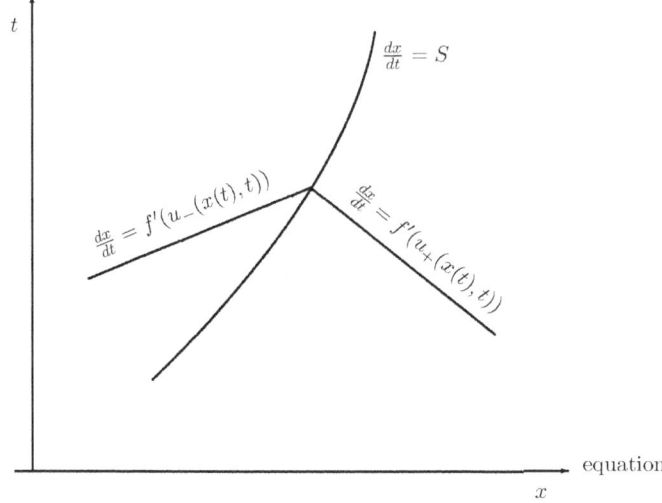

Figure 2.5. A jump discontinuity satisfying the entropy condition.

with respect to the medium ahead of it, but subsonic with respect to the medium behind it."

Example 2.16 In the case of the Burgers' equation (Example 2.12), the "forward-facing step" [Figure 2.3(a)] generates a discontinuity that moves at the speed $S = \frac{1}{2}$ and satisfies the entropy condition. In contrast, the "rarefaction shock," which results from the "backward-facing step" [Figure 2.3(b)], and which also moves at the same speed, is clearly in violation of the entropy condition.

Remark 2.17 Given Equation (2.22) and initial data as in Figure 2.3(b), the reader can easily verify that the function

$$u(x, t) = \begin{cases} 1, & x > \frac{3}{4}t, \\ \frac{1}{2}, & \frac{1}{4}t < x < \frac{3}{4}t, \\ 0, & x < \frac{1}{4}t, \end{cases}$$

is also a weak solution consisting of two consecutive rarefaction shocks, whereas the solution depicted in Figure 2.3(b) consists of a single one. Continuing in the same fashion, a weak solution to this initial data can be constructed as a sequence of arbitrarily many "small rarefaction shocks," with an ever decreasing jump per shock. In fact, when the number of shocks goes to infinity this procedure produces the correct "centered rarefaction wave" solution (shown in Figure 2.4), which conforms to the entropy condition.

For the general theory of existence and uniqueness of "entropy-satisfying weak solutions" to the conservation law (2.1), (2.2) we refer the reader to the books by Smoller [103], Godlewski and Raviart [54], Hörmander [63], and Serre [101].

The Riemann Problem

As we shall see in Section 2.2, the elementary solutions discussed in the preceding subsection are fundamental in the numerical treatment of conservation laws. It is therefore useful to formalize this situation as follows.

Definition 2.18 (Riemann problem) The "Riemann problem" for the conservation law (2.1) is the IVP subject to the initial data (2.25), where u_L, u_R are constants.

We can summarize this discussion by saying that the "entropy-satisfying" solution to the Riemann problem for (2.1), when $f(u)$ is strictly convex, consists of a shock wave (moving at speed $S = \frac{f(u_R)-f(u_L)}{u_R-u_L}$) when $u_L > u_R$, whereas for $u_L < u_R$ it is a centered rarefaction wave. The solution is self-similar; namely, it depends only on the ratio x/t. For future reference we denote this solution by $R(\frac{x}{t}; u_L, u_R)$. When the solution is a shock wave, it moves either to the left or to the right, according to the sign of $S = \frac{f(u_R)-f(u_L)}{u_R-u_L}$. When it is a centered rarefaction wave, it propagates to the right (resp. to the left) when the initial jump satisfies the additional relation $0 < f'(u_L) < f'(u_R)$ (resp. $f'(u_L) < f'(u_R) < 0$). However, in the case of a rarefaction wave there is a third possibility, which we refer to as the "sonic case," where $f'(u_L) \leq 0 \leq f'(u_R)$. Here the line $x = 0$ is contained within the rarefaction fan, coinciding with the characteristic line that moves at zero speed. It then carries the value $u = u_{min}$, i.e., the value of u at the minimum point of $f(u)$ (where $f'(u_{min}) = 0$).

In all these cases the following corollary is evident.

Corollary 2.19 *The range of values attained by the Riemann solution $R(\frac{x}{t};$ $u_L, u_R)$ is contained in $[u_L, u_R]$ [rarefaction as in (2.26)] or consists of the values u_L, u_R (shock). In particular, the waves issuing from $x = 0$ move at speeds that do not exceed (in absolute value) $\max\left(|f'(u_L)|, |f'(u_R)|\right)$.*

This is in fact a special case of a general property of solutions to scalar conservation laws, as stated in the following theorem.

Theorem 2.20 *Let $f(u)$ be continuously differentiable (not necessarily convex), and let $u_0 \in L^\infty(\mathbb{R})$. Then the initial value problem (2.1), (2.2) has a unique weak solution $u(x, t)$ satisfying the entropy condition. This solution is defined for all $x \in \mathbb{R}$, $t \geq 0$ and has the following properties:*

(i) *Maximum–Minimum Principle. For every $t \geq 0$,*

$$\sup_{x \in \mathbb{R}} u(x, t) \leq \sup_{x \in \mathbb{R}} u_0(x), \qquad \inf_{x \in \mathbb{R}} u(x, t) \geq \inf_{x \in \mathbb{R}} u_0(x).$$

(ii) *Continuity in $L^1_{\text{loc}}(\mathbb{R})$. For every fixed $X > 0$, the function $t \to u(\cdot, t)$ is continuous from $[0, \infty)$ into $L^1(-X, X)$. In particular, the initial condition (2.2) is satisfied in the sense that, for all $X > 0$,*

$$\int_{-X}^{X} |u(x, t) - u_0(x)| \, dx \to 0 \quad as \quad t \to 0.$$

(iii) *L^1 Contraction Property. If the initial data $u_0, v_0 \in L^1(\mathbb{R}) \cap L^\infty(\mathbb{R})$ and $u(x, t), v(x, t)$ are the corresponding weak solutions, then, for every $t \geq 0$,*

$$\int_{\mathbb{R}} |u(x, t) - v(x, t)| \, dx \leq \int_{\mathbb{R}} |u_0(x) - v_0(x)| \, dx.$$

For a proof of this theorem we refer to Hörmander [63] and Godlewski and Raviart [54].

For the purpose of conservation law schemes, in particular the GRP scheme considered in the following chapters, we shall make frequent use of the value $u^* = R(0; u_L, u_R)$, that is, the (constant) value of the solution to a Riemann problem along the line $x = 0$. Clearly

$$u^* = \begin{cases} u_L & \text{for a right wave,} \\ u_R & \text{for a left wave,} \\ u_{\min} & \text{if } x = 0 \text{ is a sonic point,} \end{cases} \tag{2.28}$$

where it is assumed that $f(u)$ is strictly convex.

Observe that $f'(u^*) > 0$ (resp. $f'(u^*) < 0$) if the wave moves to the right (resp. to the left), whereas $f'(u^*) = 0$ at a sonic point.

2.2 Basic Concepts of Numerical Approximation

The basic idea in the construction of approximate solutions to (2.1), (2.2) is that of *time discretization*. One defines a sequence of times $0 = t_0 < t_1 < t_2 < \cdots$ and seeks functions $U^n(x)$, $n = 0, 1, 2, \ldots$, that should approximate the exact solution $u(x, t_n)$ at these times. For simplicity we shall henceforth assume a fixed time step $\Delta t > 0$, so that $t_n = n\Delta t$, $n = 0, 1, 2, \ldots$. A suitable discrete analog of Equation (2.1) is used to determine $U^{n+1}(x)$, assuming that $U^j(x)$, $0 \leq j \leq n$, are already known. In this setting, time derivatives may be approximated by expressions like $\Delta t^{-1} \left[U^{n+1}(x) - U^n(x) \right]$. Thus we need to find expressions to approximate the action of x-derivatives on $U^n(x)$. If $U^n(x)$ is indeed known for all $x \in \mathbb{R}$, we may think again of $\Delta x^{-1} [U^n(x + \Delta x) - U^n(x)]$, where $\Delta x > 0$ is some fixed increment, as approximating $\frac{d}{dx} U^n(x)$. However, in our context such an approximation is not realistic, since in general $U^n(x)$ will be a discontinuous and nonmonotonic function. So we restrict ourselves to a discretized version of this idea. For simplicity, let us take equally spaced points $x_j = j\Delta x$, $-\infty < j < \infty$. The function $U^n(x)$ is then represented by its values at the *grid points* $U_j^n = U(x_j, t_n)$. The way in which the values $\left\{ U_j^n \right\}_{j=-\infty}^{\infty}$ are interpolated in determining the values $U^n(x)$, $x \notin \left\{ x_j \right\}_{j=-\infty}^{\infty}$, is fundamental for the class of high-resolution methods discussed in this monograph (as well as many others). We shall come back to this discussion later on. However, to illustrate the basic principles of the underlying *finite-difference method*, let us first consider (as we did in Section 2.1) the case of the linear equation $u_t + au_x = 0$.

Here are four different ways in which $\left\{ U_j^{n+1} \right\}_{j=-\infty}^{\infty}$ can be derived from the known values $\left\{ U_j^n \right\}_j$:

(a) The "backward"-difference scheme

$$\frac{U_j^{n+1} - U_j^n}{\Delta t} = -a \frac{U_j^n - U_{j-1}^n}{\Delta x}. \tag{2.29}$$

(b) The "forward"-difference scheme

$$\frac{U_j^{n+1} - U_j^n}{\Delta t} = -a \frac{U_{j+1}^n - U_j^n}{\Delta x}. \tag{2.30}$$

(c) The Lax–Friedrichs scheme

$$\frac{U_j^{n+1} - \frac{1}{2} \left[U_{j+1}^n + U_{j-1}^n \right]}{\Delta t} = -a \frac{U_{j+1}^n - U_{j-1}^n}{2\Delta x}. \tag{2.31}$$

(d) The Lax–Wendroff scheme

$$\frac{U_j^{n+1} - U_j^n}{\Delta t} = -a \frac{U_{j+1}^n - U_{j-1}^n}{2\Delta x} + a^2 \frac{U_{j+1}^n - 2U_j^n + U_{j-1}^n}{\Delta x^2} \frac{\Delta t}{2}. \qquad (2.32)$$

Observe that in the first three schemes the left-hand side approximates the derivative u_t while the right-hand side approximates $-au_x$. In the Lax–Wendroff scheme the left-hand side approximates $u_t + \frac{1}{2}\Delta t\, u_{tt}$, while the right-hand side approximates $-au_x + \frac{1}{2}\Delta t\, a^2 u_{xx}$. Note that for smooth solutions the relation $u_{tt} = a^2 u_{xx}$ follows from $u_t = -au_x$ by differentiation with respect to t. In other words, all four schemes are "consistent" with the equation $u_t + au_x = 0$. To make this notion (as well as that of "accuracy" of the scheme) precise, we shall now introduce some widely used conventions.

First, we set $k = \Delta t$ and take $\Delta x = \frac{\Delta t}{\lambda}$, where $\lambda > 0$ is given (and fixed). Thus, k is the only "small parameter" in the scheme. Next, we use the symbolic writing $U^{n+1} = H_k(U^n)$ to denote any of these schemes. Here U^n, U^{n+1} are the sequences $\{U_j^n\}_{j=-\infty}^{\infty}$, $\{U_j^{n+1}\}_{j=-\infty}^{\infty}$, respectively, and H_k is the associated *linear* operator. In other words, we observe that the sequence U_j^n is transformed linearly to U_j^{n+1}, and we denote by H_k the corresponding map. The map H_k differs from one scheme to the other and depends on the parameter k. Now let $u(x, t)$ be the exact solution to the equation, and let $u^n = \{u(x_j, t_n)\}_{j=-\infty}^{\infty}$ be the sequence of its values at the spatial grid points for $t = t_n$. Clearly, it is generally not true that $u^{n+1} = H_k(u^n)$, for any of the preceding schemes. However, "consistency" means here that this last equality is approximately satisfied, whereas "accuracy" measures the error involved in this approximation. As a matter of fact, the definition of accuracy is not limited to the linear case at hand. In Section 3.1 we shall refer to it in the context of discrete schemes approximating solutions of the general conservation law (2.1). In this case the discrete algorithm $U^{n+1} = H_k(U^n)$ is in general not linear (replacing U^n by $2U^n$, for example, will not produce a result of $2U^{n+1}$). However, we are still interested in measuring the "truncation error" committed by H_k, when applied to the exact solution $u^n = \{u(x_j, t_n)\}_{j=-\infty}^{\infty}$. We therefore formulate our definition of accuracy in this more general setting ($f(u) = au$ will take us back to the present case).

Definition 2.21 Let $u(x, t)$ be a smooth solution to the conservation law $u_t + f(u)_x = 0$ and let $U^{n+1} = H_k(U^n)$ be an approximating scheme. We say that H_k is accurate of order $p \geq 1$ if, with $u^n = \{u(x_j, t_n)\}_{j=-\infty}^{\infty}$,

$$u^{n+1} - H_k(u^n) = O\left(k^{p+1}\right), \qquad k \to 0. \qquad (2.33)$$

Remark 2.22 Observe that the notion of *consistency* is built into (2.33) in the following way:

Define $F_k(u^n) = u^n - H_k(u^n)$. Then (2.33) can be rewritten as

$$u^{n+1} - u^n + F_k(u^n) = O\left(k^{p+1}\right) \tag{2.34}$$

and dividing by k we have

$$\frac{u^{n+1} - u^n}{k} + \frac{1}{k}F_k(u^n) = O\left(k^p\right), \quad k \to 0. \tag{2.35}$$

Since $u(x, t)$ is an exact solution, $\frac{u^{n+1}-u^n}{k}$ approximates u_t (at $t = t_n$), and therefore $\frac{1}{k}F_k(u^n)$ should be an approximation for $f(u)_x$ (at $t = t_n$), as $k \to 0$. This last conclusion is commonly referred to as the "consistency" of the scheme $H_k(u^n) = u^n - F_k(u^n)$ with the differential equation. Suppose that $u(x, t)$ is a smooth function satisfying (2.33) and assume that

$$-u^n + H_k(u^n) + kau_x(x_j, t_n) = O\left(k^{p+1}\right) \quad \text{as} \quad k \to 0. \tag{2.36}$$

Inserting (2.34) into (2.36) we obtain

$$\frac{u^{n+1} - u^n}{k} + au_x(x_j, t_n) = O\left(k^p\right) \quad \text{as} \quad k \to 0, \tag{2.37}$$

so that by letting $k \to 0$ we get $u_t + au_x = 0$.

Example 2.23 For the backward-difference scheme (2.29) we replace U_j^n by $u(x, t)$, U_j^{n+1} by $u(x, t + k)$, and U_{j-1}^n by $u(x - \Delta x, t)$. Assuming u to be a smooth solution of $u_t + au_x = 0$ and using Taylor expansion we get

$$\begin{aligned}
u(x, t + k) &= u(x, t) + ku_t(x, t) + O(k^2) \\
&= u(x, t) - kau_x(x, t) + O(k^2) \\
&= u(x, t) - \lambda a\left[u(x, t) - u(x - \Delta x, t)\right] + O(k^2) + O(\Delta x^2) \\
&= H_k u(x, t) + O(k^2),
\end{aligned}$$

where

$$H_k\, u(x, t) = (1 - \lambda a)u(x, t) + \lambda au(x - \Delta x, t), \tag{2.38}$$

and we have absorbed $O(\Delta x^2)$ into $O(k^2)$ in view of $\Delta x = \lambda^{-1}k$.

We conclude that the scheme is first-order accurate [$p = 1$ in (2.33)]. The reader can prove similarly that both the forward-difference and the

Lax–Friedrichs schemes are of first-order accuracy, whereas the Lax–Wendroff scheme is of second-order accuracy.

Convergence

The major question that poses itself in connection with discretized schemes is that of convergence (as $k \to 0$) of the "approximate solution" $\{U_j^n\}$ to a (weak) solution of the differential equation. To illustrate the situation, we take as before the linear equation

$$u_t + au_x = 0, \quad u(x, 0) = u_0(x). \tag{2.39}$$

Example 2.24 (Non convergence for large $\lambda > 0$) Assume $a > 0$ and take the backward-difference scheme (2.29). Using (2.38) the reader can verify easily (say, by induction) that

$$U_j^n = \sum_{l=0}^n \binom{n}{l} (\lambda a)^l (1 - \lambda a)^{n-l} U_{j-l}^0. \tag{2.40}$$

The initial values $\{U_j^0\}_{j=-\infty}^\infty$ are computed from the given initial function $u_0(x)$. A common choice is to define U_j^0 as the average of $u_0(x)$ over the interval of size Δx centered at x_j, that is,

$$U_j^0 = \frac{1}{\Delta x} \int_{x_{j-\frac{1}{2}}}^{x_{j+\frac{1}{2}}} u_0(x)\,dx, \quad x_r = r\Delta x. \tag{2.41}$$

If we take the initial step function

$$u_0(x) = \begin{cases} 1, & x \le 0, \\ 0, & x > 0, \end{cases} \tag{2.42}$$

we have by Example 2.4 the weak solution

$$u(x, t) = u_0(x - at). \tag{2.43}$$

For the corresponding approximation we get from (2.41)

$$U_j^0 = \begin{cases} 1, & j < 0, \\ \frac{1}{2}, & j = 0, \\ 0, & j > 0, \end{cases} \tag{2.44}$$

and from (2.40), along with simple facts about the binomial coefficients, we get

$$\sum_{j=0}^{n} U_j^n = \sum_{j=0}^{n} \sum_{l=0}^{n} \binom{n}{l} (\lambda a)^l (1 - \lambda a)^{n-l} U_{j-l}^0$$

$$= \frac{1}{2} \sum_{j=0}^{n} \binom{n}{j} (\lambda a)^j (1 - \lambda a)^{n-j} + \sum_{j=0}^{n} \sum_{l=j+1}^{n} \binom{n}{l} (\lambda a)^l (1 - \lambda a)^{n-l}$$

$$= \frac{1}{2} + \sum_{l=0}^{n} l \binom{n}{l} (\lambda a)^l (1 - \lambda a)^{n-l} = \frac{1}{2} + \lambda n a,$$

so that

$$\Delta x \sum_{j=0}^{n} U_j^n = \frac{\Delta x}{2} + nka. \tag{2.45}$$

In analogy with the interpretation of U_j^0, we think of U_j^n as approximating the mean value of $u(x, nk)$ in the interval $(x_{j-\frac{1}{2}}, x_{j+\frac{1}{2}})$. Thus, the sum in (2.45) should be compared with the integral

$$\int_{-\frac{\Delta x}{2}}^{(n+\frac{1}{2})\Delta x} u(x, nk) \, dx = \int_{-\frac{\Delta x}{2}}^{(n+\frac{1}{2})\Delta x} u_0(x - ank) \, dx$$

$$= \int_{-\frac{\Delta x}{2} - ank}^{(n+\frac{1}{2})\Delta x - ank} u_0(x) \, dx.$$

Now fix $t > 0$ and take $nk = t$. As $k \to 0$ and because $\lambda = \frac{k}{\Delta x}$ is fixed, we have $\Delta x \to 0$ and the limit of the last integral can be evaluated as

$$\int_{-\frac{\Delta x}{2} - at}^{\frac{\Delta x}{2} + t(\frac{1}{\lambda} - a)} u_0(x) \, dx \to \begin{cases} at, & \lambda^{-1} \geq a, \\ \frac{1}{\lambda} t, & \lambda^{-1} < a. \end{cases} \tag{2.46}$$

However, from (2.45), as $k \to 0$,

$$\Delta x \sum_{j=0}^{n} U_j^n \to at. \tag{2.47}$$

We conclude that

$$\left| \int_{-\frac{\Delta x}{2}}^{(n+\frac{1}{2})\Delta x} u(x, t) \, dx - \Delta x \sum_{j=0}^{n} U_j^n \right| \to 0, \qquad \text{as} \quad k \to 0 \tag{2.48}$$

(with $nk = t$ and $\lambda = \frac{k}{\Delta x}$) if and only if

$$\lambda a \le 1. \tag{2.49}$$

Clearly, convergence "in the mean" is a very reasonable way of requesting that the sequence $\{U_j^n\}_{j=-\infty}^{\infty}$ should approximate the exact solution $u(x, nk)$. Since the step function (2.42) represents only one possible initial datum, we can only derive a necessary condition for convergence from the foregoing discussion. First, we formalize the convergence in the mean as follows.

Definition 2.25 Fix $T > 0$. We say that $\{U_j^n\}_{j=-\infty}^{\infty}$ converges to the solution $u(x, t)$ in $L_{\text{loc}}^1(\mathbb{R})$ at fixed time t if, for any $[\alpha, \beta] \in \mathbb{R}$,

$$\lim_{\substack{k \to 0 \\ (nk=t)}} \sum_{j=[\frac{\alpha}{\Delta x}]}^{[\frac{\beta}{\Delta x}]} \int_{x_{j-1/2}}^{x_{j+1/2}} \left| U_j^n - u(x, t) \right| \, dx = 0 \tag{2.50}$$

for any $0 \le t \le T$.[2] ($[z]$ is the integer satisfying $z - 1 < [z] \le z$ for any $z \in \mathbb{R}$.)

Our conclusion (2.48) can now be restated as follows.

Corollary 2.26 *Let $a > 0$. Then (2.49) is a necessary condition for the backward-difference scheme to converge in $L_{\text{loc}}^1(\mathbb{R})$ to the solution of (2.39).*

Definition 2.27 The condition (2.49) is called the CFL (Courant–Friedrichs–Lewy) condition associated with Equation (2.39) and scheme (2.29).

Since $\lambda = \frac{k}{\Delta x}$, the CFL condition can be written as

$$k \le \frac{\Delta x}{a}. \tag{2.51}$$

It therefore forces a necessary restriction on the size of the time step $k = \Delta t$, relative to the cell size Δx, for convergence to take place. Later in this section we shall have a geometric interpretation of this condition, in terms of the characteristic lines of the equation. Note that the condition refers not only to the equation but also to the particular scheme used to approximate it. Although it plays a fundamental role in the theory of linear equations, it serves only as a guideline in the nonlinear case (via linearization). Because our primary objective here is the treatment of the nonlinear case, we shall make little use

---

[2] $L_{\text{loc}}^1(\mathbb{R})$ consists of functions that are integrable over any finite interval.

of the general theory related to the CFL condition, and we refer the reader to Richtmyer and Morton [96] for a thorough discussion of this topic.

As we shall see throughout this monograph, the backward-difference scheme (2.29) (for $a > 0$!) plays a fundamental role in the development of accurate high-resolution schemes. The first step in this development is taken in the following theorem, proving the sufficiency of the CFL condition for convergence in $L^1_{loc}(\mathbb{R})$. Some knowledge of the binomial distribution is needed in the proof, which the reader may skip on first reading.

Theorem 2.28 *Consider the equation $u_t + au_x = 0, a > 0$, and assume that the initial function $u_0(x) = u(x, 0)$ is uniformly bounded. Then, under the CFL condition (2.49) the backward-difference scheme (2.29) converges in $L^1_{loc}(\mathbb{R})$, that is, in the sense of Definition 2.25.*

Proof In view of (2.6), (2.40), and (2.41) we can write, with $nk = t$,

$$
\sum_{j=[\frac{\alpha}{\Delta x}]}^{[\frac{\beta}{\Delta x}]} \int_{x_{j-1/2}}^{x_{j+1/2}} \left| U_j^n - u(x, t) \right| dx
$$

$$
= \sum_{j=[\frac{\alpha}{\Delta x}]}^{[\frac{\beta}{\Delta x}]} \int_{x_{j-1/2}}^{x_{j+1/2}} \left| \sum_{l=0}^{n} \binom{n}{l} (\lambda a)^l (1 - \lambda a)^{n-l} \left[U_{j-l}^0 - u_0(x - at) \right] \right| dx
$$

$$
\leq \sum_{l=0}^{n} \binom{n}{l} (\lambda a)^l (1 - \lambda a)^{n-l} p_l, \tag{2.52}
$$

where

$$
p_l = \sum_{j=[\frac{\alpha}{\Delta x}]}^{[\frac{\beta}{\Delta x}]} \int_{x_{j-1/2}}^{x_{j+1/2}} \left| U_{j-l}^0 - u_0(x - at) \right| dx, \tag{2.53}
$$

and where we have used the identity

$$
\sum_{l=0}^{n} \binom{n}{l} (\lambda a)^l (1 - \lambda a)^{n-l} = 1.
$$

We note that $n\Delta x = \frac{nk}{\lambda} = \frac{t}{\lambda} \leq \frac{T}{\lambda}$, so that the numbers $p_l, 0 \leq l \leq n$, are uniformly bounded by

$$
|p_l| \leq 2 \int_{\alpha - \frac{T}{\lambda} - \Delta x}^{\beta + \Delta x} |u_0(x)| dx, \qquad 0 \leq l \leq n.
$$

Recall that the "Law of Large Numbers" (see Feller [45, Vol. I, Section VI-4]) states that the binomial distribution $b_{n,l} = \binom{n}{l}(\lambda a)^l(1-\lambda a)^{n-l}$ is "concentrated around $l = \lambda an$," or, more precisely, that for any $\epsilon > 0$

$$\lim_{n\to\infty} \sum_{|l-\lambda an|>n\epsilon} b_{n,l}\, p_l = 0. \tag{2.54}$$

Thus, going back to (2.52), (2.53), we obtain for any $\epsilon > 0$

$$\lim_{\substack{k\to 0\\(nk=t)}} \sum_{j=[\frac{\alpha}{\Delta x}]}^{[\frac{\beta}{\Delta x}]} \int_{x_{j-1/2}}^{x_{j+1/2}} \left|U_j^n - u(x,t)\right|\,dx \leq \lim_{\substack{n\to\infty\\(nk=t)}} \sum_{|l-\lambda an|\leq n\epsilon} b_{n,l}\, p_l. \tag{2.55}$$

But if $|l - \lambda an| \leq n\epsilon$, we have $|(j-l)\Delta x - (x_j - at)| \leq n\epsilon\Delta x \leq \frac{T}{\lambda}\epsilon$, so

$$p_l \leq \sum_{j=[\frac{\alpha}{\Delta x}]}^{[\frac{\beta}{\Delta x}]} \int_{x_{j-l-1/2}}^{x_{j-l+1/2}} \left|U_{j-l}^0 - u_0(x)\right|\,dx$$

$$+ \sup_{0<h\leq\frac{T}{\lambda}\epsilon} \int_{\alpha-\frac{T}{\lambda}}^{\beta} |u_0(y+h) - u_0(y)|\,dy. \tag{2.56}$$

Given $\delta > 0$ we can choose $\epsilon > 0$ sufficiently small so that the second term on the right-hand side of (2.56) is smaller than $\frac{\delta}{2}$. (This follows from elementary properties of functions in $L^1(\mathbb{R})$; simply approximate u_0 in L^1 by a smooth function.) As for the first term in the right-hand side of (2.56), recall from (2.41) that U_{j-l}^0 is the average value of $u_0(x)$ over $[x_{j-l-1/2}, x_{j-l+1/2}]$. Thus, if $u_0(x)$ is smooth, $U_{j-l}^0 = u_0(y_{j-l})$ for some y_{j-l} in the interval and, since

$$u_0(y_{j-l}) - u_0(x) = \int_x^{y_{j-l}} u_0'(\xi)\,d\xi$$

we get

$$\sum_{j=[\frac{\alpha}{\Delta x}]}^{[\frac{\beta}{\Delta x}]} \int_{x_{j-l-1/2}}^{x_{j-l+1/2}} \left|U_{j-l}^0 - u_0(x)\right|\,dx \leq \Delta x \int_{\alpha-\frac{T}{\lambda}}^{\beta} |u_0'(y)|\,dy,$$

which is smaller than $\frac{\delta}{2}$ (for $0 \leq l \leq n$) if Δx is small. If u_0 is not smooth, it can be approximated (in L^1) by a smooth function, so that the same result holds.

We conclude that $\sup_{|l-\lambda an|<n\epsilon} p_l$ can be made arbitrarily small by taking ϵ, Δx sufficiently small. From (2.55) we now get

$$\lim_{\substack{k\to 0\\(nk=t)}} \sum_{j=[\frac{\alpha}{\Delta x}]}^{[\frac{\beta}{\Delta x}]} \int_{x_{j-1/2}}^{x_{j+1/2}} \left|U_j^n - u(x,t)\right|\,dx = 0,$$

which proves our theorem.   □

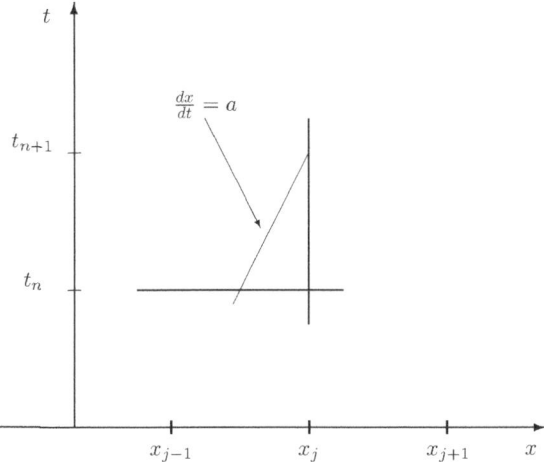

Figure 2.6. Geometric (characteristic) interpretation of the backward-difference scheme, $u_t + au_x = 0$. The CFL condition implies that the intersection is between x_{j-1} and x_j.

The backward-difference scheme has a simple geometric interpretation. Consider the grid (x_j, t_n) as in Figure 2.6.

As mentioned earlier, the approximating values $\{U_j^n\}$ are associated with the points (x_j, t_n). If the CFL condition (2.49) holds, then the characteristic line $x'(t) = a$, issuing from (x_j, t_{n+1}), intersects the line $t = t_n$ at the point $\bar{x} = \lambda a x_{j-1} + (1 - \lambda a)x_j \in [x_{j-1}, x_j]$. If we use the linear interpolation

$$U^n(\bar{x}) = \lambda a U_{j-1}^n + (1 - \lambda a)U_j^n \tag{2.57}$$

then the backward-difference scheme (2.29) states simply that

$$U_j^{n+1} = U^n(\bar{x}),$$

which just expresses the fact that the corresponding exact solution is constant along a characteristic line. We can summarize this discussion as follows.

Summary 2.29 (The backward-difference scheme as *exact* solution of *approximate* initial data) *The values $\{U_j^{n+1}\}_{j=-\infty}^{\infty}$ as obtained by the scheme (2.29) ($a > 0$), subject to the CFL condition (2.49), are the exact values $\tilde{u}(x_j, t_{n+1})$, where $\tilde{u}(x, t)$ satisfies the equation $\tilde{u}_t + a\tilde{u}_x = 0$, subject to the initial condition $\tilde{u}(x, t_n) = U^n(x)$. The function $U^n(x)$ is the piecewise linear (continuous) function obtained by interpolating the values $\{U_j^n\}_{j=-\infty}^{\infty}$ at the grid points $\{(x_j, t_n)\}_{j=-\infty}^{\infty}$.*

Definition 2.30 (Upwinding) We say that the backward-difference scheme (2.29), with $a > 0$, is an "upwind scheme," meaning that the values $\left\{U_j^{n+1}\right\}_{j=-\infty}^{\infty}$ are obtained from $\left\{U_j^n\right\}_{j=-\infty}^{\infty}$ by following the characteristic lines of the equation.

We now suggest yet another interpretation of the backward-difference scheme. This one, as in the preceding discussion, will also be based on an exact solution of the equation, subject to approximate initial data. However, now we take $U^n(x)$ as the *piecewise-constant* function defined by

$$U^n(x) = U_j^n, \qquad x_{j-1/2} < x < x_{j+1/2}, \qquad -\infty < j < \infty. \qquad (2.58)$$

We can make the following claim.

Claim 2.31 *Solve the equation* $\tilde{u}_t + a\tilde{u}_x = 0$, *subject to the initial condition* $\tilde{u}(x, t_n) = U^n(x)$ *as in (2.58). Then the values* U_j^{n+1}, *as determined by the backward-difference scheme (2.29), satisfy*

$$U_j^{n+1} = \frac{1}{\Delta x} \int_{x_{j-1/2}}^{x_{j+1/2}} \tilde{u}(x, t_{n+1})\,dx, \qquad (2.59)$$

provided the CFL condition (2.49) holds.

Proof The CFL condition implies that the "moving step" solution (2.6) satisfies

$$\tilde{u}(x_{j+1/2}, t) = U_j^n, \qquad t_n \le t \le t_{n+1}, \qquad -\infty < j < \infty. \qquad (2.60)$$

It follows from the balance equation (2.3) that

$$\int_{x_{j-1/2}}^{x_{j+1/2}} \tilde{u}(x, t_{n+1})\,dx = U_j^n\,\Delta x - a\left[U_j^n - U_{j-1}^n\right]k,$$

and, by $k = \lambda \Delta x$,

$$\frac{1}{\Delta x} \int_{x_{j-1/2}}^{x_{j+1/2}} \tilde{u}(x, t_{n+1})\,dx = (1 - \lambda a)U_j^n + \lambda a U_{j-1}^n = U_j^{n+1}. \qquad (2.61)$$

□

Observe that although $\tilde{u}(x, t)$ (in Summary 2.29) and $\tilde{\tilde{u}}(x, t)$ (in Claim 2.31) satisfy the same differential equation, they are actually different since the initial data $U^n(x)$, used to interpolate the discrete values $\{U_j^n\}_{j=-\infty}^{\infty}$, are different for the two cases. In the case of $\tilde{u}(x, t)$ the initial function $U^n(x)$, and hence $\tilde{u}(x, t)$, are *continuous*, and U_j^{n+1} is taken as the approximate (pointwise) "upwind" value. In contrast, in the case of $\tilde{\tilde{u}}(x, t)$, the initial function $U^n(x)$ is piecewise constant, and hence is in general *discontinuous*, and the value U_j^{n+1} is taken as the average of the ensuing solution $\tilde{\tilde{u}}(x, t_{n+1})$ over $(x_{j-1/2}, x_{j+1/2})$.

Recall that, for the nonlinear conservation law $u_t + f(u)_x = 0$, the solution can develop discontinuities even when subject to very smooth initial data. In this case, therefore, the pointwise upwinding approach expressed by $\tilde{u}(x, t)$, based on continuous interpolation, does not seem appropriate. In contrast, the "averaging" approach, based on the balance equation (2.3) applied to piecewise-constant initial data, can be readily generalized to the nonlinear case. It is this approach, first suggested by Godunov [56], that will serve as the basis of the GRP method discussed in the next chapter.

Remark 2.32 Note that none of the schemes (2.30)–(2.32) (i.e., the forward-difference, Lax–Friedrichs, and Lax–Wendroff schemes) are amenable to an interpretation based on characteristic values ("upwinding" as in Definition 2.30) or averaging in the sense of Godunov (as in Claim 2.31). Nonetheless, all these schemes [including (2.29)] are "conservative" in the sense that

$$\sum_{j=-\infty}^{\infty} U_j^{n+1} = \sum_{j=-\infty}^{\infty} U_j^n$$

(when the values U_j^n vanish sufficiently fast as $|j| \to \infty$). This is of course consistent with the conservation property (2.4). However, in this monograph we shall not make much use of this conservation property.

Remark 2.33 We refer the reader to Godlewski and Raviart [54] for a proof of the following: The Lax–Friedrichs scheme (2.31) and the Lax–Wendroff scheme (2.32) converge to the solution of $u_t + au_x = 0 \, (a > 0)$, in the sense of Definition 2.25, subject to the CFL condition (2.49); however, the forward-difference scheme (2.30) fails to converge, no matter how small $\lambda = \frac{k}{\Delta x}$ is.

Except for a few examples, we shall not make use of these schemes in the present monograph.

3

The GRP Method for Scalar Conservation Laws

This chapter introduces the GRP method in the context of the scalar conservation law $u_t + f(u)_x = 0$. We start in Section 3.1 with the classical first-order (conservative) "Godunov Scheme," which leads naturally to its second-order GRP extension. Section 3.2 contains a number of numerical (one-dimensional) examples, for linear and nonlinear equations, illustrating the improved resolution obtained by the GRP method. In Section 3.3 we extend the GRP methodology to the two-dimensional scalar conservation law $u_t + f(u)_x + g(u)_y = 0$. Analytical and numerical results are compared for simple as well as complex wave interactions.

3.1 From Godunov to the GRP Method

In this section we discuss the GRP method, aimed at a high-resolution numerical approximation of the solution to a conservation law of the form (2.1). We always assume that $f(u)$ is strictly convex: $f''(u) \geq \mu > 0$. It is shown that this method is a natural analytic extension to the Godunov (upwind) scheme. This latter scheme has been extensively studied in Section 2.2, in the context of the linear convection equation. We start here by studying this scheme in the nonlinear case.

As in Section 2.2, we take a uniform spatial grid $x_j = j\Delta x$, $-\infty < j < \infty$, and uniformly spaced time levels $t_{n+1} = t_n + k$, $t_0 = 0$. As before, we refer to the interval $(x_{j-1/2}, x_{j+1/2})$ as "cell j" and to $x_{j\pm1/2}$ as its "cell boundaries." Given the approximating functions $U^1(x), \ldots, U^n(x)$, a numerical scheme consists in constructing $U^{n+1}(x)$, approximating $u(x, t_{n+1})$. The functions $U^n(x)$ are piecewise constant (Godunov) or piecewise linear (GRP). Their averages over cell j are denoted by U_j^n.

Our starting point is the balance equation (2.3), to be used over the rectangle $[x_{j-1/2}, x_{j+1/2}] \times [t_n, t_{n+1}]$. Since U_j^n, U_j^{n+1} are supposed to be the average values of the approximating function over "bottom" and "top" respectively, the discrete version of (2.3) should be

$$U_j^{n+1} = U_j^n - \lambda \left(f_{j+1/2}^{n+1/2} - f_{j-1/2}^{n+1/2} \right), \tag{3.1}$$

where $\lambda = \frac{k}{\Delta x}$ and $f_{j+1/2}^{n+1/2}$ is an approximation for the average flux $\frac{1}{k} \int_{t_n}^{t_{n+1}} f\left(u(x_{j+1/2}, t)\right) dt$.

Definition 3.1 The term $f_{j+1/2}^{n+1/2}$ is called the "numerical flux" at the boundary $x_{j+1/2}$ over the time interval $[t_n, t_{n+1}]$.

Clearly, once the numerical fluxes are known, the numerical scheme is fully determined. We refer the reader to the books by LeVeque [81] and Godlewski and Raviart [54] for surveys of many existing schemes.

In this monograph we adapt the approach suggested by Godunov [56] as mentioned at the end of Section 2.2. In the present nonlinear case it can be described as follows: Take the function $U^n(x)$ as piecewise constant, with

$$U^n(x) = U_j^n, \qquad x_{j-1/2} < x < x_{j+1/2}. \tag{3.2}$$

Let $\tilde{u}(x, t)$ be the weak solution to (2.1) for $t \geq t_n$, subject to initial data $U^n(x)$ at $t = t_n$. Now evaluate the numerical flux as

$$f_{j+1/2}^{G, n+1/2} = \frac{1}{k} \int_{t_n}^{t_{n+1}} f\left(\tilde{u}(x_{j+1/2}, t)\right) dt, \qquad -\infty < j < \infty, \tag{3.3}$$

where the superscript "G" refers to the fact that this is the numerical flux associated with the Godunov method.

The main idea in the application of (3.1) (with $f_{j+1/2}^{n+1/2} = f_{j+1/2}^{G, n+1/2}$) is that if k is *sufficiently small* then

$$\tilde{u}(x_{j+1/2}, t) = constant, \qquad t \in [t_n, t_{n+1}], \tag{3.4}$$

so that (3.3) is easily evaluated. The reader should observe that this is in full agreement with the linear case [$f(u) = au$, $a > 0$] as discussed in Claim 2.31, where $f_{j+1/2}^{G, n+1/2} = aU_j^n$.

To give a more precise meaning to (3.4), we set

$$M_n = \sup_{-\infty < j < \infty} \left| U_j^n \right| < \infty, \tag{3.5}$$

and let $k = \Delta t$ satisfy the "CFL condition" (compare Definition 2.27)

$$k \cdot \max_{|u| \leq M_n} \left| f'(u) \right| < \Delta x. \tag{3.6}$$

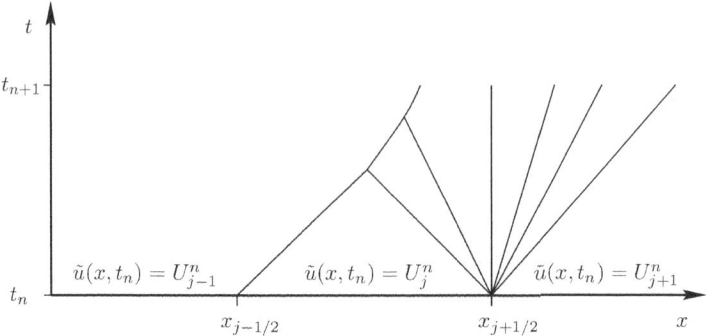

Figure 3.1. Wave pattern for piecewise-constant initial data.

Observe that near every cell boundary $x_{j+1/2}$ the solution $\tilde{u}(x, t)$ is a "Riemann solution" $R\left(\frac{x-x_{j+1/2}}{t-t_n}; U_j^n, U_{j+1}^n\right)$ (see Definition 2.18) associated with the initial data U_j^n, U_{j+1}^n (see Figure 3.1). It follows from Corollary 2.19 that the speeds of all waves emanating from the points $x_{j+1/2}$, $-\infty < j < \infty$, are bounded by

$$S_n = \max_{|u| < M_n} \left| f'(u) \right|. \tag{3.7}$$

The CFL condition (3.6) therefore entails the following important conclusion.

Corollary 3.2 (Validity of "local Riemann solutions") *Under the CFL condition (3.6), a wave issuing from $(x_{j+1/2}, t_n)$ does not reach any other cell boundary $(x_{l+1/2}, t)$ within the time interval $[t_n, t_{n+1}]$.*

Remark 3.3 Note that in the case $f(u) = au$ the CFL condition (3.6) coincides with (2.49).

Remark 3.4 Observe that (see Figure 3.1) the waves issuing from neighboring cell boundaries $x_{j\pm1/2}$ may interact during the time interval $[t_n, t_{n+1}]$. However, since the speeds of all resulting waves are still bounded by S_n [Equation (3.7)], they do not reach the opposite cell boundary.

The preceding remark and Corollary 3.2 imply, in particular, that when (3.6) is fulfilled, the solution $\tilde{u}(x_{j+1/2}, t)$ satisfies, for every $-\infty < j < \infty$,

$$\tilde{u}(x_{j+1/2}, t) = R(0; U_j^n, U_{j+1}^n), \qquad t_n \leq t \leq t_{n+1}, \tag{3.8}$$

so that the numerical flux is given by

$$f_{j+1/2}^{G, n+1/2} = f\left(R(0; U_j^n, U_{j+1}^n)\right), \tag{3.9}$$

and the balance equation (2.3) yields

$$\frac{1}{\Delta x} \int_{x_{j-1/2}}^{x_{j+1/2}} \tilde{u}(x, t_{n+1})dt = U_j^n - \lambda \left(f_{j+1/2}^{G,n+1/2} - f_{j-1/2}^{G,n+1/2} \right). \qquad (3.10)$$

In the linear case $f(u) = au, a > 0$, we get $f_{j+1/2}^{G,n+1/2} = aU_j^n$, so that Equation (3.1) yields the upwind scheme as in Definition 2.30 and Claim 2.31.

Definition 3.5 (The Godunov Scheme) The scheme given by

$$U_j^{n+1} = U_j^n - \lambda \left[f \left(R(0; U_j^n, U_{j+1}^n) \right) - f \left(R(0; U_{j-1}^n, U_j^n) \right) \right],$$

$$U_j^0 = \frac{1}{\Delta x} \int_{x_{j-1/2}}^{x_{j+1/2}} \tilde{u}_0(x)\, dx, \qquad (3.11)$$

$$\lambda = \frac{k}{\Delta x}$$

is called the "Godunov scheme" for the approximation of the conservation law (2.1).

It is important to note that by (3.10), (3.11) the value U_j^{n+1} is the average, over cell j, of the exact solution $\tilde{u}(x, t_{n+1})$, subject to the initial data $U^n(x)$ at $t = t_n$ [as in (3.2)]. In fact, this property singles out the Godunov scheme: It is fully determined by the requirement that the sequence of updated values $\{U_j^{n+1}\}_{j=-\infty}^{\infty}$ consists of the averaged (over computational cells) of the exact solution $\tilde{u}(x, t_{n+1})$. This is true because the balance equation (2.3) of the conservation law (2.1) is satisfied by $\tilde{u}(x, t)$.

The first and most fundamental question to be asked about the Godunov scheme (as well as any other approximating scheme) concerns its convergence to the exact weak solution of (2.1), (2.2). In Example 2.24 and Theorem 2.28 we have seen that, in the linear case $f(u) = au$, the CFL condition (3.6) is a necessary and sufficient condition for convergence in $L^1_{\text{loc}}(\mathbb{R})$ (see Definition 2.25). As observed already in Section 2.2, the idea of convergence in $L^1_{\text{loc}}(\mathbb{R})$ is very reasonable, especially when dealing with discontinuous solutions. It allows for phenomena common to numerical approximation, such as oscillations or "spurious waves," as long as they tend to zero *in the mean* as the grid is refined ($k = \Delta t \to 0$), over any fixed finite interval.

Considering the convergence properties of the Godunov scheme in the case of a nonlinear flux function $f(u)$, we can cite the following theorem.

Theorem 3.6 *Let $u_0(x) \in L^1(\mathbb{R}) \cap L^\infty(\mathbb{R})$ and assume further that $u_0(x)$ is a function of finite total variation. Let $u(x, t)$ be the unique entropy solution*

to (2.1), (2.2), and let $\{U_j^n\}_{\substack{-\infty<j<\infty \\ n\geq 0}}$ be obtained by the Godunov scheme (3.11).
Then, under the CFL condition (3.6), $\{U_j^n\}$ converges to $u(x,t)$ in $L_{loc}^1(\mathbb{R})$ (see
Definition 2.25).

The proof of this theorem is presented in Appendix B, where we recall the
notion of total variation of a function and its discrete approximation. This notion
is of basic significance in the study of scalar conservation laws.

An important ingredient in the proof of Theorem 3.6 is that the Godunov
scheme satisfies a "maximum principle," in analogy with the case of the exact
solution (see Theorem 2.20). In our case it has also a practical significance, as it
permits us to replace in the CFL condition (3.6) the (usually unknown) maximal
value S_n by the initial value S_0, which is easily available in most typical cases.
We therefore state and prove the maximum principle as follows.

Claim 3.7 ("Maximum principle for the Godunov scheme")   *Given   the*
scheme (3.11), and using the notation (3.5), we have

$$M_{n+1} \leq M_n \leq \cdots \leq M_0. \tag{3.12}$$

Proof According to (3.1) and (3.3) U_j^{n+1} is an average (over cell j) of the
exact solution $\tilde{u}(x, t_{n+1})$, subject to initial data $\tilde{u}(x, t_n) = U^n(x)$. Thus, by the
maximum principle for an exact solution (Theorem 2.20),

$$M_{n+1} = \sup_j \left|U_j^{n+1}\right| \leq \sup_x |\tilde{u}(x, t_{n+1})| \leq \sup_x |\tilde{u}(x, t_n)| = \sup_j \left|U_j^n\right| = M_n.$$
$$\tag{3.13}$$

□

We shall not develop here the mathematical theory of numerical schemes.
The reader is referred to Godlewski and Raviart [54] for an extensive survey.
We emphasize that there are almost no results concerning the convergence of
schemes approximating *systems* of equations, and this is particularly true for the
system of equations governing compressible inviscid flow, at which this monograph is primarily aimed. Furthermore, even in the case of a *scalar* conservation
law in one space dimension [such as (2.1)], the existing proofs pertain mostly
to the case of a first-order scheme (such as the Godunov scheme), whereas we
are concerned with the GRP method, which is of second-order accuracy.

This method may be introduced as follows: We consider again the balance
equation (2.3). However, we now assume that at $t = t_n$ the initial distribution
is *linear* in each cell j. This seemingly modest modification of (3.2), proposed by van Leer [112], proved to be the most crucial ingredient in all further

developments. Retaining the notation U_j^n for the average cell values, we there-fore assume that

$$U^n(x) = U_j^n + (x - x_j)s_j^n, \qquad x_{j-1/2} < x < x_{j+1/2}, \tag{3.14}$$

where s_j^n is the slope of the linear segment $U^n(x)$ in cell j. Note that at cell boundaries $x_{j+1/2}$ we have in general a jump discontinuity in the values of $U^n(x)$ (namely, between $U_j^n + \frac{\Delta x}{2} s_j^n$ and $U_{j+1}^n - \frac{\Delta x}{2} s_{j+1}^n$) and also in the values of the slopes (s_j^n, s_{j+1}^n).

Let $\tilde{u}(x, t)$, $t_n \le t \le t_{n+1}$, be the weak solution to (2.1), subject to initial data $\tilde{u}(x, t_n) = U^n(x)$ [as in (3.14)]. The values $\tilde{u}(x_{j+1/2}, t)$ at cell boundaries now depend on t, even for $t - t_n$ small, in contrast to the previous (Godunov) case, as given in (3.8). This is of course because now $U^n(x)$ is not constant on either side of $x_{j+1/2}$, so we cannot expect a Riemann solution there. It follows that in the present case the difference scheme (3.1) can only be written with numerical fluxes $f_{j+1/2}^{n+1/2}$ that are only *approximately* equal to $\frac{1}{k} \int_{t_n}^{t_{n+1}} f\left(\tilde{u}(x_{j+1/2}, t)\right) dt$. Specifically, we assume now that the numerical fluxes $f_{j+1/2}^{n+1/2}$ satisfy

$$f_{j+1/2}^{n+1/2} - f_{j-1/2}^{n+1/2} = \frac{1}{k} \int_{t_n}^{t_{n+1}} \left[f\left(\tilde{u}(x_{j+1/2}, t)\right) - f\left(\tilde{u}(x_{j-1/2}, t)\right) \right] dt + O\left(k^3\right),$$

$$-\infty < j < \infty. \tag{3.15}$$

We now define the new averages U_j^{n+1}, $-\infty < j < \infty$, by

$$U_j^{n+1} = U_j^n - \lambda \left(f_{j+1/2}^{n+1/2} - f_{j-1/2}^{n+1/2} \right). \tag{3.16}$$

Combining (3.14) and (3.15) with the balance equation (2.3) for $\tilde{u}(x, t)$ over $[x_{j-1/2}, x_{j+1/2}] \times [t_n, t_{n+1}]$ we get, from (3.16),

$$\begin{aligned} U_j^{n+1} &= \frac{1}{\Delta x} \int_{x_{j-1/2}}^{x_{j+1/2}} \tilde{u}(x, t_n)\, dx \\ &\quad - \frac{1}{\Delta x} \int_{t_n}^{t_{n+1}} \left[f\left(\tilde{u}(x_{j+1/2}, t)\right) - f\left(\tilde{u}(x_{j-1/2}, t)\right) \right] dt + O\left(k^3\right) \\ &= \frac{1}{\Delta x} \int_{x_{j-1/2}}^{x_{j+1/2}} \tilde{u}(x, t_{n+1})\, dx + O\left(k^3\right), \qquad -\infty < j < \infty. \end{aligned} \tag{3.17}$$

Note that for the special initial data (3.2) Equations (3.15), (3.17) were satis-fied with no truncation error [see (3.11)]. In terms of Definition 2.21 we now conclude that the scheme (3.16) is of *second-order accuracy* ($p = 2$).

The foregoing derivation was based on the hypothesis (3.15). To study its validity we prove the following claim.

Claim 3.8 *Let $\tilde{u}(x, t)$ be smooth in $x \in [x_{j-1/2}, x_{j+1/2}]$ and $t \geq t_n$. Then (3.15) is satisfied with*

$$f_{j+1/2}^{n+1/2} = f\left(\tilde{u}(x_{j+1/2}, t_n)\right) + \frac{k}{2}\frac{\partial}{\partial t}f\left(\tilde{u}(x_{j+1/2}, t_n)\right) \tag{3.18}$$

[namely, $f_{j+1/2}^{n+1/2} = $ the linear approximation (in t) of $f\left(\tilde{u}(x_{j+1/2}, t)\right)$ evaluated at the midpoint $t_{n+1/2} = t_n + \frac{k}{2}$].

Proof This is a direct consequence of Taylor's theorem and the fact that $\lambda = k/\Delta x = constant$. To simplify notation, we introduce the functions

$$g_{j+1/2}(t) = f\left(\tilde{u}(x_{j+1/2}, t)\right), \qquad -\infty < j < \infty.$$

Writing

$$g_{j+1/2}(t) = g_{j+1/2}(t_{n+1/2}) + g'_{j+1/2}(t_{n+1/2})(t - t_{n+1/2})$$
$$+ \frac{1}{2}g''_{j+1/2}(t_{n+1/2})(t - t_{n+1/2})^2 + O\left(k^3\right), \qquad t_n \leq t \leq t_{n+1},$$

we obtain by integration

$$\int_{t_n}^{t_{n+1}} \left[g_{j+1/2}(t) - g_{j-1/2}(t)\right] dt = \left[g_{j+1/2}(t_{n+1/2}) - g_{j-1/2}(t_{n+1/2})\right] k$$
$$+ \frac{1}{24}\left[g''_{j+1/2}(t_{n+1/2}) - g''_{j-1/2}(t_{n+1/2})\right] k^3$$
$$+ O\left(k^4\right). \tag{3.19}$$

However, $g''_{j+1/2}(t_{n+1/2}) - g''_{j-1/2}(t_{n+1/2}) = O(k)$ and by (3.18)

$$g_{j+1/2}(t_{n+1/2}) - g_{j-1/2}(t_{n+1/2}) = f_{j+1/2}^{n+1/2} - f_{j-1/2}^{n+1/2}$$
$$+ \frac{1}{8}\left[g''_{j+1/2}(t_n) - g''_{j-1/2}(t_n)\right] k^2 + O\left(k^3\right)$$
$$= f_{j+1/2}^{n+1/2} - f_{j-1/2}^{n+1/2} + O\left(k^3\right).$$

Inserting these relations in (3.19) yields (3.15)   □

Remark 3.9 It is clear from the proof that we could replace (3.18) by any other expression that approximates, up to $O(k^2)$, the value $f\left(\tilde{u}(x_{j+1/2}, t_{n+1/2})\right)$, such as

$$f_{j+1/2}^{n+1/2} = f\left(\tilde{u}(x_{j+1/2}, t_n) + \frac{k}{2}\frac{\partial \tilde{u}}{\partial t}(x_{j+1/2}, t_n)\right). \tag{3.20}$$

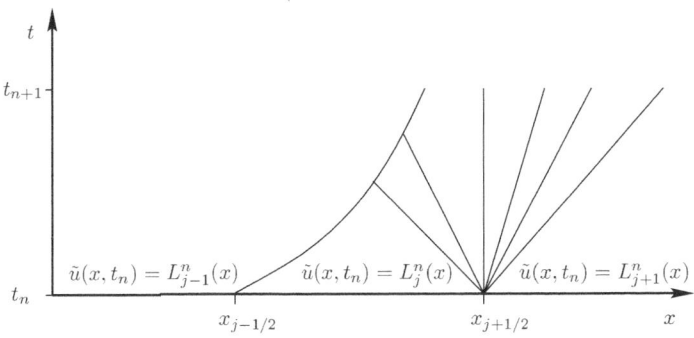

Figure 3.2. Wave pattern for the GRP algorithm. $L_j^n(x) = U_j^n + (x - x_j)s_j^n$.

Although Claim 3.8 is of a formal value (as the solution $\tilde{u}(x, t)$ is generally not smooth), it provides the guideline to the construction of the GRP numerical fluxes. Because of the fundamental importance of this construction in the present monograph, we shall first list the technical steps, then follow with a detailed discussion.

Construction 3.10 (GRP algorithm) *Given the piecewise-linear distribution $U^n(x)$ (3.14) and $\Delta t = k$ such that $\lambda = \frac{k}{\Delta x}$ satisfies the CFL condition (3.6) [with $M_n = \sup_{x \in \mathbb{R}} |U^n(x)|$], construct $U^{n+1}(x)$ (which should approximate $\tilde{u}(x, t_n + k)$) as follows.*

Step 1. At every cell boundary $x_{j+1/2}$ evaluate $U^n(x)$ on the two sides by

$$U_{j+1/2, \pm}^n = \lim_{\delta \to 0^+} U^n(x_{j+1/2} \pm \delta) = \begin{cases} U_{j+1}^n - \frac{\Delta x}{2} s_{j+1}^n, & ''+'', \\ U_j^n + \frac{\Delta x}{2} s_j^n, & ''-''. \end{cases}$$

Then determine the Riemann solution

$$U_{j+1/2}^n = R\left(0; U_{j+1/2, -}^n, U_{j+1/2, +}^n\right). \tag{3.21}$$

Note that, as in (2.28),

$$U_{j+1/2}^n = \begin{cases} U_{j+1/2, -}^n & wave\ moves\ right,\ f'\left(U_{j+1/2}^n\right) > 0, \\ U_{j+1/2, +}^n & wave\ moves\ left,\ \ f'\left(U_{j+1/2}^n\right) < 0, \\ u_{\min} & if\ x_{j+1/2}\ is\ a\ sonic\ point, \\ & f'\left(U_{j+1/2, -}^n\right) \le 0 \le f'\left(U_{j+1/2, +}^n\right). \end{cases}$$

$$\tag{3.22}$$

Step 2. *Determine the instantaneous time derivatives $\frac{\partial \tilde{u}}{\partial t}(x_{j+1/2}, t_n)$ by*

$$\frac{\partial \tilde{u}}{\partial t}(x_{j+1/2}, t_n) = \begin{cases} -f'\left(U^n_{j+1/2}\right) s^n_j & \text{if } f'\left(U^n_{j+1/2}\right) > 0, \\ -f'\left(U^n_{j+1/2}\right) s^n_{j+1} & \text{if } f'\left(U^n_{j+1/2}\right) < 0, \\ 0 & \text{if } U^n_{j+1/2} = u_{\min}. \end{cases}$$

(3.23)

Then compute the approximate solution and numerical flux [see (3.18)] at the midpoint $(x_{j+1/2}, t_{n+1/2})$ by

$$U^{n+1/2}_{j+1/2} = U^n_{j+1/2} + \frac{k}{2}\frac{\partial \tilde{u}}{\partial t}(x_{j+1/2}, t_n),$$

(3.24)

$$f^{n+1/2}_{j+1/2} = f\left(U^n_{j+1/2}\right) + \frac{k}{2} f'\left(U^n_{j+1/2}\right)\frac{\partial \tilde{u}}{\partial t}(x_{j+1/2}, t_n).$$

Step 3. *Evaluate the new cell averages as in (3.16),*

$$U^{n+1}_j = U^n_j - \lambda\left(f^{n+1/2}_{j+1/2} - f^{n+1/2}_{j-1/2}\right), \qquad -\infty < j < \infty, \quad (3.25)$$

and the new slopes by

$$U^{n+1}_{j+1/2} = U^n_{j+1/2} + k\frac{\partial \tilde{u}}{\partial t}(x_{j+1/2}, t_n), \qquad -\infty < j < \infty,$$

(3.26)

$$s^{n+1}_j = \frac{1}{\Delta x}\left(U^{n+1}_{j+1/2} - U^{n+1}_{j-1/2}\right).$$

The construction of the algorithm is not yet complete. We shall later supplement it [see (3.28)] by a suitable "slope limiter," which ensures certain monotonicity properties of the new profile $U^{n+1}(x)$ (see Figure 3.4 below). However, we shall first make a few comments concerning this algorithm.

The basic hypothesis underlying the GRP construction is that the wave pattern associated with the solution $\tilde{u}(x, t)$ can be fully determined (for sufficiently small $k = \Delta t$) by the Riemann solutions of Step 1 [see (3.21)]. Of course, as has already been observed, a shock wave issuing from $x_{j+1/2}$ will not be (in general) self-similar. In other words, its trajectory will not be of constant slope. This is in contrast to the characteristic lines (comprising a centered rarefaction wave), which in the *scalar* case are always straight lines. However, the assumption here is that at each cell boundary $x_{j+1/2}$ the solution $\tilde{u}(x, t)$ consists of a single wave (shock for $U^n_{j+1/2,-} > U^n_{j+1/2,+}$, centered rarefaction otherwise), where instantaneous features at $x = x_{j+1/2}$, $t = t_n$ (i.e., slopes of a shock trajectory or head and tail characteristics of a rarefaction) are completely determined by the Riemann solution $R\left(\frac{x-x_{j+1/2}}{t-t_n}; U^n_{j+1/2,-}, U^n_{j+1/2,+}\right)$. Also, the solution $\tilde{u}$ on the two sides of the shock, or inside and outside a centered rarefaction wave, is smooth, with a jump discontinuity across a shock trajectory or jump discontinuities of the derivatives across the head and tail characteristics of a centered rarefaction wave [compare the discussion following (2.26)].

In fact, in the present case (a scalar conservation law with a strictly convex flux function) this assumption can be proved for the unique entropy solution. Observe that by the maximum principle (Theorem 2.20), the CFL condition implies, as in the case of the Godunov scheme, that a wave issuing from a cell boundary $x_{j+1/2}$ is limited (for $t \in [t_n, t_n + k]$) to the neighboring cells j, $j + 1$, not reaching their opposite boundaries $x_{j+3/2}$, $x_{j-1/2}$. In particular, the solution $\tilde{u}(x_{j+1/2}, t)$, $t_n \leq t \leq t_{n+1}$, is not affected by the discontinuities at $x_{l+1/2}$, $l \neq j$, and is therefore a smooth function of t. Its derivative $\frac{\partial \tilde{u}}{\partial t}(x_{j+1/2}, t_n)$ should be interpreted as the limiting value of $\frac{\partial \tilde{u}}{\partial t}(x_{j+1/2}, t)$, $t > t_n$, as $t \to t_n$. If the wave moves to the right, the segment $(x_{j+1/2}, t)$, $t_n < t < t_{n+1}$, is contained, along with (x, t_n), $x_j < x < x_{j+1/2}$, in the same "domain of smoothness" of $\tilde{u}(x, t)$; hence $\tilde{u}(x, t)$ is a classical solution there, satisfying $\tilde{u}_t(x, t) = -f'(\tilde{u}(x, t)) \tilde{u}_x(x, t)$. A similar consideration applies to the case where the wave moves to the left. If $x_{j+1/2}$ is a sonic point, the line $x = x_{j+1/2}$ is characteristic, carrying the constant value $\tilde{u}(x_{j+1/2}, t) = u_{\min}$. We obtain therefore all three cases of Equation (3.23).

The evaluation of the numerical fluxes (3.24) follows the second-order approximation given by (3.20), where $\tilde{u}(x_{j+1/2}, t_n) = U^n_{j+1/2}$ is the limiting value (as $t \to t_n$) of $\tilde{u}(x_{j+1/2}, t)$, $t > t_n$. The same linear approximation of $\tilde{u}(x_{j+1/2}, t)$ serves to determine the new slopes in (3.26).

Remark 3.11 (Accuracy of the slope computation) As in Claim 3.8, assume that $\tilde{u}(x_{j+1/2}, t)$ is smooth in $[x_{j-1/2}, x_{j+1/2}] \times [t_n, t_{n+1}]$. Then

$$U^{n+1}_{j\pm1/2} = \tilde{u}(x_{j\pm1/2}, t_{n+1}) - \frac{1}{2}\tilde{u}_{tt}(x_{j\pm1/2}, t_n) \cdot k^2 + O\left(k^3\right)$$

$$= \tilde{u}(x_j, t_{n+1}) \pm \tilde{u}_x(x_j, t_{n+1}) \cdot \frac{\Delta x}{2}$$

$$+ \frac{1}{8}\tilde{u}_{xx}(x_j, t_{n+1}) \cdot \Delta x^2 - \frac{1}{2}\tilde{u}_{tt}(x_{j\pm1/2}, t_n) \cdot k^2 + O\left(k^3\right).$$

Thus

$$s_j^{n+1} = \frac{1}{\Delta x} \left(U_{j+1/2}^{n+1/2} - U_{j-1/2}^{n+1/2} \right) = \tilde{u}_x(x_j, t_{n+1}) + O\left(k^2\right),$$

which is, naturally, less accurate than the computation of the cell averages U_j^{n+1} [see (3.15) and (3.16)].

Remark 3.12 (Zero slopes in GRP computation)   Observe that when the slopes s_j^n are set to zero for all cells j and at every time level t_n, the GRP computational scheme naturally reduces to the Godunov scheme.

Remark 3.13 (Stationary shocks)   If the Riemann solution $R\left(\frac{x-x_{j+1/2}}{t-t_n}\right)$; $U_{j+1/2,-}^n, U_{j+1/2,+}^n$) yields a stationary shock along $x = x_{j+1/2}$, it means (by the Rankine–Hugoniot jump condition) that $f\left(U_{j+1/2,-}^n\right) = f\left(U_{j+1/2,+}^n\right)$, $U_{j+1/2,-}^n > U_{j+1/2,+}^n$. The shock speed is given by

$$\sigma(t) = \frac{f\left(\tilde{u}^+(x(t), t)\right) - f\left(\tilde{u}^-(x(t), t)\right)}{\tilde{u}^+(x(t), t) - \tilde{u}^-(x(t), t)},$$

where $x(t)$ is the shock trajectory $[x(t_n) = x_{j+1/2}, x'(t) = \sigma(t)]$ and $\tilde{u}^-$ (resp. $\tilde{u}^+$) is the value behind (resp. ahead of) the shock, $u^{\pm}(x(t_n), t_n) = U_{j+1/2,\pm}^n$. Thus,

$$\sigma'(t)\big|_{t=t_n} = \frac{-f'\left(U_{j+1/2,+}^n\right)^2 s_{j+1}^n + f'\left(U_{j+1/2,-}^n\right)^2 s_j^n}{U_{j+1/2,+}^n - U_{j+1/2,-}^n},$$

and the value of $\frac{\partial \tilde{u}}{\partial t}(x_{j+1/2}, t_n)$ is determined according to whether $\pm\sigma'(t_n) > 0$.

The last technical step in the description of the GRP algorithm is concerned with a modification of the slope s_j^{n+1}. In the language common to numerical schemes, it is a "postprocessing" step applied to the new results $\left\{U_j^{n+1}, s_j^{n+1}\right\}_{-\infty<j<\infty}$.

It is a *basic rule* in all GRP calculations (also in the case of systems later on) that the new averages U_j^{n+1}, as determined by (3.25), are *never* modified. Their values are obtained by the approximate implementation of the balance equation, which is viewed here as the basis of our methodology. However, the slopes are less accurately computed, using a discrete differentiation procedure (3.26). We can illustrate the need for a "postprocessing intervention" in their values by the following example (see Figure 3.3).

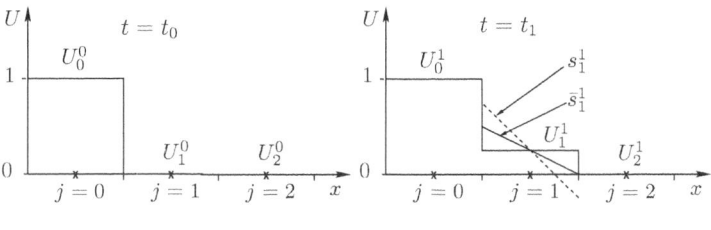

(a) Initial data (unit step function).   (b) Integration by one time step $\lambda = \frac{1}{2}$.

Figure 3.3. First GRP time-integration cycle of a moving step.

Example 3.14 Let the initial data be $U_j^0 = 1$ (resp. $U_j^0 = 0$) for $j \le 0$ (resp. $j > 0$), and let $f(u) = \frac{1}{2}u^2$, so that the solution is a shock wave moving at speed $\frac{1}{2}$ (see Example 2.12), $\tilde{u}(x, t) = u_0(x - \frac{1}{2}t)$. If we use a time step $k = \Delta t$, it is easy to see that

$$
U_j^1 = \begin{cases} 1, & j \le 0, \\ \frac{1}{2}\lambda, & j = 1, \\ 0, & j > 1. \end{cases}
$$

The corresponding computed slopes s_j^1 satisfy

$$
s_j^1 = \begin{cases} 0, & j \le 0, \\ -\frac{1}{\Delta x}, & j = 1, \\ 0, & j > 1. \end{cases}
$$

Thus, if we are to retain all slopes s_j^1, the approximating function $U^1(x)$ in the cell $j = 1$ should be

$$
U^1(x) = \frac{1}{2}\lambda - \frac{x - \Delta x}{\Delta x} = 1 + \frac{1}{2}\lambda - \frac{x}{\Delta x}, \qquad \frac{\Delta x}{2} < x < \frac{3}{2}\Delta x. \quad (3.27)
$$

Thus, $U^1(\frac{3}{2}\Delta x-) = \frac{1}{2}(\lambda - 1)$, which is negative. This is in contradiction with the "moving step" character of the exact (weak) solution.

From the mathematical point of view, the modification of the slopes $\{s_j^n\}_{j=-\infty}^{\infty}$ is needed for the control of the total variation of the approximating solution, in analogy with the total variation properties of the exact (weak) solution. We refer the reader to Godlewski and Raviart [54] for details.

The modification of the slopes used in our GRP methodology is implemented as follows.

Construction 3.15 (GRP "slope limiter") *Given the computed slopes s_j^{n+1} [as in (3.26)], set the final slope values $\bar{s}_j^{n+1}$ to be*

$$\bar{s}_j^{n+1} = \frac{1}{\Delta x} \, \text{minmod} \left[2(U_{j+1}^{n+1} - U_j^{n+1}), \, 2(U_j^{n+1} - U_{j-1}^{n+1}), \, \Delta x \, s_j^{n+1} \right],$$

$$(3.28)$$

where, for any three real numbers a, b, c,

$$\text{minmod}\,[a, b, c] = \begin{cases} \sigma \min\left(|a|, |b|, |c|\right), & \text{if } \sigma = \text{sgn}\,(a) \\ & \quad\quad = \text{sgn}\,(b) = \text{sgn}\,(c), \\ 0, & \text{otherwise.} \end{cases}$$

Geometrically speaking, our limiter reflects the minimal change (of s_j^{n+1}) needed to obtain the following "five-point monotonicity" (see Figure 3.4): If $\{U_{j-1}^{n+1}, U_j^{n+1}, U_{j+1}^{n+1}\}$ form a monotonic sequence, then so do the five values $\{U_{j-1}^{n+1}, U_j^{n+1} - \frac{\Delta x}{2} \bar{s}_j^{n+1}, U_j^{n+1}, U_j^{n+1} + \frac{\Delta x}{2} \bar{s}_j^{n+1}, U_{j+1}^{n+1}\}$. If $U_j^{n+1} \geq \max\left(U_{j\pm1}^{n+1}\right)$ or $U_j^{n+1} \leq \min\left(U_{j\pm1}^{n+1}\right)$ we set $\bar{s}_j^{n+1} = 0$. Thus, at extremal points the slopes are set to zero, whereas elsewhere it is ensured that (in the case of $U_{j-1}^{n+1} \leq U_j^{n+1} \leq U_{j+1}^{n+1}$)

$$U_j^{n+1} \geq \max(U_{j-1/2,\pm}^{n+1}), \qquad U_j^{n+1} \leq \min(U_{j+1/2,\pm}^{n+1}).$$

Remark 3.16 (Convergence of the GRP Scheme)   The convergence of the first-order Godunov scheme was stated in Theorem 3.6. At the time of writing

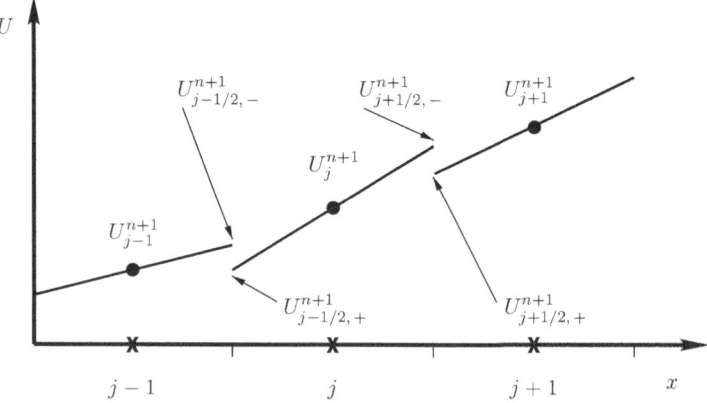

Figure 3.4. The "slope limiter" is a five-point rule.

this monograph, a similar convergence result has not yet been established for the GRP scheme. The main obstacle for a convergence proof lies in the rather weak slope limiter as given in (3.28). It allows for a nonmonotonic behavior of the set of value

$$U_{j-1}^{n+1}, U_{j-1/2,-}^{n+1}, U_{j-1/2,+}^{n+1}, U_{j}^{n+1}, U_{j+1/2,-}^{n+1}, U_{j+1/2,+}^{n+1}, U_{j+1}^{n+1} \qquad (3.29)$$

(see Figure 3.4). Replacing Equation (3.28) by the more restrictive limiter

$$\bar{s}_j^{n+1} = \frac{1}{\Delta x} \operatorname{minmod} \left[(U_{j+1}^{n+1} - U_j^{n+1}), (U_j^{n+1} - U_{j-1}^{n+1}), \Delta x \, s_j^{n+1} \right] \qquad (3.30)$$

is easily seen to lead to the monotonicity of (3.29). However, virtually all GRP computations presented in this monograph have been performed using the algorithm (3.28). Numerical experience has shown that it leads to sharper resolution of discontinuities [i.e., there is more "dissipation" built into (3.30)].

We refer to [18], [57], [79], [85], [95], [113], and [121], where the convergence properties of various upwind second-order schemes are considered. In all these works additional hypotheses are imposed on the schemes (in terms of their numerical fluxes and monotonicity algorithms). In particular, the monotonicity of (3.29) is always assumed. Note also the negative result in [120], concerning the nonconvergence of certain Godunov-type second-order schemes (at least for some initial data).

3.2 1-D Sample Problems

In this section we present numerical solutions to scalar conservation laws, linear and nonlinear, in one space dimension. The initial data considered are sufficiently simple, so that the exact solutions can be computed and compared to the finite-difference approximations. Two pairs of schemes were chosen for the sample problems, one pair of first-order schemes and one pair of second-order schemes. The idea is to demonstrate the difference between the first-order Godunov scheme (3.11) and its natural second-order extension – the GRP scheme (Construction 3.10). Then, for the sake of comparison, we use another typical scheme in each class. We selected the (first-order) Lax–Friedrichs scheme and the (second-order) Lax–Wendroff scheme.

3.2.1 The Linear Conservation Law

The equation to be considered here is

$$u_t + u_x = 0, \qquad u(x, 0) = u_0(x). \qquad (3.31)$$

According to (2.6), the exact solution is given by the "traveling wave" $u(x, t) = u_0(x - t)$.

First-Order Schemes

We shall use the following pair of first-order schemes:

(a) The *Godunov scheme*, which in this case (since $a > 0$) is identical to the backward-difference scheme (2.29), as explained in Claim 2.31.
(b) The *Lax–Friedrichs (LF) scheme*, which in this case is given by (2.31).

In all the computations of this section we take constant (fixed) space and time steps, Δx and $k = \Delta t$, respectively. Their ratio $\lambda = \frac{k}{\Delta x}$ satisfies the CFL condition (2.51), namely, $\lambda \leq 1$.

Two initial profiles $u_0(x)$ are considered, the first having smooth periodic data, and the second having step function data. These problems have been chosen for two reasons: (i) One of them has smooth data, and the other has discontinuous data (i.e., only a weak solution exists). (ii) Both problems are defined on $\mathbb{R}$, yet can be solved numerically on some finite interval $x_1 < x < x_2$, producing the same finite-difference solution that would have been obtained on an unbounded interval of x. The smooth initial data are

$$u_0(x) = \sin^4(\pi x). \tag{3.32}$$

This is a periodic function with a period $L = 1$, so that at time $t = 1$, $u_0(x)$ has propagated exactly through one period. The numerical solution is performed with periodic boundary conditions. Figure 3.5 shows the results of such a computation, using a coarse grid of $\Delta x = 1/9$ and a refined grid with $\Delta x = 1/17$. The constant ratios are $\lambda = 0.7500$ and $\lambda = 0.7391$, respectively (corresponding to

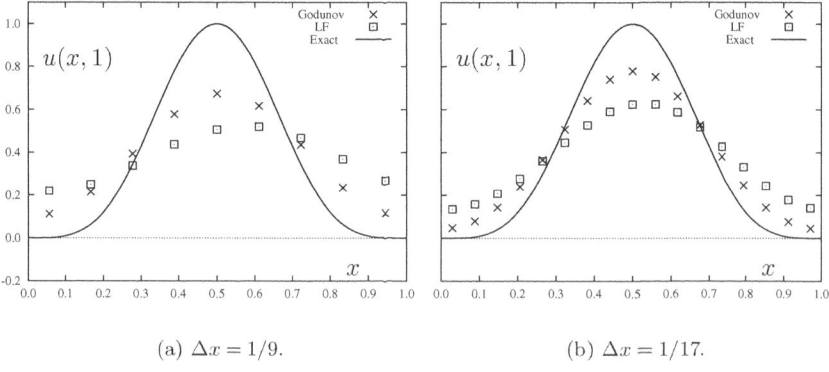

(a) $\Delta x = 1/9$.   (b) $\Delta x = 1/17$.

Figure 3.5. First-order integration of $u_t + u_x = 0$, with initial data $u_0(x) = \sin^4(\pi x)$.

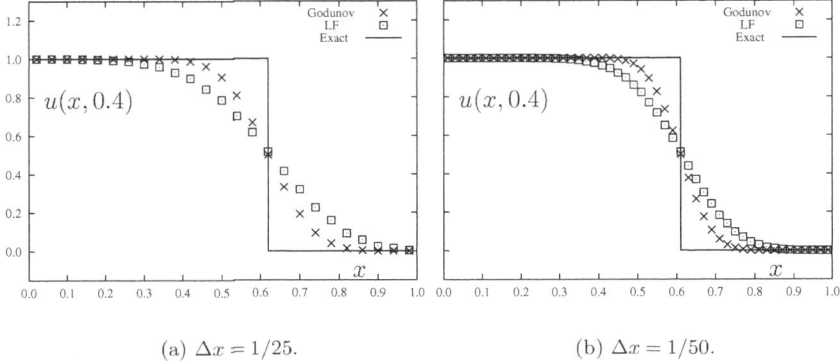

(a) $\Delta x = 1/25.$ (b) $\Delta x = 1/50.$

Figure 3.6. First-order integration of $u_t + u_x = 0$, with unit step-function initial data.

integration by 12 and 23 time steps, respectively). This example was calculated by Liu and Tadmor [87] using their second-order centered difference scheme.

As is evident from Figure 3.5, both finite-difference approximations are rather far from the exact solution, with a smaller error in the finer grid computation. Furthermore, the Godunov scheme clearly produces more accurate results than the LF scheme.[1] This can be interpreted as indicating that although both schemes are first-order accurate, the Godunov solution is less "smearing" than the LF one and therefore more accurate.

For the step function case, the function is $u_0(x) = 1$ for $x < x_0$ and $u_0(x) = 0$ for $x > x_0$ (see Example 2.4). The numerical integration is performed in the range $0 < x < 1$ until a time $t = 0.4$, with $\lambda = 0.5$. The boundary conditions for the time interval $0 < t < 0.4$ are $u(0) = 1$, $u(1) = 0$. Two grids were used, a coarse grid with $\Delta x = 0.04$ (20 time steps) and $x_0 = 0.22$ and a fine grid with $\Delta x = 0.02$ (40 time steps) and $x_0 = 0.21$. Referring to the exact solution, we note that the discontinuity is positioned at a mid-cell point at the initial time, as well as at the final time. The datum in cell $x_j = x_0$ is $U_j^0 = 0.5$, in accordance with Definition 3.5 of the Godunov scheme. Again, we observe in Figure 3.6 that the Godunov scheme produces more accurate results than the LF scheme. It is also noted that both coarse and fine grid solutions seem to approximate the moving step quite accurately in the mean; i.e., the numerical values are symmetrically distributed about the step, and, moreover, the sharp step is "spread" over about 8 cells in the first grid and over about 12 cells in the second. The width of the "step-spreading" appears to be proportional to $\sqrt{N}$, where N is the number of time integration cycles. This spreading effect is

---

[1] We refer the reader to Godlewski and Raviart [54, Chapter 3.2] for a discussion of "numerical viscosity." This viscosity is "maximal" in the case of the LF scheme.

typical of a linear conservation law. In the case of a nonlinear conservation law, a moving shock discontinuity is usually spread over a constant number of cells. Such a finite-difference approximation is referred to as a "captured shock."

We recall that for the Godunov scheme the "convergence in the mean" has been proved in Theorem 2.28. A similar result also can be proved for the LF scheme (see Godlewski and Raviart [54]).

Second-Order Schemes

Turning to second-order schemes, our primary interest is GRP, but for comparison we also consider the Lax–Wendroff (LW) scheme (2.32). It is readily verified from Equation (2.32) that if $u_0(x)$ is of compact support in $\mathbb{R}$, then $\sum_j U_j^{n+1} = \sum_j U_j^n$. This means that the LW scheme is conservative, although not upwind (see Remark 2.32).

The GRP scheme, by contrast, is both upwind and conservative. It is given by adapting Construction 3.10 to the case $f(u) = u$. Thus, the Riemann solution is simply a moving step solution, so that

$$U_{j+1/2}^n = R\left(0; U_{j+1/2,-}^n, U_{j+1/2,+}^n\right) = U_{j+1/2,-}^n. \tag{3.33}$$

It follows from Equation (3.23) that

$$\frac{\partial \tilde{u}}{\partial t}(x_{j+1/2}, t_n) = -s_j^n = -\frac{1}{\Delta x}\left(U_{j+1/2,-}^n - U_{j-1/2,+}^n\right); \tag{3.34}$$

hence, as in Equation (3.24),

$$U_{j+1/2}^{n+1/2} = U_{j+1/2}^n - \frac{k}{2}s_j^n, \tag{3.35}$$

$$f_{j+1/2}^{n+1/2} = U_{j+1/2}^{n+1/2}. \tag{3.36}$$

The resulting GRP scheme is, as in Equation (3.25),

$$U_j^{n+1} = U_j^n - \lambda(U_{j+1/2}^{n+1/2} - U_{j-1/2}^{n+1/2}). \tag{3.37}$$

Finally, the new slopes s_j^{n+1} are obtained as follows [see Eq. (3.26)]:

$$U_{j+1/2}^{n+1} = U_{j+1/2}^n - ks_j^n, \tag{3.38}$$

$$s_j^{n+1} = \frac{1}{\Delta x}\left(U_{j+1/2}^{n+1} - U_{j-1/2}^{n+1}\right). \tag{3.39}$$

The slopes s_j^{n+1} are further subjected to the monotonicity algorithm given by Construction 3.15.

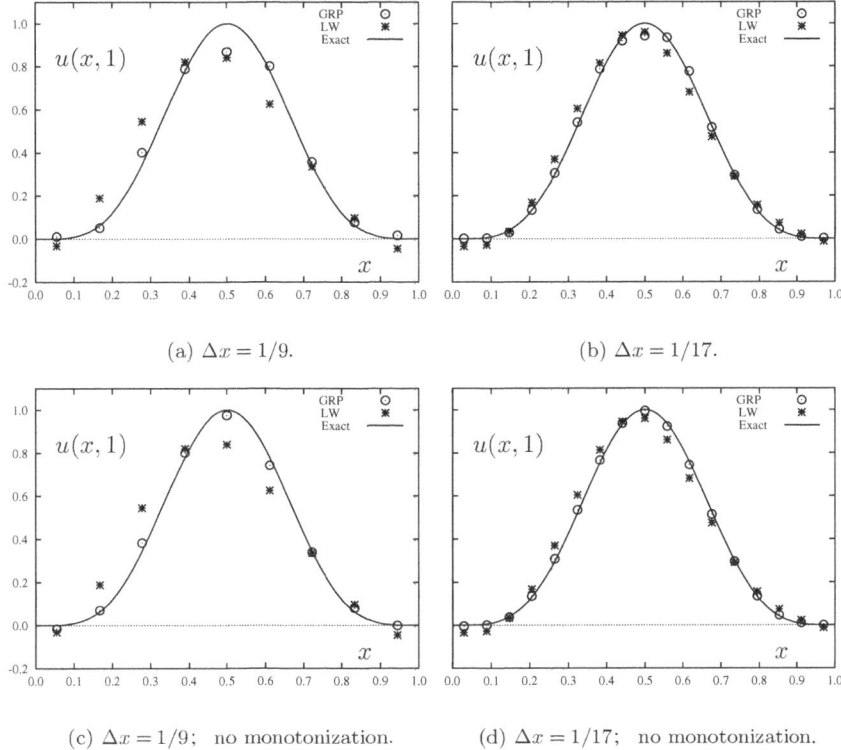

(a) $\Delta x = 1/9$.

(b) $\Delta x = 1/17$.

(c) $\Delta x = 1/9$;   no monotonization.

(d) $\Delta x = 1/17$;   no monotonization.

Figure 3.7. Second-order integration of $u_t + u_x = 0$, with initial data $u_0(x) = \sin^4(\pi x)$.

The sample problems considered here are the same two problems previously used for the first-order schemes, including the same grids and time step specifications.

The second-order results for the periodic case are given in Figure 3.7, where a comparison between GRP and LW schemes is shown. We notice a significant improvement relative to the first-order results in Figure 3.5, and it is also evident that the convergence with grid refinement is faster in the second-order case than in the first-order one. On the whole, the GRP values are closer to the exact solution than are the LW values. Furthermore, the LW results have a significant phase-shift error, whereas the GRP results do not.

How does the monotonization algorithm affect the GRP results? In Figures 3.7(a) and 3.7(b) we show the GRP results that were subject to the slope limiter given in Construction 3.15. The LW scheme, however, does not include any monotonization or slope limiting algorithm. For comparison, we therefore repeated the GRP computation without applying the monotonization

algorithm, and the results are shown in Figures 3.7(c) and 3.7(d). Clearly, the
GRP points near the peak (where slope limiting is most effective) are now
higher, indicating that indeed the limiting algorithm is required to suppress
peak-forming tendencies. We also note on Figures 3.7(c) and 3.7(d) that some
GRP and LW points have $u < 0$. These "undershoot" values are in violation of
the Maximum–Minimum Principle (Theorem 2.20). Slope limiting eliminates
such violation by a second-order scheme and is hence mandatory to comply
with the Maximum–Minimum Principle. We note that the Godunov scheme
is in agreement with that principle (as stated in Claim 3.7), and the results in
Figure 3.5 are evidence to that property.

We now turn to the step-function problem, identical to that considered in
the first-order scheme. In particular, we use the same λ, Δx, and final time.
As is clearly visible in Figures 3.8(a) and 3.8(b), the "shock-captured" solution
obtained here is similar to that of the first-order scheme already discussed

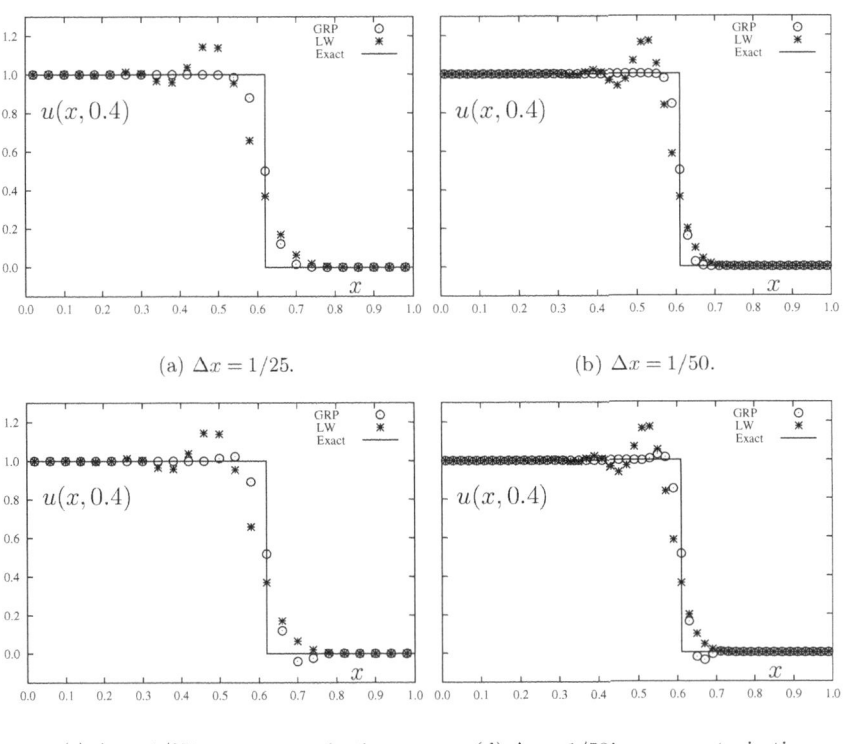

(a) $\Delta x = 1/25$.   (b) $\Delta x = 1/50$.

(c) $\Delta x = 1/25$; no monotonization.   (d) $\Delta x = 1/50$; no monotonization.

Figure 3.8. Second-order integration of $u_t + u_x = 0$, with unit step-function initial
data.

(Figure 3.6). However, the discontinuity is more sharply resolved by the second-order schemes, with the sharpest (and most accurate) results obtained by the GRP. Here the jump in U_j^n is spread over about three cells, both in the coarse-grid and in the fine-grid computations.

Again, for comparison we repeated the two cases without the GRP monotonization constraint, and the results are shown in Figures 3.8(c) and 3.8(d). The GRP now produces some "overshoot" and "undershoot" values near the step. The indispensability of monotonization has thus been amply demonstrated, and in subsequent GRP computations we shall no longer consider the nonmonotonization option.

It is also interesting to notice the nature of the LW solution. No monotonization is applied in this scheme, and indeed the numerical solution develops pronounced oscillations behind the shock, which is also a typical feature for this scheme when it is extended to the fluid dynamical equations.

3.2.2 The Burgers Nonlinear Conservation Law

Here we consider the Burgers [22] equation,

$$u_t + (\tfrac{1}{2}u^2)_x = 0 , \qquad u(x,0) = u_0(x). \tag{3.40}$$

As explained in Section 2.1, in the case of smooth initial data the solution to this equation is obtained by the invariance of $u(x,t)$ along characteristic lines [see Equation (2.10) and the subsequent discussion]. When characteristic lines intersect, a smooth solution no longer exists, and from that time on only a (weak) solution, with shocks that obey the jump condition (2.21), is possible. In the case of the Burgers equation [see Example 2.12 and Equation (2.22)] the characteristic speed is $\frac{dx}{dt} = u$, and the speed of a shock wave is given by $S = \frac{1}{2}(u_L + u_R)$, where the left and right values at the shock discontinuity u_L, u_R must obey the inequality $u_L \geq u_R$.

Two initial value problems are considered. The first has the smooth periodic data

$$u_0(x) = \sin(2\pi x), \tag{3.41}$$

and the second is a moving step problem (or a Riemann problem), having the initial data $u_0(x) = 1$ for $x \leq x_0$ and $u_0(x) = 0$ for $x > x_0$. Both problems have exact solutions, which, for the simple initial data considered here, are readily calculated by using the previously mentioned characteristics construction. Both problems are defined on $\mathbb{R}$, yet, with appropriate boundary conditions, they can be solved numerically on some bounded interval $[x_1, x_2]$ of $\mathbb{R}$, yielding the same finite-difference solution that would have been obtained on $\mathbb{R}$. We note that Liu and Tadmor [87] have considered a similar IVP, with the (periodic) initial data $u_0(x) = 1 + \frac{1}{2}\sin(\pi x)$. The reader is also referred to Yang and Przekwas [122] for a survey of many other schemes applied to the Burgers equation.

First-Order Computation

Here we use the same two first-order schemes (Godunov and Lax–Friedrichs) previously considered in the context of the linear sample problems. The Godunov scheme is given by (3.11). The Lax–Friedrichs scheme, however, for a general flux function $f(u)$ (see Smoller [103], Godlewski and Raviart [54], and LeVeque [81]) is given by

$$U_j^{n+1} = \frac{1}{2}\left(U_j^n + U_{j-1}^n\right) - \frac{\lambda}{2}\left(f(U_{j+1}^n) - f(U_{j-1}^n)\right), \tag{3.42}$$

where the Burgers equation scheme is obtained by taking $f(u) = \frac{1}{2}u^2$. In our computations (both first and second order) we take fixed values for k and Δx, so that the ratio $\lambda = \frac{k}{\Delta x}$ satisfies the CFL condition (3.6). In fact, we take k such that the left-hand side in (3.6) is approximately equal to $\frac{1}{2}\Delta x$.

In the case of smooth initial data, the equation is solved in the domain $[0,1]$ with periodic boundary conditions. The computational cell size is $\Delta x = \frac{1}{22}$. The results are displayed as a time sequence in Figure 3.9, which compares the finite-difference solutions with the exact solution obtained by the method of characteristics. Prior to shock formation, in Figure 3.9(b), the solution is smooth and displays the expected steepening in the interval where $\partial_x u(x, t) < 0$. The smooth solution breaks down at the moment $t = 1/2\pi$, where the slope at $x = 0.5$ becomes unbounded, as readily derived by taking the limit

$$\lim_{\epsilon \to 0} \frac{\epsilon}{\sin(2\pi(0.5 - \epsilon))} = \lim_{\epsilon \to 0} \frac{\epsilon}{2\pi\epsilon} = \frac{1}{2\pi},$$

which corresponds to the point where the characteristic line emanating from $(x, t) = (0.5 - \epsilon, 0)$ intersects the line $x = 0.5$. The solution at the breakdown time is shown in Figure 3.9(c). Beginning at this time the jump discontinuity at $x = 0.5$ gradually increases, reaching a maximal value (between $u = -1$ and $u = 1$) at $t = 0.25$ [see Figure 3.9(d)]. This is the moment at which the characteristic lines emanating from the extremal points $(x, t) = (0.50 \pm 0.25, 0)$ reach the discontinuity point $(x, t) = (0.50, 0.25)$. We also observe that by the jump condition for the Burgers equation the speed of propagation of a shock discontinuity $[u_L, u_R]$ is $S = 0.5(u_L + u_R)$, which vanishes owing to the symmetry of (u, t) about $x = 0.5$. The shock discontinuity at $x = 0.5$ is thus a standing shock.

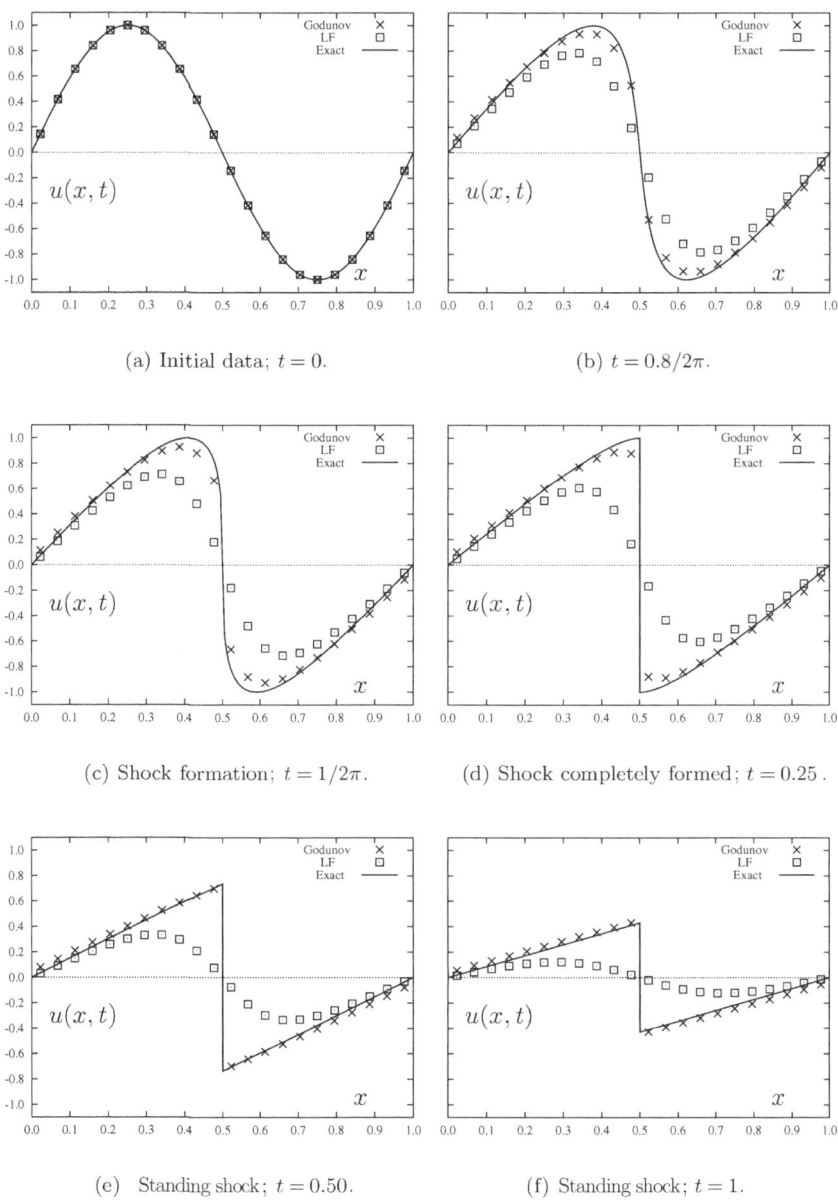

(a) Initial data; $t = 0$.

(b) $t = 0.8/2\pi$.

(c) Shock formation; $t = 1/2\pi$.

(d) Shock completely formed; $t = 0.25$.

(e) Standing shock; $t = 0.50$.

(f) Standing shock; $t = 1$.

Figure 3.9. First-order integration of $u_t + (\frac{1}{2}u^2)_x = 0$, with initial data $u_0(x) = \sin(2\pi x)$.

At later times ($t > 0.25$), the jump $[u]$ at the shock discontinuity decreases progressively from its maximal value of $[u] = u_L - u_R = 2$, as clearly visible in Figures 3.9(d)–3.9(f). Can this observation be explained by theoretical consideration? Indeed, it can be explained using the concepts of energy and dissipation. Let the "energy measure" of $u(x, t)$ be the finite integral $E(t) = \int_a^{a+1} \frac{1}{2} [u(x, t)]^2 \, dx$, where $0 \leq a \leq 1$ is a constant. The periodicity of $u(x, t)$ in x implies that $E(t)$ is independent of a. Now, multiply the Burgers equation by $u(x, t)$, obtaining $(\frac{1}{2}u^2)_t + (\frac{1}{3}u^3)_x = 0$, and integrate the resulting equation over an interval $[a, a + 1]$. Interchanging the order of x integration and time derivative, we obtain $\partial_t E(t) = 0$ [since $u(x, t)$ is periodic in x]. This means that the Burgers equation preserves the energy measure over time, which seems to agree with the finite-difference solution *prior* to the shock formation. After shock formation, however, this result is clearly in disagreement with both the exact and the numerical solutions. Indeed, this equation may not be integrated over an interval containing a shock discontinuity, while disregarding the jump in $u(x, t)$ there. A correct way to perform the integration when a shock is present is to choose $a = -0.5$, so that the integration will extend from the right side of the shock at $x = -0.5$ to the left side of the shock at $x = 0.5$. The resulting rate of dissipation is then given by

$$\partial_t E(t) = -\frac{2}{3} \left[u_L^3 - u_R^3 \right], \tag{3.43}$$

and since $u_L = -u_R > 0$ the energy $E(t)$ is decreasing in time, as is also evident by observing the evolution of the solution from $t = 0.25$ [Figure 3.9(d)] to $t = 1$ [Figure 3.9(f)]. This clearly demonstrates the dissipation effect of a shock wave in the solution to the Burgers equation.[2]

We now turn to the step-function example, the initial data being $u_0(x) = 1$ for $x < x_0$ and $u_0(x) = 0$ for $x > x_0$ (see Example 2.4), where x_0 is the initial position of the discontinuity. The numerical integration is performed in the range $0 < x < 1$, with boundary conditions $u(0) = 1$, $u(1) = 0$. Two grids were used: (i) a coarse grid with $\Delta x = 0.04$ and an initial (mid-cell) position $x_0 = 0.22$ and (ii) a fine grid with $\Delta x = 0.02$ and an initial (mid-cell) position $x_0 = 0.11$. In both grids we chose a constant $\lambda = k/\Delta x$, having the value $\lambda = 0.5$, and the integration was performed to time $t = 0.8$, so that the step propagates through the same distance as in the linear case.

As is clearly observed in Figure 3.10, the Godunov scheme approximates the moving step to a considerably higher level of accuracy and resolution than

---

[2] See Smoller [103] for a more general discussion of the L^2 decay of solutions to the scalar conservation law (2.1).

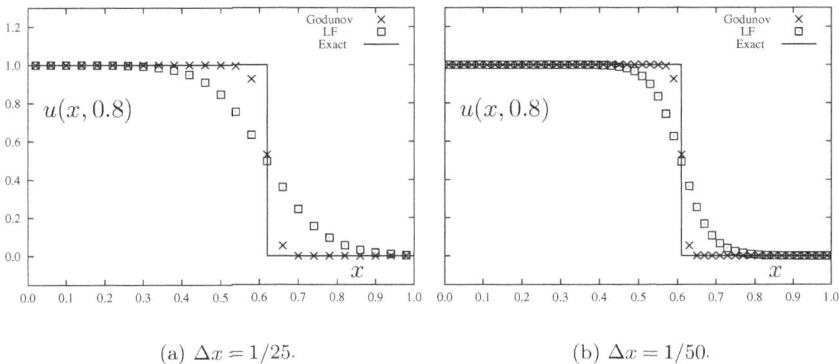

(a) $\Delta x = 1/25$.   (b) $\Delta x = 1/50$.

Figure 3.10. First-order integration of $u_t + (\frac{1}{2}u^2)_x = 0$, with unit step-function initial data.

the LF scheme. In particular, the Godunov scheme captures the shock over about three cells, versus nine cells in the LF scheme. It is also noted that these cell numbers are virtually unchanged by grid refinement (for constant time). Comparing this feature to the linear case (Figure 3.6), where the "discontinuity spreading" increases with grid refinement, we interpret this as a "stabilization" effect typical of shock capturing in the nonlinear case. As will be shown in subsequent chapters, an analogous relation exists in the case of the fluid dynamical equations between a captured contact discontinuity ("linearly degenerate" wave) and a captured shock wave.

Second-Order Computation

Our primary interest here is the GRP scheme given by Construction 3.10 while taking $f(u) = \frac{1}{2}u^2$. The slopes s_j^{n+1} are further subjected to the monotonicity algorithm given by Construction 3.15. For comparison we take the Lax–Wendroff scheme, which in the case of Equation (2.1) generalizes (2.32) as

$$U_j^{n+1} = U_j^n - \frac{\lambda}{2}\left[f_{j+1}^n - f_{j-1}^n\right] + \frac{\lambda^2}{2}\left[g_{j+1/2}^n - g_{j-1/2}^n\right], \tag{3.44}$$

where $\lambda = \frac{k}{\Delta x}$. The first-order term (in λ) in (3.44) approximates $k\,(U_t)_j^n$ with

$$f_{j\pm 1}^n = f(U_{j\pm 1}^n),$$

and the second-order term approximates $\frac{k^2}{2}(U_{tt})_j^n$. The second time-derivative is based on the identity $u_{tt} = g(u)_x$, $g(u) = f'(u)\,f(u)_x$, obtained by differentiating the scalar conservation law $u_t + f(u)_x = 0$. Its finite-difference approximation is then given by

$$(U_{tt})_j^n = \left[g_{j+1/2}^n - g_{j-1/2}^n \right] / \Delta x,$$

$$g_{j\pm1/2}^n = \pm f' \left(\frac{1}{2}(U_j^n + U_{j\pm1}^n) \right) \left[f_{j\pm1}^n - f_j^n \right]. \tag{3.45}$$

For the reader's convenience we recall the GRP algorithm (Construction 3.10) in the special case $f(u) = \frac{1}{2}u^2$.

(I) Given the values

$$U_{j+1/2,\pm}^n = \begin{cases} U_{j+1}^n - \frac{\Delta x}{2} s_{j+1}^n, & \text{``}+\text{''}, \\ U_j^n + \frac{\Delta x}{2} s_j^n, & \text{``}-\text{''}, \end{cases} \tag{3.46}$$

we let $U_{j+1/2}^n$ be the solution to the corresponding Riemann problem. Explicitly, it is given by

$$U_{j+1/2}^n = \begin{cases} \max\left(|U_{j+1/2,-}^n|, |U_{j+1/2,+}^n| \right) \cdot \operatorname{sgn}\left(U_{j+1/2,-}^n + U_{j+1/2,+}^n \right), \\ \qquad \text{if} \quad U_{j+1/2,-}^n > U_{j+1/2,+}^n, \\ \min\left(|U_{j+1/2,-}^n|, |U_{j+1/2,+}^n| \right) \cdot \operatorname{sgn}\left(U_{j+1/2,-}^n \right), \\ \qquad \text{if} \quad \operatorname{sgn}\left(U_{j+1/2,-}^n \right) \cdot \operatorname{sgn}\left(U_{j+1/2,+}^n \right) \geq 0, \\ 0, \qquad \text{if} \quad U_{j+1/2,-}^n < 0 < U_{j+1/2,+}^n, \end{cases} \tag{3.47}$$

and of course $U_{j+1/2}^n = U_{j+1/2,-}^n$ if $U_{j+1/2,-}^n = U_{j+1/2,+}^n$.

(II) The instantaneous time derivatives $\left(\frac{\partial \tilde{u}}{\partial t} \right)_{j+1/2}^n = \frac{\partial \tilde{u}}{\partial t}(x_{j+1/2}, t_n)$ are now given by

$$\left(\frac{\partial \tilde{u}}{\partial t} \right)_{j+1/2}^n = \begin{cases} -U_{j+1/2}^n \cdot s_j^n, & \text{if} \quad U_{j+1/2}^n > 0, \\ -U_{j+1/2}^n \cdot s_{j+1}^n, & \text{if} \quad U_{j+1/2}^n < 0, \\ 0, & \text{if} \quad U_{j+1/2}^n = 0, \end{cases} \tag{3.48}$$

and the numerical fluxes are given by

$$f_{j+1/2}^{n+1/2} = f\left(U_{j+1/2}^n \right) + \frac{k}{2} U_{j+1/2}^n \cdot \left(\frac{\partial \tilde{u}}{\partial t} \right)_{j+1/2}^n,$$

so that

$$U_j^{n+1} = U_j^n - \lambda \left(f_{j+1/2}^{n+1/2} - f_{j-1/2}^{n+1/2} \right).$$

(III) The new values at the cell boundaries are given by

$$U_{j+1/2}^{n+1} = U_{j+1/2}^n + k \left(\frac{\partial \tilde{u}}{\partial t} \right)_{j+1/2}^n,$$

and the new slopes are updated by

$$s_j^{n+1} = \frac{1}{\Delta x} \left(U_{j+1/2}^{n+1} - U_{j-1/2}^{n+1} \right).$$

(IV) Finally, the computed slopes s_j^{n+1} are modified by the slope limiter (3.28).

The same two sample problems are considered here as for the first-order schemes already discussed, with identical data, boundary conditions, grids, and final times.

The time sequence results for the periodic case are shown in Figure 3.11. How do the second-order results compare to the first-order ones (Figure 3.9)? The GRP approximation is generally very close to the exact solution, which is a considerable improvement relative to the Godunov approximation. The LW scheme produces fairly accurate results, although less accurate than those of the GRP. As in the smooth linear case (Figure 3.7), the LW scheme is characterized by overshoots near extremal points, notably in Figures 3.11(c) and 3.11(d). It is particularly interesting to compare the GRP formation of the "N-wave" [Figures 3.11(d)–3.11(f)] with the corresponding Godunov results [Figures 3.9(d)–3.9(f)]. The GRP points are considerably closer to the exact solution than those of the Godunov scheme, notably near the shock. For still higher order schemes (not considered in this monograph), we refer to Liu and Tadmor [87, Section 5], where a similar periodic case (for the Burgers equation) was computed by a third-order, nonoscillatory, central scheme. They used 80 mesh points (per period), whereas only 22 were used in the present GRP calculation.

Turning to the step-function example, having the same data as in the first-order case, we show the results in Figure 3.12 at $t = 0.8$. As in the linear case (Figure 3.8), the LW scheme produces significant oscillations behind the step, indicating that this feature of the scheme is not suppressed by the nonlinearity of the scalar conservation law. By comparing the GRP and the Godunov approximations (Figure 3.12), it is evident that the improvement in accuracy takes place only near the shock discontinuity. The GRP results are a near-perfect approximation to the step, with only (the inevitable) single point representing the average value of 0.5 at the mid-cell $x = 0.62$ or $x = 0.61$, where the exact jump is positioned. This demonstrates the high-resolution feature of GRP, which also characterizes the fluid dynamical GRP scheme.

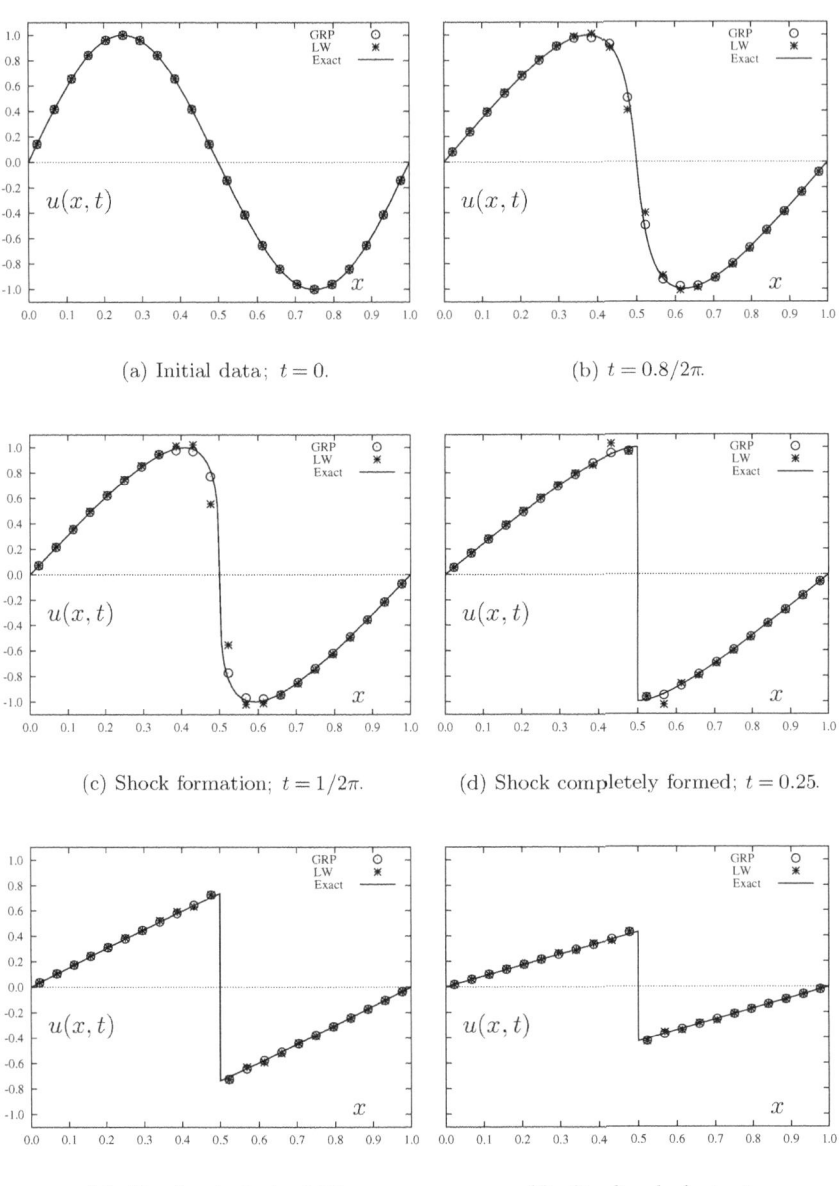

Figure 3.11. Second-order integration of $u_t + (\frac{1}{2}u^2)_x = 0$, with initial data $u_0(x) = \sin(2\pi x)$.

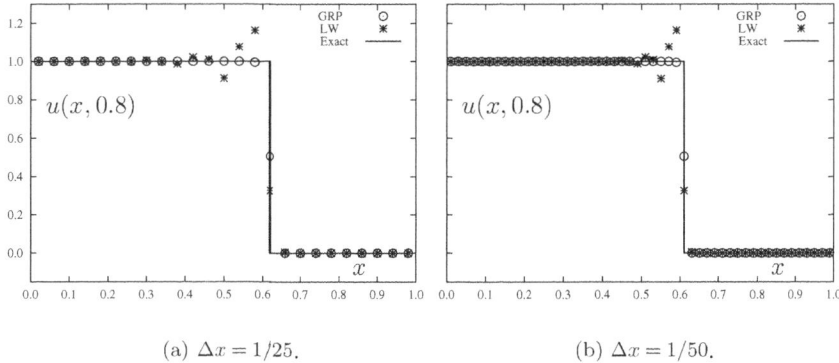

(a) $\Delta x = 1/25$. (b) $\Delta x = 1/50$.

Figure 3.12. Second-order integration of $u_t + (\frac{1}{2}u^2)_x = 0$, with unit step-function initial data.

3.3 2-D Sample Problems

Consider the two-dimensional extension of the (1-D) scalar conservation law (2.1), leading to the following Cauchy problem for $u = u(x, y, t)$:

$$\frac{\partial}{\partial t}u + \frac{\partial}{\partial x}f(u) + \frac{\partial}{\partial y}g(u) = 0, \qquad (x, y) \in \mathbb{R}^2, t > 0, \qquad (3.49)$$

$$u(x, y, 0) = u_0(x, y), \qquad (x, y) \in \mathbb{R}^2. \qquad (3.50)$$

The theory for this IVP is as complete as for the 1-D case (see Theorem 2.20). Briefly, if $f(u)$ and $g(u)$ are continuously differentiable and u_0 is uniformly bounded then there exists a unique "entropy" solution for all $t > 0$. This solution satisfies the Maximum–Minimum Principle and also the properties of L^1_{loc} continuity and L^1 contraction, as expressed in Theorem 2.20 (ii) and (iii) (with an interval of integration replaced by a rectangle). We refer the reader to Hörmander [63] and to Godlewski and Raviart [54, Section 2.5] for full details of the proof (as well as a definition of "entropy" in the multidimensional scalar case).

The important point in our examples is the "self-similarity" of solutions, which is a consequence of uniqueness; if the initial function u_0 is constant along rays emanating from the origin [i.e., depends only on the direction $\arctan(\frac{y}{x})$], then the solution $u(x, y, t)$ is "self-similar" in the sense that it depends only on $(\frac{x}{t}, \frac{y}{t})$. In the 1-D case this corresponds to the Riemann problem (see Definition 2.18 and subsequent discussion). However, in the 2-D case this restriction still allows a great variety of initial value problems; taking u_0 to be constant in (finitely many) sectors in the (x, y) plane, with vertices at $(0, 0)$, always leads to a self-similar solution.

Studying Equation (3.49) is of interest because it serves as a model for the 2-D fluid dynamical case. In particular, the 2-D setting allows for a variety of wave interactions, as we shall see in this section. A finite-difference approximation of the 2-D equation (3.49) is obtained by using a sequence of 1-D conservation law schemes, integrating the equation alternately in the x or y directions. This approach is commonly referred to as the "operator-splitting" method. It enables us to apply the methodology introduced in Section 3.1 to the 2-D case. In Part II we shall discuss its extension to multidimensional flow problems. Here we demonstrate that this approach correctly produces 2-D solutions to (3.49), (3.50), by applying a sequence of (the 1-D) Godunov or GRP schemes to a "split" form of (3.49).

We start this section with an outline of the operator-splitting method and then proceed with a detailed study of three sample equations, obtained by specifying the flux functions and a variety of initial data in (3.49), (3.50). The first equation is the linear conservation law $f(u) = g(u) = u$, with an "oblique step" initial data $u_0(x, y) = h(x + y + 1)$, where $h(w)$ is the Heaviside function, defined as

$$h(w) = \begin{cases} 1 & \text{for} \quad w < 0, \\ 0 & \text{for} \quad w > 0. \end{cases}$$

In this case, the IVP [(3.49), (3.50)] becomes "one dimensional" when transformed by rotating (x, y) through an angle of $\pi/4$, enabling a comparison with the corresponding 1-D case.

The second equation is the nonlinear, two-dimensional, Burgers equation, $f(u) = g(u) = u^2/2$, where we first take the same initial data as in the previous linear case. As before, this leads to a situation where the one-dimensional equation is "rotated" through an angle of $\pi/4$ and we can compare the results with those of the corresponding 1-D case. Next we consider four different cases with initial data that are constant in sectors of the (x, y) plane. None of these sectors, however, is a half-plane. Thus, although the solutions are "self-similar" [depending on $(\frac{x}{t}, \frac{y}{t})$] they are certainly not "one dimensional" as in the rotated case. The initial discontinuities (lying along straight rays emanating from the origin) give rise to waves that interact with each other, leading to complex wave structures. These structures are then studied, both analytically and numerically.

The third equation, along with its special (sectorial) initial data, is due to Guckenheimer [59], with fluxes given by $f(u) = u^2/2$, $g(u) = u^3/3$. The self-similar solution (which can be obtained analytically) is surprisingly rich in wave interactions. These wave interactions are analogous to phenomena encountered in compressible fluid dynamics (Mach stems, triple points, formation of contact discontinuities, etc.). This problem, for which full analytic details are available, constitutes a significant and valuable test case in the study of multidimensional numerical schemes.

The Operator-Splitting Method

Consider the following system obtained by "splitting" (3.49) into the two one-dimensional equations:

$$\frac{\partial}{\partial t} u + \frac{\partial}{\partial x} f(u) = 0, \tag{3.51i}$$

$$\frac{\partial}{\partial t} u + \frac{\partial}{\partial y} g(u) = 0. \tag{3.51ii}$$

Loosely speaking, the system (3.51) is taken to mean that the evolution of an initial state u_0 by (3.49) over a short time interval Δt can be approximated by evolving u_0 first subject to (3.51i) (over time Δt), obtaining a state u_1, then evolving u_1 in accordance with (3.51ii) again over time Δt. Let $H^x_{\Delta t}$, $H^y_{\Delta t}$, and $H_{\Delta t}$ denote finite-difference approximation operators for the integration by a time step Δt of (3.51i), (3.51ii), and (3.49), respectively (compare Definition 2.21). Then, as shown by Strang [105], the operator sequence

$$H_{\Delta t} = H^x_{\Delta t/2} \, H^y_{\Delta t} \, H^x_{\Delta t/2} \tag{3.52}$$

is a second-order-accurate finite-difference approximation to (3.49), provided this is true for the 1-D operators $H^x_{\Delta t}$, $H^y_{\Delta t}$. We remark, however, that Strang's analytic arguments are valid only for a smooth solution $u(x, y, t)$. Thus, when $u(x, y, t)$ contains discontinuities, the accuracy of "shock-capturing" finite-difference solutions reflects the applicability and accuracy of the operator-splitting method. We refer to Section 7.2 for a more detailed discussion of this topic.

Two additional topics to consider here are the issue of stability and the numerical boundary conditions. These are given in the following remarks.

Remark 3.17 (The CFL condition) This condition [see Definition 2.27 and Equation (3.6)] is necessary for numerical stability, and we adapt it to the present (two-dimensional) case by requiring that it holds separately with respect to each of the split one-dimensional equations (3.51) and to the corresponding finite-difference scheme. Thus, the 2-D CFL condition is

$$\mu^{x,n}_{\text{CFL}} = \max_{i,j} \left\{ f'(U^n_{i,j}) \right\} \Delta t / \Delta x < 1, \tag{3.53}$$

$$\mu^{y,n}_{\text{CFL}} = \max_{i,j} \left\{ g'(U^n_{i,j}) \right\} \Delta t / \Delta y < 1, \tag{3.54}$$

$$\mu_{\text{CFL}} = \max_n \left\{ \mu^{x,n}_{\text{CFL}}, \mu^{y,n}_{\text{CFL}} \right\}, \tag{3.55}$$

where (i, j) denotes a cell of dimensions $\Delta x \times \Delta y$ in the (x, y) plane, centered at $(i \Delta x, j \Delta y)$. Then $U_{i,j}^n$ is the discrete value approximating $u(x, y, t)$ in that cell at time level $t = t_n = n \Delta t$. In practice, the CFL ratio μ_{CFL} is computed according to (3.53), (3.54) at each time level and then the maximum value over all time levels is retained. In simple examples such as those considered here, however, the value of μ_{CFL} is readily determined a priori, owing to the maximum-minimum principle.

Remark 3.18 (Treatment of the numerical boundary conditions) The Cauchy problem (3.49), (3.50) is solved in the entire (x, y) plane. Our finite-difference approximation, by contrast, is naturally computed in a finite domain, so that suitable boundary conditions have to be imposed on its boundaries. In the simple examples considered here, where the exact solution is known, the exact solution is prescribed on the domain boundary. However, even then the finite-difference integration in boundary cells is not perfectly accurate, so that we exclude the boundary zones from the presented results by showing them in a smaller subdomain.

The Linear Conservation Law

Here we consider (3.49), (3.50), with $f(u) = g(u) = u$ and initial data $u_0(x, y) = h(x + y + 1)$. Let us now perform a transformation to Cartesian coordinates (ξ, η) obtained by a $\pi/4$ rotation of (x, y):

$$\begin{aligned} \xi &= x/\sqrt{2} + y/\sqrt{2}, \\ \eta &= -x/\sqrt{2} + y/\sqrt{2}. \end{aligned} \tag{3.56}$$

The transformed equation [with $v(\xi, \eta, t) = u(x, y, t)$] is

$$\frac{\partial}{\partial t} v + \sqrt{2} \frac{\partial}{\partial \xi} v = 0, \tag{3.57}$$

$$v(\xi, \eta, 0) = v_0(\xi, \eta) = h(\sqrt{2}\, \xi + 1). \tag{3.58}$$

Clearly, (3.57), (3.58) is a Cauchy problem in one space dimension, with a moving step solution (with speed of propagation $\sqrt{2}$). For the numerical computation we use the domain $[-2 \leq x \leq 2, -2 \leq y \leq 2]$, which is divided into a grid of 52×52 square cells. A constant time step $\Delta t = 0.025$ is used, corresponding to $\mu_{\text{CFL}} = 0.325$. The integration is performed to final time $t = 1$; i.e., the step moves through a distance $\Delta \xi = \sqrt{2}$. The initial conditions were $U_{i,j}^0 = 0$ in cells (i, j) with center point $\xi_{i,j} > -\sqrt{2}/2$ and $U_{i,j}^0 = 1$ when $\xi_{i,j} < -\sqrt{2}/2$. In the cells having $\xi_{i,j} = -\sqrt{2}/2$ we set $U_{i,j}^0 = \frac{1}{2}$, since they

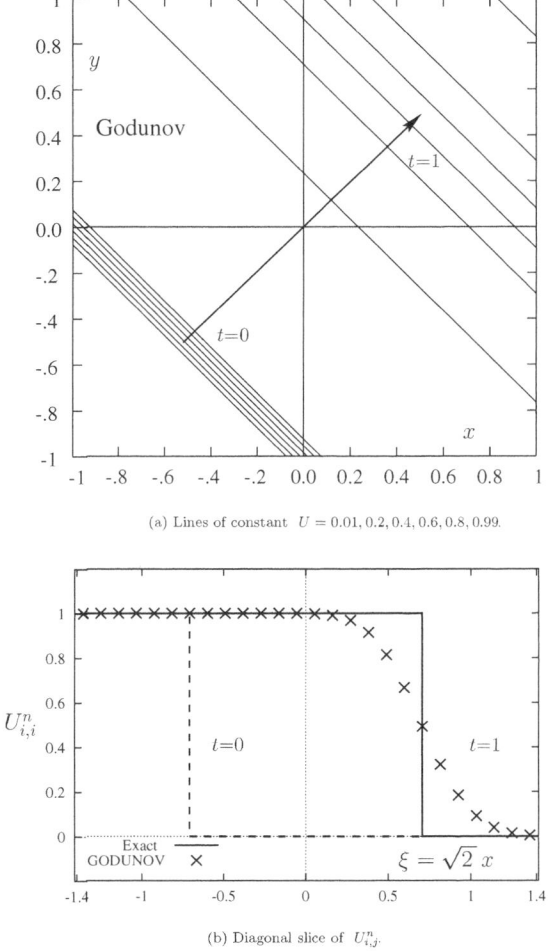

(a) Lines of constant $U = 0.01, 0.2, 0.4, 0.6, 0.8, 0.99$.

(b) Diagonal slice of $U_{i,j}^n$.

Figure 3.13. First-order integration (Godunov) of $u_t + u_x + u_y = 0$. Initial data $u_0(x, y) = h(x + y + 1)$. $\Delta x = \Delta y = \frac{1}{13}$; $\Delta t = 0.025$.

are (diagonally) bisected by the discontinuity. Two schemes are used, the (first-order) Godunov scheme and the (second-order) GRP scheme. Both schemes are applied according to the split sequence (3.52). The numerical results at time level $t = t_n$ are denoted by $U_{i,j}^n$.

The results obtained by the Godunov scheme are presented in two graphical formats (see Figure 3.13). The first is a U-level map, obtained by linear interpolation between cell-centered values (the cell-averages $U_{i,j}^n$). These lines are shown both at the initial time $t = 0$ and at the final time $t = 1$ in Figure 3.13(a). They are shown in a half-size subdomain, thereby avoiding the effects of

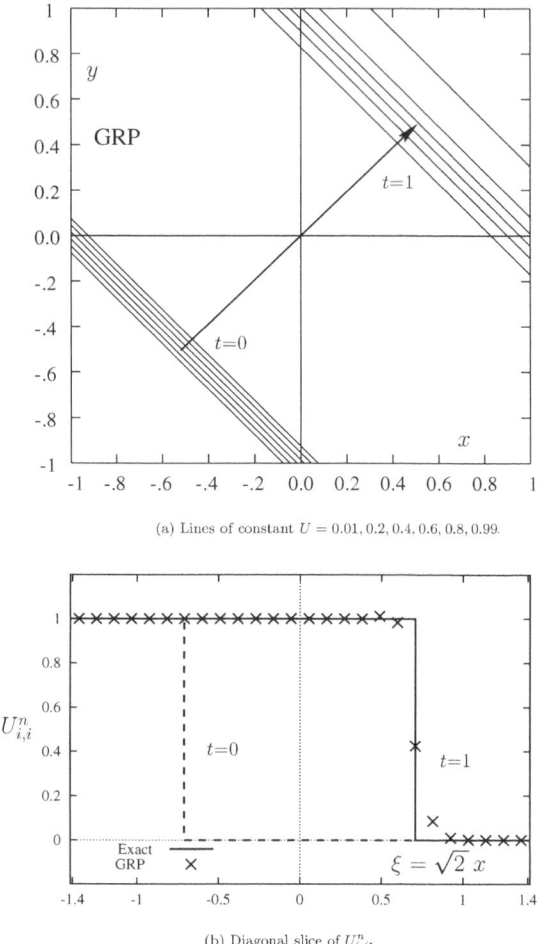

(a) Lines of constant $U = 0.01, 0.2, 0.4, 0.6, 0.8, 0.99$.

(b) Diagonal slice of $U_{i,j}^n$.

Figure 3.14. Second-order integration (GRP) of $u_t + u_x + u_y = 0$. Initial data $u_0(x, y) = h(x + y + 1)$. $\Delta x = \Delta y = \frac{1}{13}$; $\Delta t = 0.025$.

numerical boundary conditions (see Remark 3.18). The second presentation is a one-dimensional distribution along the diagonal $x = y$, where we show the distributions at $t = 0$ and at $t = 1$ [Figure 3.13(b)]. The results obtained by the GRP scheme are presented in an identical format (see Figure 3.14).

In the first-order solution (Figure 3.13) the discontinuity seems to propagate at the correct speed, but it is captured at a rather low resolution. The "diagonal slice" distribution in Figure 3.13(b) shows that the discontinuity is spread over about nine cells. The second-order solution (GRP) (see Figure 3.14) shows a definite improvement in resolution. Referring to the distribution in Figure 3.14(b), we see that the discontinuity is spread over three cells. This is a significantly enhanced resolution relative to the first-order scheme.

How do these results compare with the 1-D moving step in Section 3.2? Consider the case with $\Delta x = 1/25$ in Figure 3.6(a) and Figure 3.8(a). In those the step moves through a distance of 10 computational cells, while in the present 2-D case the step traverses 13 cells along the diagonal. In the corresponding 1-D and 2-D cases, the discontinuity is spread over the same number of cells: 9 in the Godunov solution and 3 in the GRP solution. This is a significant observation, as it indicates that (at least in this simple case where the solution is one dimensional in the ξ coordinate) the quality of "discontinuity capturing" is not degraded by the 2-D operator splitting.

The Nonlinear Burgers Equation

First we consider (3.49), (3.50), with the flux functions $f(u) = g(u) = u^2/2$ and the same initial data as in the previous case, $u_0(x, y) = h(x + y + 1)$. Using the transformation (3.56), we obtain [with $v(\xi, \eta, t) = u(x, y, t)$]

$$\frac{\partial}{\partial t} v + \sqrt{2} \frac{\partial}{\partial \xi} (v^2/2) = 0, \tag{3.59}$$

$$v(\xi, \eta, 0) = v_0(\xi, \eta) = h(\sqrt{2}\xi + 1), \tag{3.60}$$

so the speed of propagation of the oblique shock in this case is $\frac{d\xi}{dt} = \sqrt{2}/2$, i.e., half the speed of the moving step in the linear case. The final time here is taken as $t = 2$, twice that of the previous case. Thus the shock propagates through the same distance of 13 computational cells. In all other aspects the present case is computed in the same way as the linear one. In particular, the same spatial domain (and subdomain) is retained, with the same mesh size $\Delta x = \Delta y = \frac{1}{13}$, time step $\Delta t = 0.025$, and $\mu_{\mathrm{CFL}} = 0.325$.

The results are presented in Figure 3.15 for the Godunov scheme and in Figure 3.16 for the GRP scheme. The most conspicuous feature is that the shock here is spread over about 1 cell, as opposed to 9 or 3 cells in the linear case. Comparing this observation to the corresponding 1-D problems in Figure 3.10 and Figure 3.12, we find that the shock is resolved over the same thickness in both the 1-D and the 2-D cases. As in the previous linear problem, this finding confirms that our operator splitting preserves the same level of resolution as that of the 1-D scheme used in the split sequence (3.52). Also, as already noted in the 1-D case, here too the nonlinear shock discontinuity is more sharply captured than the linear discontinuity.

Next, using the same flux functions $f(u) = g(u) = u^2/2$, we study four different cases, in all of which the initial data are constant in sectors of the (x, y) plane. Thus, $u(x, y, 0) = u_0(\theta)$, where $\theta = \arctan(\frac{y}{x})$ is the angle between the

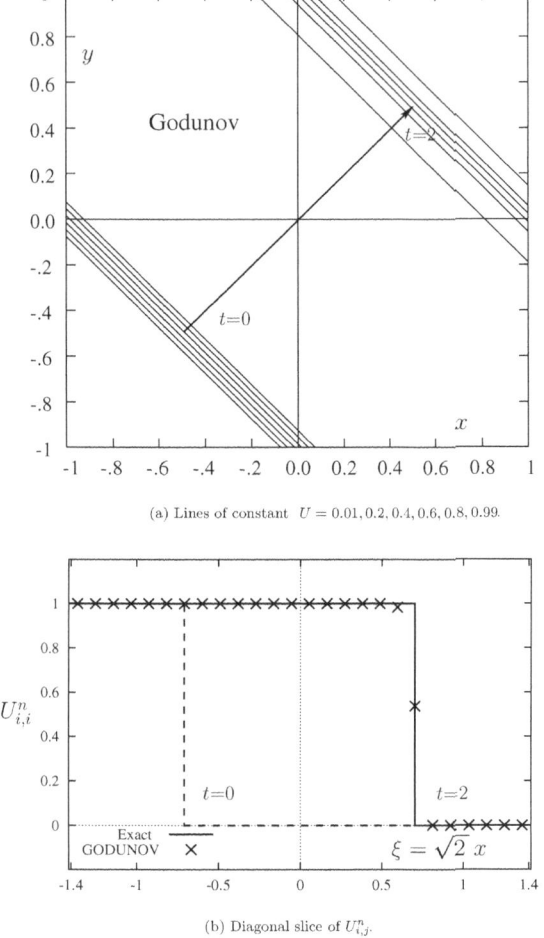

(a) Lines of constant $U = 0.01, 0.2, 0.4, 0.6, 0.8, 0.99.$

(b) Diagonal slice of $U_{i,j}^{n}$.

Figure 3.15. First-order integration (Godunov) of $u_t + (u^2/2)_x + (u^2/2)_y = 0.$ Initial data $u_0(x, y) = h(x + y + 1).$ $\Delta x = \Delta y = \frac{1}{13};$ $\Delta t = 0.025.$

vector (x, y) and the positive x-axis. The exact (self-similar) solutions of these cases (at time $t = 1$) are shown schematically in Figure 3.17.

Case A

As a first case, we take the initial data

$$u_0(\theta) = \begin{cases} 1, & 0 < \theta < \frac{\pi}{2}, \\ -1, & \text{otherwise,} \end{cases} \tag{3.61}$$

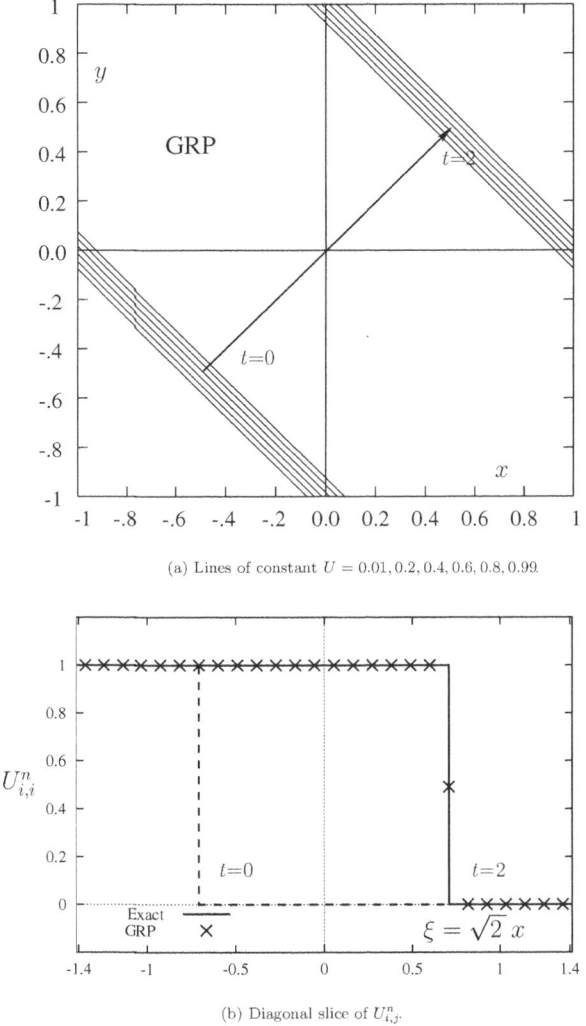

(a) Lines of constant $U = 0.01, 0.2, 0.4, 0.6, 0.8, 0.99$.

(b) Diagonal slice of $U_{i,j}^n$.

Figure 3.16. Second-order integration (GRP) of $u_t + (u^2/2)_x + (u^2/2)_y = 0$. Initial data $u_0(x, y) = h(x + y + 1)$. $\Delta x = \Delta y = \frac{1}{13}$; $\Delta t = 0.025$.

which produces the two-dimensional rarefaction wave pattern shown in Figure 3.17(a). The initial data (3.61) clearly imply that outside of a large disk the solution consists of two rarefaction waves, emanating from $\{x > 0, \, y = 0\}$ and $\{x = 0, \, y > 0\}$. It can be shown (Li, Yang, and Zhang [83], Ben-Artzi, Falcovitz, and Li [14]) that these waves do not interact inside the disk. The

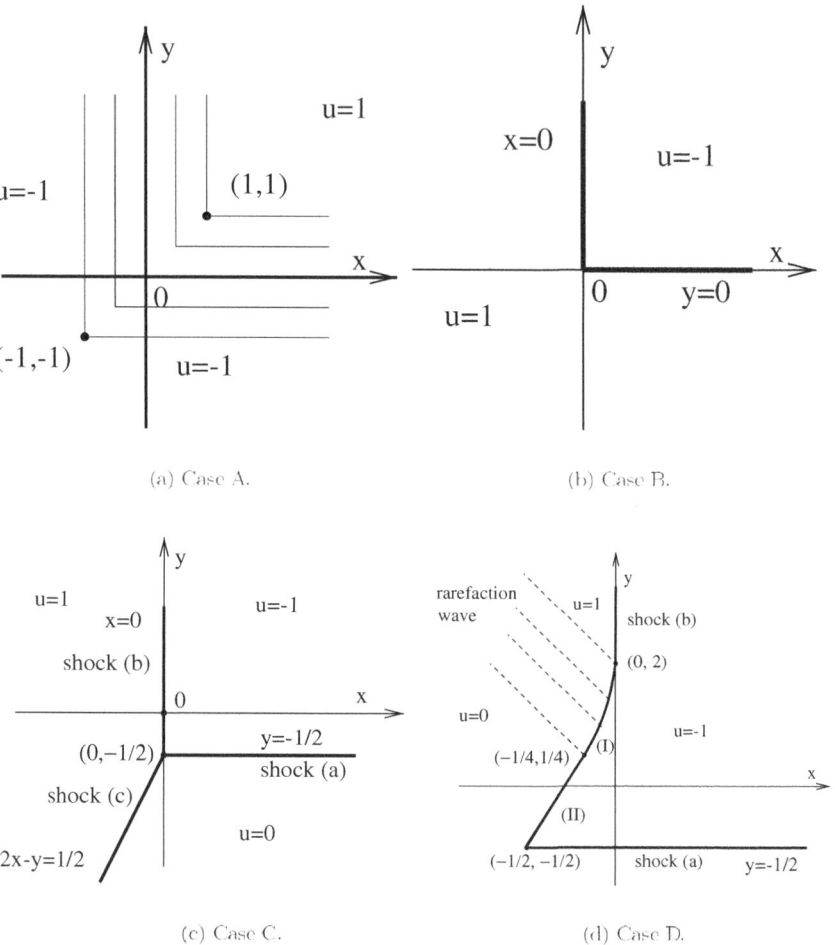

Figure 3.17. Exact solutions at $t = 1$ of 2-D Burgers equation with sectorial initial data.

entropy solution at $t = 1$ is therefore given by

$$u(x, y, 1) = \begin{cases} 1, & x > 1 \text{ and } y > 1, \\ -1, & x < -1 \text{ or } y < -1, \\ y, & x > y \text{ and } -1 < y < 1, \\ x, & y > x \text{ and } -1 < x < 1. \end{cases} \tag{3.62}$$

The finite-difference computation was performed in the (x, y) domain $[-3, 3] \times [-3, 3]$, which was divided into a grid of 600×600 (square) cells,

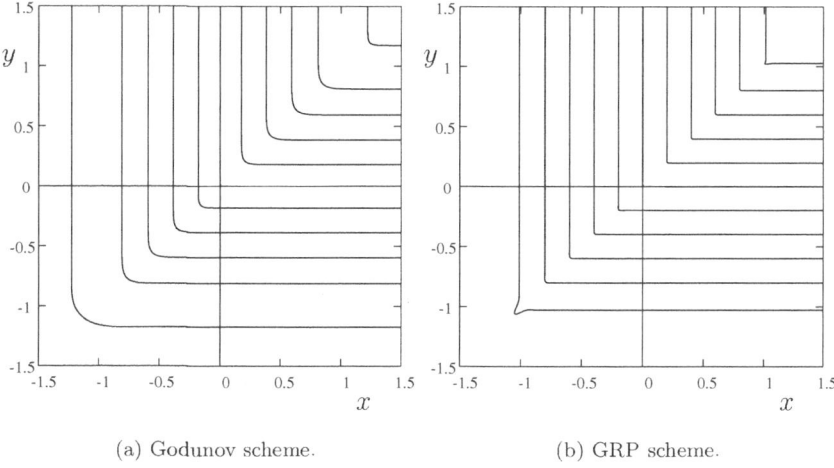

(a) Godunov scheme. (b) GRP scheme.

Figure 3.18. U-level curves for the 2-D Burgers equation at $t = 1$ for Case A.

where the exact (self-similar) solution was specified on the domain boundaries. According to Remark 3.18, the numerical solution is displayed in the subdomain $[-1.5, 1.5] \times [-1.5, 1.5]$ (i.e., a region of 300×300 cells). The integration was performed from $t = 0$ to $t = 1$, with a time step $\Delta t = 0.005$, corresponding to a CFL ratio of $\mu_{\mathrm{CFL}} = 0.5$. The computation results are shown in Figure 3.18(a) for the Godunov scheme and in Figure 3.18(b) for the GRP scheme, as U-level plots at the equally spaced levels $U_L = -1 + 0.2\,L$, $L = 0, 1, \ldots, 10$ (the head and tail values, however, were modified to ± 0.999 instead of ± 1, to enable an interpolation for the head and tail U-level lines).

Comparing the exact solution (3.62) [see Figure 3.17(a)] to the numerical solution, we consider separately the "1-D" regions away from the diagonal $x = y$ and the "corner" region near $x = y$ where the x-facing and y-facing rarefaction waves intersect. In the 1-D regions, the GRP solution is quite accurate, showing slightly displaced level lines only at the head and tail edges. The Godunov solution, however, shows a much larger "outward" displacement of the head and tail lines. This observation demonstrates the improved accuracy of the GRP scheme at the head and tail regions where the gradients (u_x or u_y) are discontinuous.

Turning to the corner region near $x = y$, we note that although the GRP solution reproduces quite well the exact pattern of the 2-D rarefaction wave [Figure 3.17(a)], the Godunov solution is more rounded at the corner points where the x-facing and y-facing (1-D) rarefaction waves intersect. This is a genuinely two-dimensional effect, demonstrating the higher "dissipativity" of the Godunov scheme.

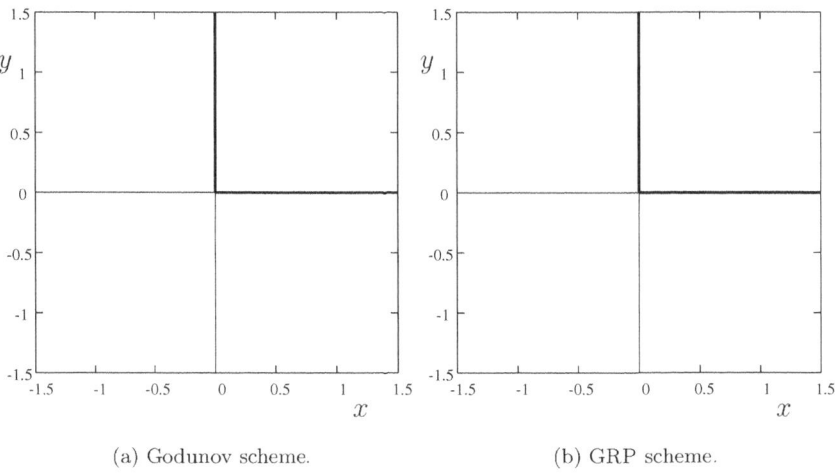

(a) Godunov scheme. (b) GRP scheme.

Figure 3.19. U-level curves for the 2-D Burgers equation at $t = 1$ for Case B.

Case B

The initial data here are obtained by reversing the sign of u_0 in the previous case:

$$u_0(\theta) = \begin{cases} -1, & 0 < \theta < \frac{\pi}{2}, \\ 1, & \text{otherwise.} \end{cases} \tag{3.63}$$

These data produce the stationary shock wave pattern shown in Figure 3.17(b). The numerical computation was conducted just as in the previous case, and the results are shown in Figure 3.19(a) for the Godunov scheme and in Figure 3.19(b) for the GRP scheme. This is a rather trivial example, since both schemes reproduce the standing shock pair accurately, and it serves to demonstrate that this known capability of the Godunov and GRP schemes is unaltered by the two-dimensional splitting.

Case C

The third case has the initial data

$$u_0(\theta) = \begin{cases} -1, & 0 < \theta < \frac{\pi}{2}, \\ 1, & \frac{\pi}{2} < \theta < \theta_0, \\ 0, & \theta_0 < \theta < 2\pi, \end{cases} \tag{3.64}$$

where $\pi < \theta_0 < \frac{3\pi}{2}$ satisfies $\tan \theta_0 = 2$ (so that the jump between 1 and 0 lies

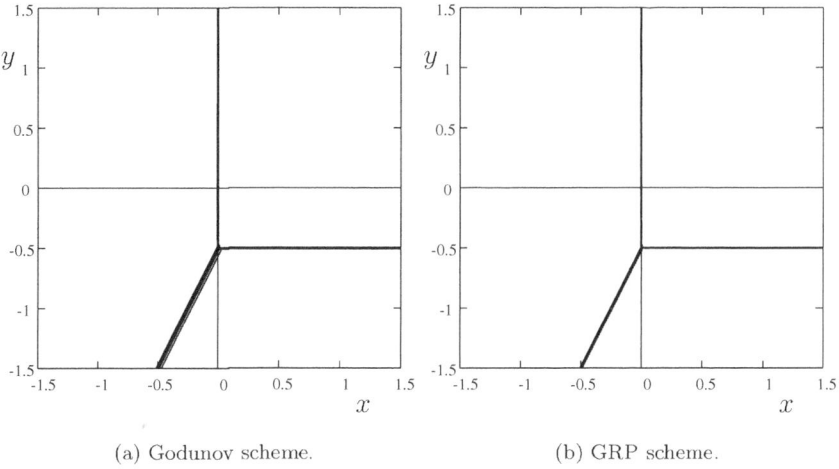

(a) Godunov scheme.   (b) GRP scheme.

Figure 3.20. U-level curves for the 2-D Burgers equation at $t = 1$ for Case C.

along $y = 2x, x < 0$). Thus, at time $t > 0$, outside of a large disk we have three shocks (in the (x, y) plane):

(a) a shock along $y = -\frac{1}{2}t$ moving at speed $-\frac{1}{2}$ (in the y direction),
(b) a standing shock along $x = 0$, $y > 0$, and
(c) a shock at $2x - y = \frac{1}{2}t$ (for x sufficiently negative).

The exact (entropy) solution is the two-dimensional shock wave pattern shown in Figure 3.17(c) (see Li, Yang, and Zhang [83] and Ben-Artzi et al. [14]). Shocks (a) and (c) "interact" with shock (b), which results in an extension of the latter to the segment $\{x = 0, -\frac{1}{2} < y < 0\}$. The numerical results are shown in Figure 3.20(a) for the Godunov scheme and in Figure 3.20(b) for the GRP scheme. Note the sharper resolution of shock (c) in the GRP solution.

Case D

The fourth case has the initial data

$$u_0(\theta) = \begin{cases} -1, & 0 < \theta < \frac{\pi}{2}, \\ 1, & \frac{\pi}{2} < \theta < \frac{3\pi}{4}, \\ 0, & \frac{3\pi}{4} < \theta < 2\pi. \end{cases} \tag{3.65}$$

As before we start by looking at the solution in the (x, y, t) frame, outside of a large disk. Clearly, as in the previous case, we have the two shocks (a) and (b) [see Figure 3.17(d)]. However, instead of the third shock we have now

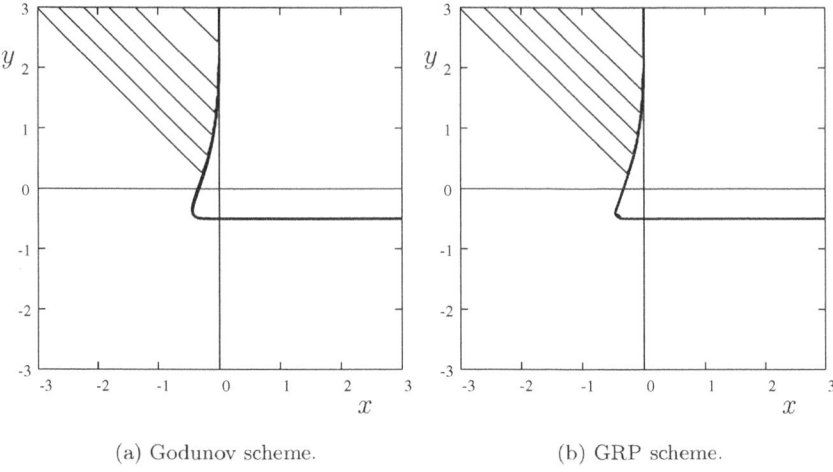

(a) Godunov scheme.   (b) GRP scheme.

Figure 3.21. Numerical solutions of 2-D Burgers equation at $t = 1$ for Case D.

a rarefaction wave that propagates parallel to the line $x + y = 0$. The exact solution at $t = 1$ is the two-dimensional wave pattern shown in Figure 3.17(d). The rarefaction wave interacts with shock (b), forming the curved branch (I) between $(0, 2)$ and $(-1/4, 1/4)$. The latter point is then connected to the point $(-1/2, -1/2)$ by a straight shock segment (II) (see Ben-Artzi, et al. [14]). The numerical results are shown in Figure 3.21(a) for the Godunov scheme and in Figure 3.21(b) for the GRP scheme. Note the rounding of the "shock wedge" at $(-1/2, -1/2)$ in both solutions.

The Guckenheimer Equation

In this case, due to Guckenheimer [59], the conservation law and the initial data are

$$\frac{\partial}{\partial t}u + \frac{\partial}{\partial x}(u^2/2) + \frac{\partial}{\partial y}(u^3/3) = 0, \tag{3.66}$$

$$u(x, y, 0) = u_0(\theta) = \begin{cases} 0 & \text{in sector} \quad 0 < \theta < \frac{3\pi}{4}, \\ 1 & \text{in sector} \quad \frac{3\pi}{4} < \theta < \frac{3\pi}{2}, \\ -1 & \text{in sector} \quad \frac{3\pi}{2} < \theta < 2\pi, \end{cases} \tag{3.67}$$

where $\theta = \arctan(\frac{y}{x})$ as before [see also Figure 3.22(a)]. Unlike the previous two equations, the flux functions here are not identical. Moreover, the function $g(u) = u^3/3$ is nonconvex, and as we shall see below, this produces a solution with a "sonic shock." Note that in this case the solution to the Riemann problem

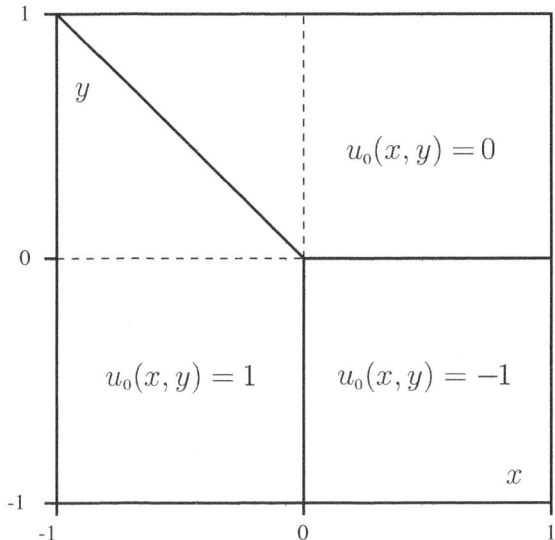

(a) Initial data for Guckenheimer structure.

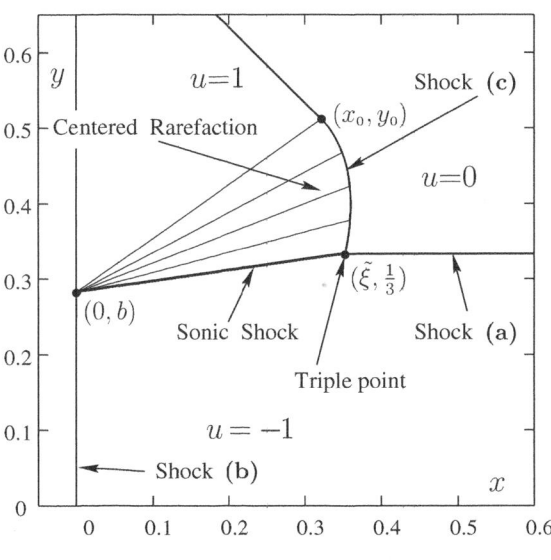

(b) Exact Guckenheimer structure at $t = 1$.

Figure 3.22. The Guckenheimer structure for $u_t + (u^2/2)_x + (u^3/3)_y = 0$.

(and likewise to the generalized Riemann problem) associated with (3.51ii) needs to be modified, taking into account the nonconvexity of $g(u)$. In the present case this modification of (3.22), (3.23) is rather simple (see Godlewski and Raviart [54, Section 2.6 and Remark 2.2 in Section 3.2]).

Consider the (self-similar) exact solution (see Guckenheimer [59], Zhang and Zhang [125], Li, Yang, and Zhang [83], and Ben-Artzi et al. [14]) of (3.66), (3.67), shown in Figure 3.22(b). Referring to this figure, we notice that outside of a large disk, the solution consists of the following three shocks:

(a) a shock emanating from the line $y = 0$ $(x > 0)$, moving at speed $1/3$ in the positive y direction (note that $g(u) = u^3/3$ is concave on $[-1, 0]$),

(b) a standing shock along $x = 0$ $(y < 0)$, and

(c) a shock emanating from the line $x + y = 0$; the self-similar analysis (Guckenheimer [59], Ben-Artzi, Falcovitz, and Li [14]) shows that at $t > 0$ this line is given by $x + y = (5/6)t$.

The interaction of these three shocks in a disk around $(0, 0)$ gives rise to a very complex wave structure. At time $t = 1$ it can be described as follows [see Figure 3.22(b)]: The shock (b) extends to a segment of the positive y-axis given by $0 \leq y \leq b$, $b = 0.2823057$. At the point $(0, b)$ it bifurcates into a CRW that has a tail characteristic coinciding with a sonic shock, across which the solution $u(x, y, 1)$ jumps from -1 to the value $\tilde{v} = 0.6087418$. Then u increases across the rarefaction from $\tilde{v}$ to 1, and it is constant along each (straight) characteristic line. The rarefaction wave modifies shock (c) in a fashion similar to that of the curved shock in the previous example of the Burgers equation (Case D). Note that the head characteristic of the CRW carries the value $u = 1$. It intersects the shock (c) at the point (x_0, y_0) given by

$$x_0 = \frac{\frac{5}{6} - b}{2 - b}, \qquad y_0 = \frac{\frac{5}{6} + \frac{b}{6}}{2 - b}.$$

The tail characteristic (sonic shock) intersects the shock (a) at the point $(\tilde{\xi}, \frac{1}{3})$, where $\tilde{\xi} = 0.3519610$. The result of the interaction between the CRW and the shock (c) leads, as already noted, to a "bending" of the latter, forming a shock branch $y = y(x)$ connecting (x_0, y_0) to $(\tilde{\xi}, \frac{1}{3})$. It can be determined by solving an ordinary differential equation (Ben-Artzi, Falcovitz, and Li [14]).

Thus, we obtain a wave pattern that includes a shock wave bifurcating into a CRW and a sonic shock that serves as a tail characteristic of the CRW. It intersects with the other two shocks at the triple point $(\tilde{\xi}, \frac{1}{3})$. This wave pattern provides for a good test of finite-difference schemes.

Two numerical tests were performed, one using the Godunov scheme and the other with the GRP scheme. The computation domain was the square

$[-1 \leq x \leq 1, \ -1 \leq y \leq 1]$, which was divided into 320×320 square cells. The time step was $\Delta t = 0.003125$ (i.e., $\mu_{\mathrm{CFL}} = 0.5$ since $\max|u| = 1$ and $f'(u) = u$, $g'(u) = u^2$), and the computation was performed to final time $t = 1$. The boundary conditions were specified by calculating the exact solution on the outer segments of boundary cells (see Remark 3.18). This is possible as long as the domain boundary is intersected only by the three shocks (a), (b), and (c), which according to Figure 3.22(b) is still true at $t = 1$.

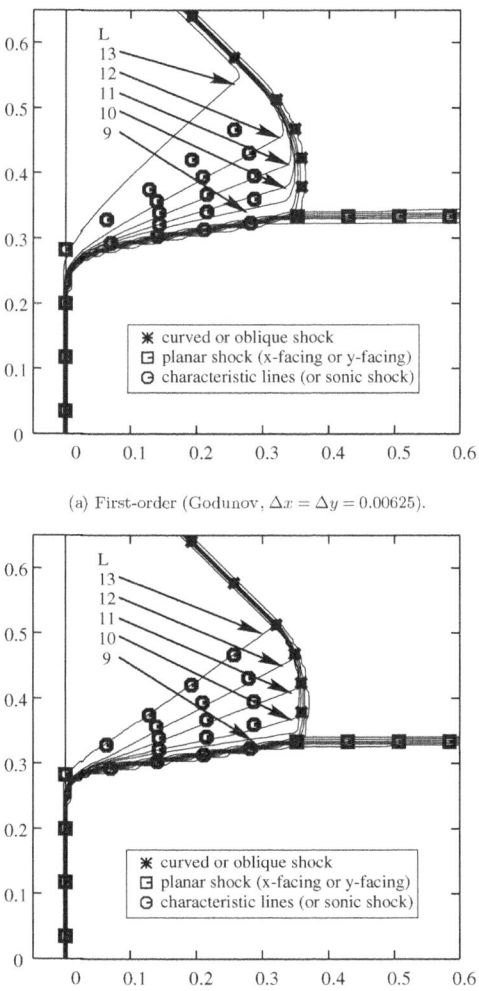

(a) First-order (Godunov, $\Delta x = \Delta y = 0.00625$).

(b) Second-order (GRP, $\Delta x = \Delta y = 0.00625$).

Figure 3.23. U-level curves for the Guckenheimer equation at $t = 1$. $u_t + (u^2/2)_x + (u^3/3)_y = 0$. Initial data in Figure 3.22(a).

The results are shown in the subdomain $[-0.05 \leq x \leq 0.60, \ 0 \leq y \leq 0.65]$ [see Figure 3.23(a) for the Godunov scheme and Figure 3.23(b) for the GRP scheme]. Recall that inside the rarefaction fan u is constant along the (straight) characteristic lines, so that numerical U-level curves approximate the fan structure. The U-level sequence

$$U_L = \begin{cases} -1 + 0.2L, & L = 0, \dots, 8, \\ 0.60874, 0.68295, 0.76366, 0.86089, 1, & L = 9, \dots, 13 \end{cases} \quad (3.68)$$

is designed to show the shock fronts and the rarefaction fan. The five levels $L = 9, \dots, 13$ correspond to the tail, head, and three inner characteristic lines of the rarefaction fan [as shown in Figure 3.22(b)]. To enable interpolation at the lowest and highest U levels, they were slightly shifted to -0.990 and 0.997, respectively. For comparison of the exact and numerical solutions, we represent the exact solution [Figure 3.22(b)] by discrete "marker points" situated on shock fronts, as shown in Figure 3.23. Additional marker points are located at points (x, y) inside the rarefaction fan, where the exact solution takes on the same values U_L, $L = 9, \dots, 13$ as given by (3.68).

Our primary observation with respect to the numerical solution is that both finite-difference schemes, applied according to the operator splitting (3.52), produce a correct approximation to this complex 2-D wave-interaction pattern (Figure 3.23). The GRP solution agrees quite well with the exact one, whereas the Godunov solution shows a nearly equal agreement for the shock fronts, but a lesser agreement in the rarefaction fan. In this centered fan, the characteristic line that coincides with the sonic shock front corresponds to a constant value of $u = \tilde{v}$, and it is one of the U-level lines plotted ($L = 9$). In the GRP solution this line is seen very near the sonic shock front [Figure 3.23(b)], whereas in the Godunov case its standoff distance is perceptibly higher [Figure 3.23(a)]. The captured sonic shock is represented by the cluster of level lines $L = 0, \dots, 9$ (since the jump across this shock is from $u = U_0$ to U_9). At the other end of the rarefaction fan, the head characteristic line is plotted with $U_{13} = 0.997$ (close to the exact value of $U_{13} = 1$, for a clear U-level interpolation). In the Godunov solution this line extends well beyond the exact solution, whereas in the GRP solution it agrees well with the exact marker points. The rarefaction fan is the only region of the solution where $u(x, y, 1)$ varies smoothly with a nonzero gradient. Hence, these observations indicate that in such regions the (second-order-accurate) GRP scheme produces considerably smaller errors than the (first-order-accurate) Godunov scheme. We observe that the bifurcation point $(0, b)$ and the triple point $(\tilde{\xi}, \frac{1}{3})$, resulting from the two-dimensional setting, are well replicated by both schemes.

4

Systems of Conservation Laws

In this chapter we review the basic facts concerning systems of conservation laws ("nonlinear hyperbolic systems"). The general theory is discussed in Section 4.1, and in Section 4.2 we specialize to the Euler system of equations of inviscid, compressible, nonisentropic flow, the subject matter of this monograph. The presentation at this stage is restricted to quasi-one-dimensional flow in the Eulerian and Lagrangian frameworks. The Riemann problem, which plays an important role in this monograph, is introduced and discussed in detail.

4.1 Nonlinear Hyperbolic Systems in One Space Dimension

A system of (in general nonlinear) conservation laws in one space dimension is given by

$$\frac{\partial}{\partial t}\mathbf{u} + \frac{\partial}{\partial x}\mathbf{F}(\mathbf{u}) = \mathbf{0}, \qquad x \in \mathbb{R}, \quad t \geq 0, \tag{4.1}$$

where

$$\mathbf{u}(x, t) = \begin{pmatrix} u_1(x, t) \\ \vdots \\ u_m(x, t) \end{pmatrix}$$

is a vector of m unknown functions and

$$\mathbf{F}(\mathbf{u}) = \begin{pmatrix} F_1(u_1, \ldots, u_m) \\ \vdots \\ F_m(u_1, \ldots, u_m) \end{pmatrix}$$

(the flux function) is a smooth vector function from $\mathbb{R}^m \to \mathbb{R}^m$.

81

In what follows we shall use boldface notation $\mathbf{v}$ to denote (real) m-vectors (or vector functions). For notational convenience we refer to $\mathbf{v}$ as a column vector so that its transpose $\mathbf{v}^T$ is a row vector $\mathbf{v}^T = (v_1, \ldots, v_m)$.

As in the scalar case (namely, $m = 1$; see Section 2.1) the equation (4.1) is supplemented by giving the initial condition

$$\mathbf{u}(x, 0) = \mathbf{u}_0(x), \qquad x \in \mathbb{R}. \tag{4.2}$$

From the discussion in Section 2.1 we know that smooth global (i.e., defined for all $t \geq 0$) solutions cannot be expected, in general, even for very regular initial data. Indeed, the main thrust of the theory was concerned with the formation of singular, discontinuous weak solutions. Clearly, these observations hold true in the case of systems. In fact, we shall see that the situation is further complicated by interaction of "discontinuity waves" associated with different "families."

Unlike the case of a scalar conservation law, where the theory of the IVP (2.1), (2.2) is fairly complete (see Theorem 2.20), for a system there are almost no theoretical results of a general character. Instead, there have been many detailed studies of special cases, related to suitably restricted initial data $\mathbf{u}_0(x)$. In what follows we shall outline these solutions, leading finally to the solution of the Riemann problem for systems. We refer the reader to Lax [75], Hörmander [63], and Evans [36] for comprehensive accounts of the existing theory.

Characteristic Curves and Centered Rarefaction Waves

We begin by studying the case of smooth solutions to the system (4.1). In this case we can use the chain rule to rewrite (4.1) as

$$\frac{\partial}{\partial t}\mathbf{u} + \mathbf{A}(\mathbf{u})\frac{\partial}{\partial x}\mathbf{u} = \mathbf{0}, \tag{4.3}$$

where

$$\mathbf{A}(\mathbf{u}) = \left(\frac{\partial F_i}{\partial u_j}\right)_{1 \leq i, j \leq m}$$

is the Jacobian matrix $\mathbf{F}'(\mathbf{u})$. Recall that in the scalar case, we saw [see Equation (2.10)] that a "distinguished" family of curves, namely, the characteristic curves, had the property of carrying constant values of the solution $u(x, t)$. Trying to look for a similar behavior in our case, we proceed as follows.

Let $\mathbf{l}^T(\mathbf{u})$ be a left eigenvector of $\mathbf{A}(\mathbf{u})$, so that

$$\mathbf{l}^T(\mathbf{u}) \cdot \mathbf{A}(\mathbf{u}) = \lambda(\mathbf{u})\, \mathbf{l}^T(\mathbf{u}).$$

We assume that $\lambda(\mathbf{u})$, $\mathbf{l}^T(\mathbf{u})$ are continuous functions of $\mathbf{u}$ (at least in some domain $D \subseteq \mathbb{R}^m$). Taking the scalar product of Equation (4.3) with $\mathbf{l}^T(\mathbf{u})$, we get

$$\mathbf{l}^T(\mathbf{u}) \cdot [\mathbf{u}_t + \lambda(\mathbf{u})\mathbf{u}_x] = 0 \tag{4.4}$$

(we assume that the solution $\mathbf{u}(x, t)$ takes values in D). We now single out certain curves.

Definition 4.1 A curve $x = x(t)$ is called characteristic for (4.1) or (4.3) [and relative to a given solution $\mathbf{u}(x, t)$] if its slope is an eigenvalue of $\mathbf{A}(\mathbf{u}(x, t))$,

$$\frac{dx}{dt} = \lambda(\mathbf{u}(x, t)). \tag{4.5}$$

Remark 4.2 We emphasize that, in contrast to the linear case, the *character-istic curves depend on the solution* $\mathbf{u}(x, t)$. This corresponds to the scalar case [see Equation (2.10) and the discussion following it], where $f'(u)$ is the only "eigenvalue."

Using the total-time-derivative notation introduced in (2.7), we can rewrite (4.4) as

$$\mathbf{l}^T(\mathbf{u}) \cdot \frac{D}{Dt}\mathbf{u} = 0 \qquad \text{along} \qquad \frac{dx}{dt} = \lambda(\mathbf{u}(x, t)), \tag{4.6}$$

which should be compared with Equation (2.10). Unlike for the scalar case, Equation (4.6) does not mean that $\mathbf{u}(x, t)$ must be constant along a characteristic curve. However, it forces limitations on the value of $\mathbf{u}$ on such a curve. In other words, if the values of $\mathbf{u}(x, t)$ are assigned along a curve $x = x(t)$, and if this curve is characteristic with respect to these values [i.e., $x'(t) = \lambda(\mathbf{u}(x, t))$, with a corresponding eigenvector $\mathbf{l}^T(\mathbf{u}(x, t))$), then (4.6) is an extra condition that must be satisfied. This condition will play an important role in the fluid dynamical context (see Section 4.2), where it is referred to as the characteristic relation.

In a somewhat more specialized situation, we can actually consider solutions that are *constant* along curves. We look at families of such curves covering domains in the (x, t) plane. Let Ω be a domain contained in the half-plane $x \in \mathbb{R}$, $t \geq 0$. Let $\phi(x, t)$ be a smooth function (i.e., at least continuously dif-ferentiable) defined in Ω, so that $\phi_x \neq 0$. Let $I \subseteq \mathbb{R}$ be the range of values taken on by ϕ in Ω. Thus the curves

$$L_c \equiv \{\phi(x, t) = c\}, \qquad c \in I,$$

form a family of smooth curves covering Ω (see Figure 4.1).

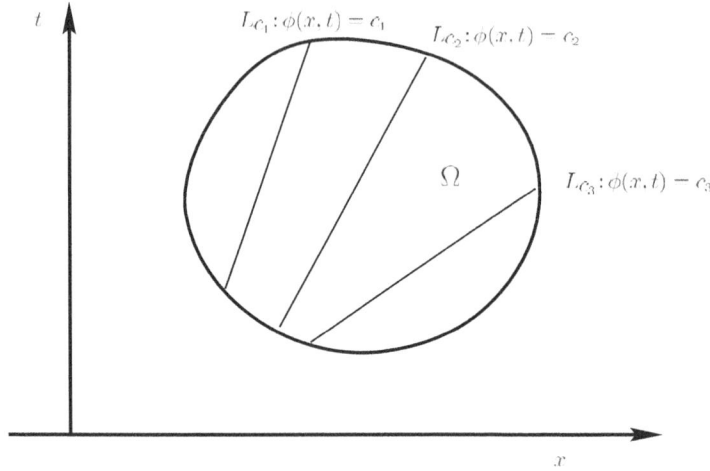

Figure 4.1. A simple wave.

A solution $\mathbf{u}(x, t)$ of (4.3), defined in Ω, is constant along any curve L_c, $c \in I$, if there exists a smooth function $\mathbf{w}(y)$, $y \in I$, such that

$$\mathbf{u}(x, t) = \mathbf{w}(\phi(x, t)), \qquad (x, t) \in \Omega. \tag{4.7}$$

Inserting this in (4.3) we get

$$\mathbf{u}_t + \mathbf{A}(\mathbf{u}) \cdot \mathbf{u}_x = \mathbf{w}'\phi_t + \mathbf{A}(\mathbf{w}(\phi(x, t))) \cdot \mathbf{w}'\phi_x = \mathbf{0} \tag{4.8}$$

(note that $\mathbf{w}$ is an m-(column) vector). We now assume further that $\mathbf{w}'(y) \neq \mathbf{0}$. It follows from (4.8) that $\mathbf{w}'(\phi(x, t))$ is an eigenvector of the matrix $\mathbf{A}(\mathbf{w}(\phi(x, t)))$, for all $(x, t) \in \Omega$. The slope $\frac{dx}{dt} = -\frac{\phi_t}{\phi_x}$ of L_c ($c = \phi(x, t)$) is the corresponding eigenvalue of $\mathbf{A}(\mathbf{w}(c))$ and hence is constant along L_c. In other words, all the curves L_c are straight lines and *characteristic* with respect to the solution $\mathbf{u}(x, t)$.

Summary 4.3 *Let $\mathbf{u}(x, t)$ be a smooth solution to (4.3) and assume that $\mathbf{u} = \mathbf{w}(c)$ along any curve $\phi(x, t) = c$ ($(x, t) \in \Omega$), where $\mathbf{w}(c)$ is a vector depending smoothly on c and such that $\frac{d}{dc}\mathbf{w}(c) \neq \mathbf{0}$. Then necessarily the curves $\phi(x, t) = c$ are straight lines and characteristic (in Ω). (Such a solution is called a "simple wave" solution.)*

Example 4.4 (A centered rarefaction wave) Consider the case that Ω is a sector, $\Omega = \{(x, t), t > 0, \mu_1 \leq \frac{x}{t} \leq \mu_2\}$, and let $\phi(x, t) = \frac{x}{t}$, so that ϕ is constant along rays emanating from the origin. A solution $\mathbf{u}(x, t)$ that is constant

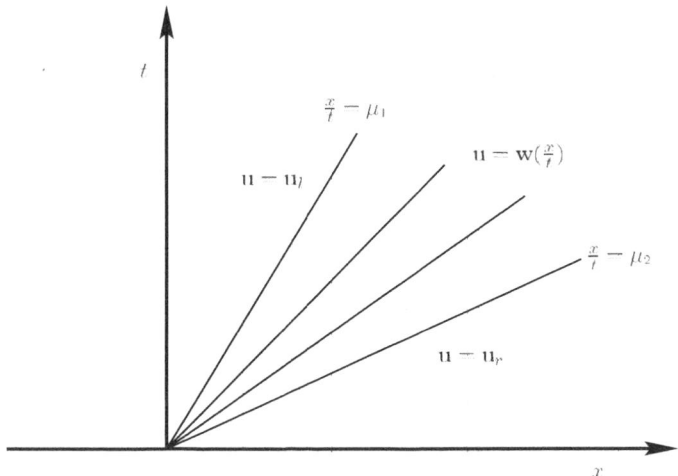

Figure 4.2. A centered rarefaction wave.

along these rays can be written as $\mathbf{u}(x, t) = \mathbf{w}(\frac{x}{t})$. Equation (4.8) becomes in this case

$$\mathbf{A}(\mathbf{w}(y))\mathbf{w}'(y) = y\mathbf{w}'(y), \qquad y = \frac{x}{t} \in [\mu_1, \mu_2]. \qquad (4.9)$$

In particular, the rays $\frac{x}{t} = y$, $\mu_1 \leq y \leq \mu_2$, are characteristic curves. For each curve the slope $y = \frac{x}{t}$ is an eigenvalue of $\mathbf{A}(\mathbf{w}(\frac{x}{t}))$.

Definition 4.5 If a solution $\mathbf{w}(y)$ to (4.9) exists (for $y \in [\mu_1, \mu_2]$) we say that the corresponding solution $\mathbf{u}(x, t) = \mathbf{w}(\frac{x}{t})$ is a centered rarefaction wave (CRW), connecting the "right state" $\mathbf{u}_r = \mathbf{w}(\mu_2)^1$ to the "left state" $\mathbf{u}_l = \mathbf{w}(\mu_1)$ (see Figure 4.2).

It is seen from (4.9) that for a CRW $\mathbf{w}'(y)$ is the eigenvector associated with the eigenvalue y. The existence of such an eigenvector, depending smoothly on y, is guaranteed if y is an isolated simple eigenvalue. We shall not go here into a more detailed study of the dependence of eigenvalues and eigenvectors on parameters; instead, we refer the reader to Evans [36, Chapter 11]. However, we cite the following definitions and fundamental facts.

Definition 4.6 (Strict hyperbolicity) The system (4.1) [or (4.3)] is *strictly hyperbolic* if, for every vector $\mathbf{v} \in \mathbb{R}^m$, the matrix $\mathbf{A}(\mathbf{v})$ has m real and distinct

---

[1] A given vector in $\mathbb{R}^m$ is often referred to as a "state," a term that is inspired by fluid dynamics (see next section).

(simple) eigenvalues

$$\lambda_1(\mathbf{v}) < \lambda_2(\mathbf{v}) < \cdots < \lambda_m(\mathbf{v}) \tag{4.10}$$

with a corresponding complete set of eigenvectors $\mathbf{r}_1(\mathbf{v}), \mathbf{r}_2(\mathbf{v}), \ldots, \mathbf{r}_m(\mathbf{v})$.

In this case, the functions $\lambda_k(\mathbf{v})$ and $\mathbf{r}_k(\mathbf{v}), k = 1, \ldots, m$, can be taken as smooth functions in $\mathbb{R}^m$. Throughout the rest of this section, we assume that our system is strictly hyperbolic.

Definition 4.7 (k-centered rarefaction wave) We say that the CRW of Definition 4.5 is a k-centered rarefaction wave if, for some $k \in \{1, \ldots, m\}$, we have [using the notation in (4.10)]

$$y = \lambda_k(\mathbf{w}(y)), \qquad \mu_1 \leq y \leq \mu_2. \tag{4.11}$$

Clearly, the existence of a k-CRW depends on the (simultaneous) solvability of Equations (4.9) and (4.11). Using the smoothness of $\lambda_k(\mathbf{v})$ and differentiating (4.11) with respect to y we get, by the chain rule,

$$1 = \frac{d}{dy}\lambda_k(\mathbf{w}(y)) = \nabla\lambda_k(\mathbf{v})\big|_{\mathbf{v}=\mathbf{w}(y)} \cdot \mathbf{w}'(y). \tag{4.12}$$

However, from (4.9) we infer that $\mathbf{w}'(y) = \alpha(y)\mathbf{r}_k(\mathbf{w}(y))$, where $\alpha(y) \neq 0$ is continuous. Inserting this in (4.12) we see that $\nabla\lambda_k \cdot \mathbf{r}_k$ cannot vanish. Hence we have the following corollary.

Corollary 4.8 *A necessary condition for the existence of a k-CRW is that the kth right eigenvector can be chosen so that*

$$\nabla\lambda_k(\mathbf{v}) \cdot \mathbf{r}_k(\mathbf{v}) = 1. \tag{4.13}$$

Remark 4.9 Condition (4.13) is known as the condition of "genuine nonlinearity" of the "kth characteristic family."

Assuming condition (4.13), suppose that a right state $\tilde{\mathbf{u}} \in \mathbb{R}^m$ is given, so that $\lambda_k(\tilde{\mathbf{u}}) = \mu, \mu_1 \leq \mu \leq \mu_2$. The system of ordinary differential equations

$$\mathbf{w}'(y) = \mathbf{r}_k(\mathbf{w}(y)), \qquad \mathbf{w}(\mu) = \tilde{\mathbf{u}}, \tag{4.14}$$

has, in view of the basic existence and uniqueness theorem for ordinary differential equations, a unique smooth solution defined for y in a small interval containing μ. By (4.12) the scalar function $\lambda_k(\mathbf{w}(y))$ is linear in this interval

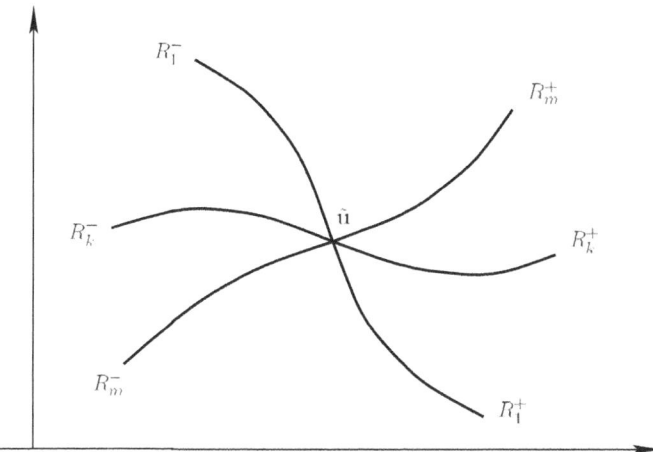

Figure 4.3. States Connected to $\tilde{\mathbf{u}}$ by a k-CRW, $\lambda_k(\tilde{\mathbf{u}}) = \mu$. (I) $\tilde{\mathbf{u}}$ is a right state: $\mathbf{w}(y) \in R_k^- \Rightarrow \lambda_k(\mathbf{w}(y)) < \mu$. (II) $\tilde{\mathbf{u}}$ is a left state: $\mathbf{w}(y) \in R_k^+ \Rightarrow \lambda_k(\mathbf{w}(y)) > \mu$.

and since $\lambda_k(\mathbf{w}(\mu)) = \mu$ we obtain the identity $\lambda_k(\mathbf{w}(y)) = y$ in the interval. In particular, we get:

Corollary 4.10 *Given a right state $\tilde{\mathbf{u}}$ such that $\lambda_k(\tilde{\mathbf{u}}) = \mu$, there exists some $\delta > 0$ such that the one-parameter family $\mathbf{w}(y)$, $\mu - \delta \leq y \leq \mu$, with $\mathbf{w}(y)$ given by (4.14), consists of all left states $\mathbf{u}_l$ that can be connected to $\tilde{\mathbf{u}}$ by a k-CRW.*

We conclude that the trajectory R_k^- in $\mathbb{R}^m$, consisting of the vectors $\mathbf{w}(y)$, $\mu - \delta \leq y \leq \mu$, contains the left states "close" to $\tilde{\mathbf{u}}$ that can be connected to $\tilde{\mathbf{u}}$ by a k-CRW. Similarly, the trajectory R_k^+ consisting of vectors $\mathbf{w}(y)$ on the "other side," $\mu \leq y \leq \mu + \delta$, represents those right states $\mathbf{u}_r$ that can be connected via a k-CRW to $\tilde{\mathbf{u}}$, which now serves as a given *left* state. We denote by R_k the union of R_k^- and R_k^+ (see Figure 4.3).

Remark 4.11 The previous discussion is of a "local" character. Only states "close" to $\tilde{\mathbf{u}}$ are considered for possible connection to $\tilde{\mathbf{u}}$ by a k-CRW. Clearly, the question of "global" connections by such waves depends on a careful study of the global properties of solutions of (4.14) and is beyond the scope of this monograph.

Weak Solutions and Jump Discontinuities

So far, we have dealt with nonlinear waves for which the solution $\mathbf{u}(x, t)$ is continuous (note, however, that for a CRW the solution is not continuous at $t = 0$). But as in the scalar case, we cannot avoid here the inclusion of weak solutions that admit discontinuities. The treatment of such solutions runs parallel to what has already been done in the scalar case.

We start by rewriting Equation (4.1) as a balance equation, in analogy with Equation (2.3). Thus, in a rectangle $x_1 \le x \le x_2, 0 \le t \le T$, we have

$$\int_{x_1}^{x_2} \mathbf{u}(x, T) \, dx - \int_{x_1}^{x_2} \mathbf{u}_0(x) \, dx = - \int_0^T \mathbf{F}(\mathbf{u}(x_2, t)) \, dt + \int_0^T \mathbf{F}(\mathbf{u}(x_1, t)) \, dt, \quad (4.15)$$

which justifies the term "flux" for $\mathbf{F}(\mathbf{u})$. Let C_0^1 be the set of test functions as in (2.12).

Definition 4.12 The bounded (vector) function $\mathbf{u}(x, t)$ is called a "weak solution" to the IVP (4.1), (4.2) if, for any test function $\phi(x, t) \in C_0^1$,

$$\int_{\mathbb{R}} \int_0^\infty (\mathbf{u}\phi_t + \mathbf{F}(\mathbf{u})\phi_x) \, dt \, dx + \int_{\mathbb{R}} \phi(x, 0)\mathbf{u}_0(x) \, dx = \mathbf{0}. \quad (4.16)$$

Basically, one can repeat the arguments in the proof of Claim 2.5 to show that if $\mathbf{u} \in C^1$ is a weak solution then it is a classical solution. Furthermore, every weak solution $\mathbf{u}$ defined in a domain Ω with smooth boundary Γ, which has finitely many jump discontinuities along smooth curves, satisfies (see Claim 2.6)

$$\oint_\Gamma -\mathbf{u}(x, t) \, dx + \mathbf{F}(\mathbf{u}(x, t)) \, dt = \mathbf{0}. \quad (4.17)$$

A case of particular interest is that of a *shock wave*, namely, a weak solution that is smooth on the two sides of a smooth curve C, across which it experiences a jump discontinuity. Following the argument leading to Corollary 2.8 we get:

Corollary 4.13 (The Rankine–Hugoniot jump condition) *Let* $C: x = x(t)$ *be a smooth trajectory traced out by a jump discontinuity of the weak solution* $\mathbf{u}(x, t)$. *Let* $\mathbf{u}_\pm(x(t), t) = \lim_{x \to x(t)\pm} \mathbf{u}(x, t)$ *be the limiting values of* $\mathbf{u}(x, t)$ *as* x *approaches* $x(t)$ *from either side of the discontinuity. Then the speed* $S = x'(t)$ *satisfies, for every* t,

$$\mathbf{F}(\mathbf{u}_+(x(t), t)) - \mathbf{F}(\mathbf{u}_-(x(t), t)) = S\big[\mathbf{u}_+(x(t), t) - \mathbf{u}_-(x(t), t)\big]. \quad (4.18)$$

The system (4.18) consists of m algebraic conditions that must be satisfied, at any time t, by the $2m$ components of $\mathbf{u}_\pm(x(t), t)$. Note that the speed $S = x'(t)$ is *scalar*, meaning that the (vector–valued) jumps $\mathbf{u}_+(x(t), t) - \mathbf{u}_-(x(t), t)$ and $\mathbf{F}(\mathbf{u}_+(x(t), t)) - \mathbf{F}(\mathbf{u}_-(x(t), t))$ must be collinear. In particular, in contrast to the scalar case, for systems with $m > 1$ *any two constant states* $\mathbf{u}_\pm$ *cannot* necessarily be connected by a shock discontinuity[2] moving at a constant speed. Indeed, given a fixed state $\tilde{\mathbf{u}} \in \mathbb{R}^m$ (which, to fix the ideas, we can take as $\mathbf{u}_+$), the states $\mathbf{u}_-$ that can be connected to it by such a discontinuity should satisfy the m equations (4.18). Since the speed S is also unknown, we have m equations for the $m + 1$ unknowns $\mathbf{u}_-$, S, which should allow, roughly speaking, a one-parameter family of solutions (at least close to $\tilde{\mathbf{u}}$). These considerations can be made more rigorous as follows.

We seek solutions to the m equations

$$\mathbf{F}(\mathbf{u}) - \mathbf{F}(\tilde{\mathbf{u}}) = S(\mathbf{u} - \tilde{\mathbf{u}}), \tag{4.19}$$

where the unknowns are $\mathbf{u}$, S. We are looking for solutions $\mathbf{u}$ close to $\tilde{\mathbf{u}}$. By Taylor's theorem, it follows from (4.19) that

$$\mathbf{A}(\tilde{\mathbf{u}}) \cdot (\mathbf{u} - \tilde{\mathbf{u}}) - S(\mathbf{u} - \tilde{\mathbf{u}}) = O\left(\|\mathbf{u} - \tilde{\mathbf{u}}\|^2\right). \tag{4.20}$$

As $\mathbf{u}$ approaches $\tilde{\mathbf{u}}$, the normalized vector $(\mathbf{u} - \tilde{\mathbf{u}})/\|\mathbf{u} - \tilde{\mathbf{u}}\|$ approaches a unit eigenvector of $\mathbf{A}(\tilde{\mathbf{u}})$, say $\mathbf{v}_k(\tilde{\mathbf{u}})$, while S approaches the corresponding eigenvalue $\lambda_k(\tilde{\mathbf{u}})$, $1 \le k \le m$. We shall therefore look for solutions $\mathbf{u}(y)$ of (4.19) having the form

$$\mathbf{u}(y) = \tilde{\mathbf{u}} + y\mathbf{r}_k(\tilde{\mathbf{u}}) + \sum_{j \ne k} \alpha_j(y)\mathbf{r}_j(\tilde{\mathbf{u}}), \tag{4.21}$$

where the eigenvectors $\mathbf{r}_i(\tilde{\mathbf{u}})$ are as in Definition 4.6, y varies in a small interval around $y = 0$, and $\alpha_j(y)$, $\alpha'_j(y)$ vanish at $y = 0$. Fixing $y \ne 0$ we let $\mathbf{v} = \sum_{j \ne k} \alpha_j(y)\mathbf{r}_j(\tilde{\mathbf{u}})$ be the unknown vector in the $(m-1)$-dimensional subspace V spanned by $\{\mathbf{r}_j(\tilde{\mathbf{u}})\}_{j \ne k}$. The unknown speed S is assumed to be close to $\lambda_k(\tilde{\mathbf{u}})$. Thus, letting

$$\Phi_y(\mathbf{v}, S) = \mathbf{F}(\tilde{\mathbf{u}} + y\mathbf{r}_k(\tilde{\mathbf{u}}) + \mathbf{v}) - \mathbf{F}(\tilde{\mathbf{u}}) - S\left[y\mathbf{r}_k(\tilde{\mathbf{u}}) + \mathbf{v}\right]$$

makes Equations (4.19) equivalent to

$$\Phi_y(\mathbf{v}, S) = \mathbf{0}. \tag{4.22}$$

---

[2] Recall that "rarefaction shocks" were excluded in the scalar case only on the basis of entropy considerations; see the discussion preceding Example 2.13.

Observe that, for a fixed y (near 0), Φ_y is a mapping from a neighborhood of $(\mathbf{0}, \lambda_k(\tilde{\mathbf{u}}))$ in $\mathbb{R}^m$ into $\mathbb{R}^m$ (recall that $\mathbf{v}$ ranges in V, an $(m-1)$-dimensional subspace). Its Jacobian matrix is given by

$$D\Phi_y(\mathbf{v}, S)_{0, \lambda_k(\tilde{\mathbf{u}})} = [\mathbf{A}_k(\tilde{\mathbf{u}} + y\mathbf{r}_k(\tilde{\mathbf{u}})) - \lambda_k(\tilde{\mathbf{u}})\mathbf{I}, \, -y\mathbf{r}_k(\tilde{\mathbf{u}})] . \qquad (4.23)$$

We have used $\mathbf{A}_k$ to denote the restriction of $\mathbf{A}$ to the $(m-1)$-dimensional subspace V. The eigenvalue $\lambda_k(\tilde{\mathbf{u}})$ is simple (by assumption of strict hyperbolicity), with a corresponding eigenvector $\mathbf{r}_k(\tilde{\mathbf{u}})$. Thus, for a sufficiently small y and S close to $\lambda_k(\tilde{\mathbf{u}})$, the restriction $\mathbf{A}_k (\tilde{\mathbf{u}} + y\mathbf{r}_k(\tilde{\mathbf{u}})) - \lambda_k(\tilde{\mathbf{u}})\mathbf{I}$ to V is of rank $(m-1)$. Indeed, $\mathbf{A}_k(\tilde{\mathbf{u}}) - \lambda_k(\tilde{\mathbf{u}})\mathbf{I}$ maps V onto V. It follows from (4.23) (and the fact that $\mathbf{r}_k(\tilde{\mathbf{u}}) \notin V$) that, for every sufficiently small $y \neq 0$,

$$\text{rank} \quad D\Phi_y(\mathbf{v}, S)_{0, \lambda_k(\tilde{\mathbf{u}})} = m,$$

so that there exists a unique solution $(\mathbf{v}, S)$, near $(\mathbf{0}, \lambda_k(\tilde{\mathbf{u}}))$, to (4.22). This solution varies smoothly with y, so that the coefficients $\alpha_j(y)$ in (4.21) are smooth functions of y near $y = 0$ and so is the shock speed $S(y)$. Indeed, it can be shown, using (4.20), (4.21), that $\alpha_j(y) = O(y^2)$; hence $\alpha'_j(0) = 0$, $j \neq k$. In view of (4.21) this is consistent with the fact that $\mathbf{r}_k(\tilde{\mathbf{u}})$ is tangent to the curve $\mathbf{u}(y)$ at $y = 0$.

We refer the reader to Evans [36, Section 11.2] for a more detailed construction of the solutions to the jump equations (4.19). The preceding argument can be repeated for $k = 1, \ldots, m$. We can therefore summarize as follows.

Summary 4.14 *Given a state $\tilde{\mathbf{u}} \in \mathbb{R}^m$, there exists a small neighborhood $C_{\tilde{\mathbf{u}}} \subseteq \mathbb{R}^m$ of $\tilde{\mathbf{u}}$, having the following property: The set*

$$U = \{\mathbf{u} \in C_{\tilde{\mathbf{u}}}, \mathbf{u} \text{ is a solution to (4.19), for some speed } S\}$$

can be represented as a union of smooth curves,

$$U = \bigcup_{k=1}^{m} P_k,$$

such that P_k passes through $\tilde{\mathbf{u}}$ and is tangent there to $\mathbf{r}_k(\tilde{\mathbf{u}})$ (see Figure 4.4).

As in the case of the CRW, we can parametrize each curve P_k, $k = 1, \ldots, m$, as

$$P_k = \{\mathbf{u}_k(y), \lambda_k(\tilde{\mathbf{u}}) - \delta < y < \lambda_k(\tilde{\mathbf{u}}) + \delta\} , \qquad (4.24)$$

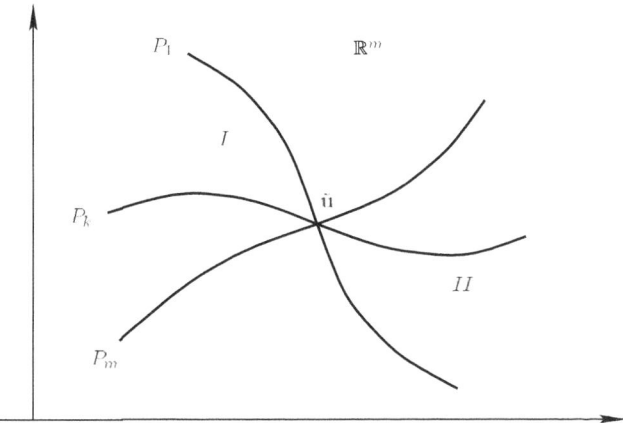

Figure 4.4. States connected to $\tilde{\mathbf{u}}$ by a jump discontinuity satisfying the Rankine–Hugoniot condition.

for some small $\delta > 0$, so that $\mathbf{u}_k(\lambda_k(\tilde{\mathbf{u}})) = \tilde{\mathbf{u}}$. Furthermore, we scale y so that

$$\mathbf{u}_k'(y)\big|_{y=\lambda_k(\tilde{\mathbf{u}})} = \mathbf{r}_k(\tilde{\mathbf{u}}). \qquad (4.25)$$

The corresponding speeds $S = S_k(y)$ are also smooth functions satisfying

$$\mathbf{F}(\mathbf{u}_k(y)) - \mathbf{F}(\tilde{\mathbf{u}}) = S_k(y)\big(\mathbf{u}_k(y) - \tilde{\mathbf{u}}\big), \qquad \lambda_k(\tilde{\mathbf{u}}) - \delta < y < \lambda_k(\tilde{\mathbf{u}}) + \delta, \qquad (4.26)$$

and such that $S_k(\lambda_k(\tilde{\mathbf{u}})) = \lambda_k(\tilde{\mathbf{u}})$.

We note that in view of (4.14) and (4.25) the trajectories R_k and P_k, $k = 1, \ldots, m$, are tangent at $\tilde{\mathbf{u}}$ [i.e., at $y = \lambda_k(\tilde{\mathbf{u}})$]. In the case of the CRW we have seen that only the part R_k^- [namely, $y < \lambda_k(\tilde{\mathbf{u}})$] represents left states that may be connected to $\tilde{\mathbf{u}}$ (as a right state) by means of a CRW. Similarly, we shall see next that only "one-half" of P_k consists of left states that can be connected to $\tilde{\mathbf{u}}$ (as a right state) by means of a moving discontinuity satisfying an additional "entropy condition."

Entropy Conditions, Shock Waves, and Contact Discontinuities

Fix $1 \le k \le m$ and consider the curve $P_k \subseteq C_{\tilde{\mathbf{u}}}$ as in (4.24). In what follows, we simplify notation (along P_k) by setting $\mathbf{A}_k(y) = \mathbf{A}(\mathbf{u}_k(y))$. We also continue to use $\mu = \lambda_k(\tilde{\mathbf{u}})$ [see (4.14)]. Differentiating Equation (4.26) with respect to y we get

$$\big(\mathbf{A}_k(y) - S_k(y)\mathbf{I}\big)\mathbf{u}_k'(y) = S_k'(y)\big(\mathbf{u}_k(y) - \tilde{\mathbf{u}}\big), \qquad (4.27)$$

and a second differentiation yields

$$\big(\mathbf{A}_k(y) - S_k(y)\mathbf{I}\big)\mathbf{u}_k''(y) + \mathbf{A}_k'(y)\mathbf{u}_k'(y) = 2S_k'(y)\mathbf{u}_k'(y) + S_k''(y)\big(\mathbf{u}_k(y) - \tilde{\mathbf{u}}\big). \tag{4.28}$$

Taking $y = \mu$ in (4.28) yields, in view of (4.25) and $S_k(\mu) = \mu$,

$$\big(\mathbf{A}(\tilde{\mathbf{u}}) - \mu\mathbf{I}\big)\mathbf{u}_k''(\mu) + \mathbf{A}_k'(\mu)\mathbf{r}_k(\tilde{\mathbf{u}}) = 2S_k'(\mu)\mathbf{r}_k(\tilde{\mathbf{u}}). \tag{4.29}$$

We shall now apply analogous treatment to the k-CRW curve R_k, which is given by Equation (4.14). By writing $\mathbf{w}_k$ for $\mathbf{w}$ and setting $\mathbf{R}_k(y) = \mathbf{r}_k(\mathbf{w}_k(y))$, Equation (4.9) takes the form [note Equation (4.11)]

$$\mathbf{A}(\mathbf{w}_k(y))\mathbf{R}_k(y) = y\mathbf{R}_k(y).$$

Differentiating this equation, we obtain

$$\big(\mathbf{A}(\mathbf{w}_k(y)) - y\mathbf{I}\big)\mathbf{R}_k'(y) + \frac{d}{dy}\left[\mathbf{A}(\mathbf{w}_k(y))\right]\mathbf{R}_k(y) = \mathbf{R}_k(y), \tag{4.30}$$

and setting $y = \mu$ in this equation leads to

$$\big(\mathbf{A}(\tilde{\mathbf{u}}) - \mu\mathbf{I}\big)\mathbf{R}_k'(\mu) + \frac{d}{dy}\left[\mathbf{A}(\mathbf{w}_k(y))\right]_{y=\mu}\mathbf{r}_k(\tilde{\mathbf{u}}) = \mathbf{r}_k(\tilde{\mathbf{u}}). \tag{4.31}$$

Subtracting (4.31) from (4.29) yields

$$\big(\mathbf{A}(\tilde{\mathbf{u}}) - \mu\mathbf{I}\big)\big(\mathbf{u}_k''(\mu) - \mathbf{R}_k'(\mu)\big) = \left[2S_k'(\mu) - 1\right]\mathbf{r}_k(\tilde{\mathbf{u}}). \tag{4.32}$$

[Note that $\mathbf{A}_k'(\mu) = \frac{d}{dy}\left[\mathbf{A}(\mathbf{w}_k(y))\right]_{y=\mu}$ since the curves $P_k : \mathbf{u}_k(y)$ and $R_k : \mathbf{w}_k(y)$ have the same tangent vector $\mathbf{r}_k(\tilde{\mathbf{u}})$ at $y = \mu$.] Let $\mathbf{l}_k^T(\tilde{\mathbf{u}})$ be the left (row) eigenvector of $\mathbf{A}(\tilde{\mathbf{u}})$, corresponding to $\lambda_k(\tilde{\mathbf{u}})$. Taking the scalar product of (4.32) with $\mathbf{l}_k^T(\tilde{\mathbf{u}})$ annihilates the left-hand side, so that, since $\mathbf{l}_k^T(\tilde{\mathbf{u}}) \cdot \mathbf{r}_k(\tilde{\mathbf{u}}) \neq 0,$[3]

$$2S_k'(\mu) = 1. \tag{4.33}$$

Remark 4.15 Observe that the equality (4.33) relies on the choice of parametrizations (4.11), (4.25) for the curves R_k, P_k, respectively. Indeed, the latter specifies the scaling only at $\lambda_k(\tilde{\mathbf{u}})$. What is important here is that the tangent vectors coincide at this point [and are equal to $\mathbf{r}_k(\tilde{\mathbf{u}})$]. Recall (Corollary 4.8) that condition (4.13) is needed for this normalization on the rarefaction side.

We can now turn to a study of the jump discontinuity as y varies in the interval $[\lambda_k(\tilde{\mathbf{u}}) - \delta, \lambda_k(\tilde{\mathbf{u}}) + \delta]$, as in (4.24). As we have seen (Summary 4.14), for every

---

[3] Note that if $j \neq k$, $\mathbf{l}_k^T(\tilde{\mathbf{u}}) \cdot \mathbf{A}(\tilde{\mathbf{u}})\mathbf{r}_j(\tilde{\mathbf{u}}) = \lambda_k(\tilde{\mathbf{u}})\mathbf{l}_k^T(\tilde{\mathbf{u}}) \cdot \mathbf{r}_j(\tilde{\mathbf{u}}) = \lambda_j(\tilde{\mathbf{u}})\mathbf{l}_k^T(\tilde{\mathbf{u}}) \cdot \mathbf{r}_j(\tilde{\mathbf{u}})$; hence $\mathbf{l}_k^T(\tilde{\mathbf{u}}) \cdot \mathbf{r}_j(\tilde{\mathbf{u}}) = 0$, $j \neq k$.

y in the interval the state $\tilde{\mathbf{u}}$, taken as a right state, can be connected to $\mathbf{u}_k(y)$ (left state) by means of a discontinuity moving at a uniform speed $S = S_k(y)$ and satisfying the Rankine–Hugoniot condition (4.26). Such a discontinuous solution is certainly a weak solution in the sense of Definition 4.12. However, in analogy with the scalar case we shall see that not all solutions are "admissible." We distinguish two cases, the first in which the kth characteristic family is genuinely nonlinear (see Remark 4.9) and the second, where in contrast to (4.13), we have identically

$$\nabla \lambda_k(\mathbf{v}) \cdot \mathbf{r}_k(\mathbf{v}) \equiv 0, \qquad \mathbf{v} \in \mathbb{R}^m. \tag{4.34}$$

Definition 4.16 When (4.34) holds identically in $\mathbb{R}^m$, we say that the kth characteristic family is *linearly degenerate*.

Proposition 4.17 *Assume that the kth characteristic family is genuinely nonlinear, satisfying identically (4.13). Then the left states $\mathbf{u}_k(y) \in P_k$, connected to the right state $\tilde{\mathbf{u}}$ by means of a discontinuity moving at speed $S_k(y)$, can be divided into two families, for sufficiently small $\delta > 0$:*

(i) $\lambda_k(\tilde{\mathbf{u}}) - \delta < y < \lambda_k(\tilde{\mathbf{u}})$, in which case

$$\lambda_k(\mathbf{u}_k(y)) < S_k(y) < \lambda_k(\tilde{\mathbf{u}}); \tag{4.35}$$

(ii) $\lambda_k(\tilde{\mathbf{u}}) < y < \lambda_k(\tilde{\mathbf{u}}) + \delta$, in which case

$$\lambda_k(\mathbf{u}_k(y)) > S_k(y) > \lambda_k(\tilde{\mathbf{u}}). \tag{4.36}$$

Proof Using (4.13), (4.25) we see that

$$\frac{d}{dy} \lambda_k(\mathbf{u}_k(y))_{y=\lambda_k(\tilde{\mathbf{u}})} = \nabla \lambda_k(\tilde{\mathbf{u}}) \cdot \mathbf{r}_k(\tilde{\mathbf{u}}) = 1,$$

whereas by (4.33), $S'_k(\lambda_k(\tilde{\mathbf{u}})) = \frac{1}{2}$. Thus, if $\delta > 0$ is sufficiently small,

$$0 < S'_k(y) < \lambda'_k(y), \qquad \lambda_k(\tilde{\mathbf{u}}) - \delta < y < \lambda_k(\tilde{\mathbf{u}}) + \delta. \tag{4.37}$$

Since $S_k(\lambda_k(\tilde{\mathbf{u}})) = \lambda_k(\tilde{\mathbf{u}})$, the inequalities (4.35), (4.36) follow directly from (4.37).   □

Since $S_k(\mu) = \mu = \lambda_k(\tilde{\mathbf{u}})$, we can use the strict hyperbolicity (Definition 4.6) and reduce $\delta > 0$ (if needed), so that (4.36) is further supplemented by

$$\lambda_{k-1}(\mathbf{u}_k(y)) < S_k(y) < \lambda_{k+1}(\tilde{\mathbf{u}}) \tag{4.38}$$

for y in the interval. The two cases in Proposition 4.17 may then be represented as in Figure 4.5. We note that in Case (i) [Figure 4.5(a)] the trajectory of the

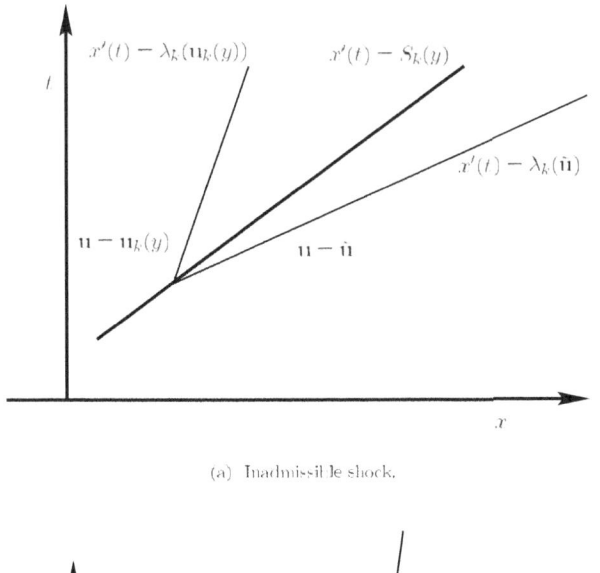

(a) Inadmissible shock.

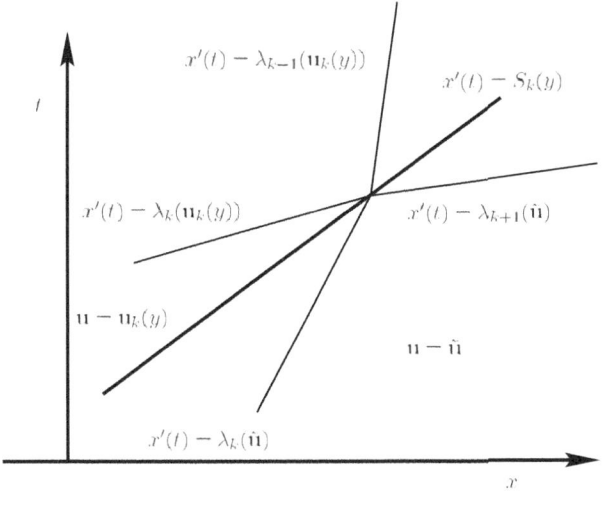

(b) Admissible shock.

Figure 4.5. Shock admissibility conditions.

discontinuity "separates" the characteristic lines of the kth family, exactly as in the scalar case of a "rarefaction shock," depicted in Figure 2.3(b).[4] However, in Case (ii) [Figure 4.5(b)] the characteristic lines (of the kth family) "run"

[4] Inspired by the fluid dynamical terminology, where the characteristic values λ_k express "sonic" speeds, we might say that in this case the shock is "supersonic" with respect to the left state $\mathbf{u}_k(y)$ but "subsonic" with respect to the right state $\hat{\mathbf{u}}$.

into the trajectory of the discontinuity as t increases. Based on our experience with the scalar case (see the discussion preceding Definition 2.15) we view only the discontinuities in Case (ii) as "admissible." As in the scalar case, they are inevitable, formed by the "forward" intersection of characteristic lines propagating in the regions where the solution is smooth.

Definition 4.18 (The Lax entropy condition for admissible shocks) The Lax entropy condition consists of the inequalities (4.36), (4.38).

Discontinuities satisfying the Lax entropy condition are called "admissible k-shocks."

Thus, admissible shocks correspond to the part

$$P_k^+ = \{\mathbf{u}_k(y), \lambda_k(\tilde{\mathbf{u}}) < y < \lambda_k(\tilde{\mathbf{u}}) + \delta\}$$

on P_k. At $y = \lambda_k(\tilde{\mathbf{u}})$, its tangent is $\mathbf{r}_k(\tilde{\mathbf{u}})$ [see (4.25)]. It therefore connects smoothly (i.e., with continuously varying slope) to the curve R_k^- [parametrized by $\lambda_k(\tilde{\mathbf{u}}) - \delta < y < \lambda_k(\tilde{\mathbf{u}})$] of the left states that can be connected to the right state $\tilde{\mathbf{u}}$ by a k-CRW (see Corollary 4.10 and Figure 4.3). We group together these two parts as follows.

Definition 4.19 (The "interaction curve") Let $\tilde{\mathbf{u}} \in \mathbb{R}^m$ be given, with $\lambda_k(\tilde{\mathbf{u}}) = \mu$, $1 \le k \le m$. The curve

$$I_k^r(\tilde{\mathbf{u}}) = \begin{cases} R_k^-, & \mu - \delta < y < \mu, \\ P_k^+, & \mu < y < \mu + \delta, \end{cases}$$

is called the "(right) kth interaction curve" for $\tilde{\mathbf{u}}$. It consists of all left states $\mathbf{u}_k(y)$, sufficiently close to $\tilde{\mathbf{u}}$, that can be connected to $\tilde{\mathbf{u}}$ by a k-CRW ($y < \mu$) or an admissible k-shock ($y > \mu$). The part P_k^+, representing the admissible shocks, is called the (kth) "Hugoniot curve" associated with $\tilde{\mathbf{u}}$ (as a right state). Similarly, $I_k^l(\tilde{\mathbf{u}})$, the "left kth interaction curve", consists of all right states sufficiently close to $\tilde{\mathbf{u}}$ that can be connected to it (as a left state) by a k-CRW or an admissible k-shock.

Remark 4.20 Observe that the part P_k^- of P_k, corresponding to values $\mu - \delta < y < \mu$, represents the *right* states $\mathbf{u}_k(y)$ that can be connected to the *left* state $\tilde{\mathbf{u}}$ by an admissible k-shock. This is clear from the inequalities (4.35).

We conclude that if the kth characteristic family is genuinely nonlinear [satisfying (4.13)], the curve $I_k^r(\tilde{\mathbf{u}})$ gives a representation of left states connected to $\tilde{\mathbf{u}}$ by a "k-wave" (CRW or admissible shock). We now turn to the linearly degenerate case, in which (4.34) is identically satisfied.

Remark 4.21 It turns out that in the conservation laws modeling fluid flow, the dichotomy "genuinely nonlinear" or "linearly degenerate" is highly prevalent. This will be seen in the discussion of compressible fluid flow (Claim 4.30).

Proposition 4.22 *Assume that the kth characteristic family, $1 \le k \le m$, is linearly degenerate, so that (4.34) is identically satisfied. Let $\tilde{\mathbf{u}} \in \mathbb{R}^m$ be given, with $\mu = \lambda_k(\tilde{\mathbf{u}})$. Then the curve P_k [see (4.24)] of states $\mathbf{u}_k(y)$ satisfying (4.26) is determined as follows.*

The function $\mathbf{u}_k(y)$ satisfies Equation (4.14), namely,

$$\mathbf{u}'_k(y) \equiv \mathbf{r}_k(\mathbf{u}_k(y)), \qquad \mu - \delta < y < \mu + \delta, \tag{4.39}$$

whereas the speed $S_k(y)$ of the discontinuity is constant and equal to the kth characteristic eigenvalue,

$$S_k(y) \equiv \mu = \lambda_k(\tilde{\mathbf{u}}), \qquad \mu - \delta < y < \mu + \delta. \tag{4.40}$$

In particular, the characteristic speed $\lambda_k(\mathbf{u}_k(y)) \equiv constant = \lambda_k(\tilde{\mathbf{u}})$ (see Figure 4.6) along P_k.

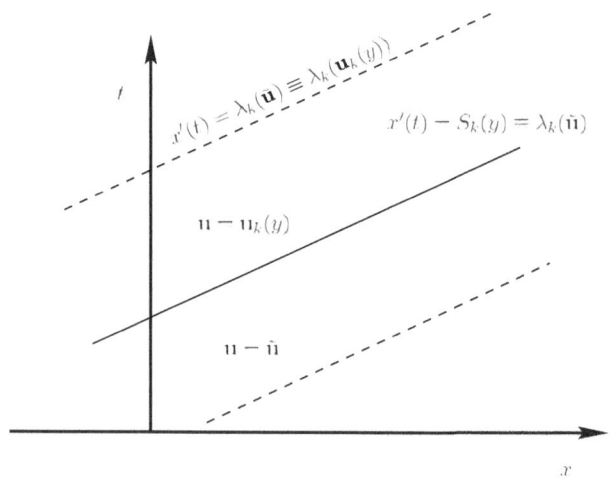

Figure 4.6. A contact discontinuity. The trajectory of the discontinuity is parallel to the characteristic lines (kth family).

Proof Let $\mathbf{u}_k(y)$ be the unique solution to (4.39), which exists if $\delta > 0$ is suffi-ciently small [see the paragraph following (4.14)]. We have

$$\frac{d}{dy}\lambda_k(\mathbf{u}_k(y)) = \nabla\lambda_k(\mathbf{v})\big|_{\mathbf{v}=\mathbf{u}_k(y)} \cdot \mathbf{u}_k'(y) = \nabla\lambda_k(\mathbf{v})\big|_{\mathbf{v}=\mathbf{u}_k(y)} \cdot \mathbf{r}_k(\mathbf{u}_k(y)) = 0,$$

(4.41)

where the last equality follows from (4.34).

Thus, $\lambda_k(\mathbf{u}_k(y)) = \mu$, $\mu - \delta < y < \mu + \delta$. Now,

$$\mathbf{F}(\mathbf{u}_k(y)) - \mathbf{F}(\tilde{\mathbf{u}}) = \int_\mu^y \frac{d}{ds}\mathbf{F}(\mathbf{u}_k(s))\,ds$$

$$= \int_\mu^y \mathbf{A}(\mathbf{u}_k(s))\mathbf{u}_k'(s)\,ds = \int_\mu^y \mathbf{A}(\mathbf{u}_k(s))\mathbf{r}_k(\mathbf{u}_k(s))\,ds \qquad (4.42)$$

$$= \int_\mu^y \lambda_k(\mathbf{u}_k(s))\mathbf{r}_k(\mathbf{u}_k(s))\,ds = \mu\int_\mu^y \mathbf{u}_k'(s)\,ds = \mu\big(\mathbf{u}_k(y) - \tilde{\mathbf{u}}\big).$$

Thus, indeed, $\mathbf{u}_k(y)$ satisfies Equation (4.26), with $S_k \equiv \mu$. By the uniqueness of the curve P_k (Summary 4.14) we conclude that the trajectory traced out by $\mathbf{u}_k(y)$ coincides with P_k. □

Definition 4.23 (A contact discontinuity) Assume that the kth characteristic field is linearly degenerate. A jump discontinuity between $\tilde{\mathbf{u}}$ and $\mathbf{u}_k(y)$, moving at speed $S_k(y) = \lambda_k(\tilde{\mathbf{u}})$, is called a k-contact discontinuity.

Remark 4.24 Observe that since the kth characteristic lines run parallel to the contact discontinuity (Figure 4.6), the Lax entropy condition (4.36), (4.38) is not satisfied. Thus, a contact discontinuity represents a weak solution satisfying the Rankine–Hugoniot jump condition (4.26), but it is not an admissible shock.

This difference is also reflected in the fact that the contact discontinuity is "symmetric," namely, $\mathbf{u}_k(y)$ and $\tilde{\mathbf{u}}$ can be interchanged as right and left states, respectively [with the same speed $S_k(y)$ for the discontinuity]. This is not the case for an admissible shock. Indeed, in the latter case, if we take $\mathbf{u}_k(y) \in P_k^+$ as the *right* state and $\tilde{\mathbf{u}}$ as the *left* state, then the jump condition (4.26) is still valid. However, the Lax entropy condition is violated [as in Figure 4.5(a)]; hence it is an inadmissible shock solution.

In the fluid dynamical context of compressible flow the contact discontinuity corresponds to a "material interface" – a discontinuity moving at "particle speed" and separating two states of different densities, but equal velocities and pressures.

Remark 4.25 Note that although Equations (4.14) and (4.39) are identical, they represent two entirely different structures. In the former, the solution $\mathbf{w}(y)$, $y = x/t$, traces out a CRW as y varies (so it is a single wave). In the latter, every y represents a state, $\mathbf{u}_k(y)$, that can be connected to $\tilde{\mathbf{u}}$ by a single wave (contact discontinuity).

The Riemann Problem

As in the scalar case (see Definition 2.18) the simplest (nontrivial) IVP for (4.1) is of particular significance.

Definition 4.26 (The Riemann problem) The "Riemann problem" (RP) for the system of conservation laws (4.1) is the IVP subject to the initial data

$$\mathbf{u}_0(x) = \mathbf{u}_L \quad (x < 0), \qquad \mathbf{u}_0(x) = \mathbf{u}_R \quad (x > 0),$$

where $\mathbf{u}_L$, $\mathbf{u}_R$ are constant states.

In Section 3.1 we have seen the fundamental role played by the (scalar) Riemann problem in the GRP approach. In the next chapter we shall see that a similar role is played by the RP in the case of compressible fluid dynamics.

There is a major difference between the solution of the RP in the scalar case and that of systems. In the former we have seen (see the discussion following Definition 2.18) that, assuming the strict convexity of $f(u)$, there exists a unique "entropy" solution for *every* two given states u_L, u_R. This cannot be repeated in the case of systems. Recall that we are always assuming that the system satisfies the hypothesis of strict hyperbolicity (Definition 4.6) and that every characteristic family is either genuinely nonlinear (Remark 4.9) or linearly degenerate (Definition 4.16). The basic reason for the difference between the scalar and the system cases (with regard to the RP) is that the interaction curves for given states $\tilde{\mathbf{u}}$ (Definition 4.19) can be defined only *locally*, namely, only for states $\mathbf{u}$ sufficiently close to $\tilde{\mathbf{u}}$. Such states (viewed as left states) can be connected to $\tilde{\mathbf{u}}$ (viewed as a fixed right state) by a fan of m waves, associated with the m characteristic families. The exact statement is as follows.

Theorem 4.27 (Solution to the RP) *Let* $\mathbf{u}_R \in \mathbb{R}^m$ *be given. Then there exists a small neighborhood* $C_{\mathbf{u}_R}$ *of* $\mathbf{u}_R$, *having the following property: For every* $\mathbf{v} \in C_{\mathbf{u}_R}$ *the Riemann problem for (4.1), with initial data* $(\mathbf{u}_L = \mathbf{v}, \mathbf{u}_R)$ *has a*

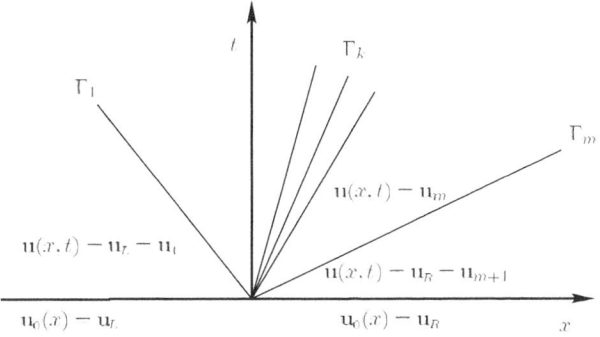

Figure 4.7. The solution to the Riemann problem.

unique solution $\mathbf{u}(x,t) = \mathbf{w}(\frac{x}{t})$ ($x \in \mathbb{R}, t > 0$). This solution consists of m waves $\Gamma_1, \Gamma_2, \ldots, \Gamma_m$. The wave $\Gamma_k (1 \leq k \leq m)$ is an admissible shock, a contact discontinuity, or a CRW associated with the kth characteristic family (see Figure 4.7).

Before proving the theorem, let us consider the structure of the solution $\mathbf{u}(x,t)$. First it is stated that it is constant along straight rays emanating from the origin. We refer to such a solution as "self-similar." This self-similarity is natural in view of the self-similarity of Equation (4.1) and the special initial data for the RP. In other words, the RP is invariant [both (4.1) and the initial data] under any rescaling $x' = \alpha x$, $t' = \alpha t$ ($\alpha > 0$) of the space and time coordinates. Second, the solution consists of sectors ("wedges") in the (x,t) half-plane ($t > 0$), in which it assumes constant values $\mathbf{u}_1, \ldots, \mathbf{u}_{m+1}$ ($\mathbf{u}_1 = \mathbf{v}$, $\mathbf{u}_{m+1} = \mathbf{u}_R$) connected by waves $\Gamma_1, \ldots, \Gamma_m$. The wave Γ_k ($1 \leq k \leq m$), connecting $\mathbf{u}_k, \mathbf{u}_{k+1}$ belongs to exactly one of the following three types:

(i) *admissible ("entropy") shock,*
   that is, a jump discontinuity moving at speed S_k such that

$$\mathbf{F}(\mathbf{u}_{k+1}) - \mathbf{F}(\mathbf{u}_k) = S_k(\mathbf{u}_{k+1} - \mathbf{u}_k)$$

   and such that the entropy conditions (4.36), (4.38) are satisfied, with

$$\mathbf{u}_{k+1} = \mathbf{u}_k(y), \mathbf{u}_k = \tilde{\mathbf{u}};$$

(ii) *contact discontinuity,*
   that is, a jump discontinuity satisfying (4.40), with

$$\mathbf{u}_{k+1} = \mathbf{u}_k(y), \mathbf{u}_k = \tilde{\mathbf{u}} \text{ and } S_k = \lambda_k(\mathbf{u}_k); \text{ or}$$

(iii) CRW $\mathbf{w}_k(y)$ as in Definition 4.5, with

$$\mu_1 \leq y \leq \mu_2, \mu_1 = \lambda_k(\mathbf{u}_k), \mu_2 = \lambda_k(\mathbf{u}_{k+1}). \tag{4.43}$$

An important observation is that any two adjacent waves Γ_k, Γ_{k+1}, $1 \leq k \leq m - 1$, do not interact with each other. Indeed, the speed of Γ_k is S_k [in cases (4.43) (i) and (ii)] or $\lambda_k(\mathbf{u}_{k+1})$, the fastest moving characteristic of the CRW. By the entropy conditions (4.36) and (4.38) we have $S_k < \min(\lambda_k(\mathbf{u}_k), \lambda_{k+1}(\mathbf{u}_{k+1}))$. The eigenvalue $\lambda_{k+1}(\mathbf{u}_{k+1})$ is the speed of the slowest moving characteristic of Γ_{k+1}, if it happens to be a CRW. If Γ_{k+1} is a contact discontinuity, then $S_{k+1} = \lambda_{k+1}(\mathbf{u}_{k+1})$ [see (4.40)]. Finally, if Γ_{k+1} is a shock wave, then the entropy condition (4.36) yields $S_{k+1} > \lambda_{k+1}(\mathbf{u}_{k+2}) > \lambda_k(\mathbf{u}_k)$. The last inequality follows from (4.10), assuming that $|\mathbf{u}_k - \mathbf{u}_{k+2}|$ is small (as all values are in the neighborhood of $\mathbf{u}_R$). We have therefore obtained a stable pattern of noninteracting self-similar waves $\Gamma_1, \ldots, \Gamma_m$, where Γ_{k+1} is faster than Γ_k, $k = 1, \ldots, m - 1$. Note that we have made use here of the two parts [(4.36), (4.38)] of the entropy condition.

We can now turn to the proof of Theorem 4.27. Geometrically speaking, the proof relies on the fact that the interaction curves $I_k^r(\tilde{\mathbf{u}})$, $k = 1, \ldots, m$ (see Definition 4.19), can serve as a "coordinate system" in a neighborhood of $\tilde{\mathbf{u}}$ in the "phase space" $\mathbb{R}^m$.

Proof (of Theorem 4.27) Given a state $\mathbf{u}_R$, we can find a small neighborhood $C_{\mathbf{u}_R} \subseteq \mathbb{R}^m$ of $\mathbf{u}_R$ having the following property:

For every $\mathbf{v} \in C_{\mathbf{u}_R}$ the interaction curves $I_k^r(\mathbf{v})$, $k = 1, \ldots, m$ (see Definition 4.19), exist and, for any $1 \leq k \leq m$, the curve $I_k^r(\mathbf{v})$ comprises all possible states (in $C_{\mathbf{u}_R}$) that can be connected as left states to the right state $\mathbf{v}$ by means of a k-CRW, an admissible k-shock, or a k-contact discontinuity (see Figure 4.8).

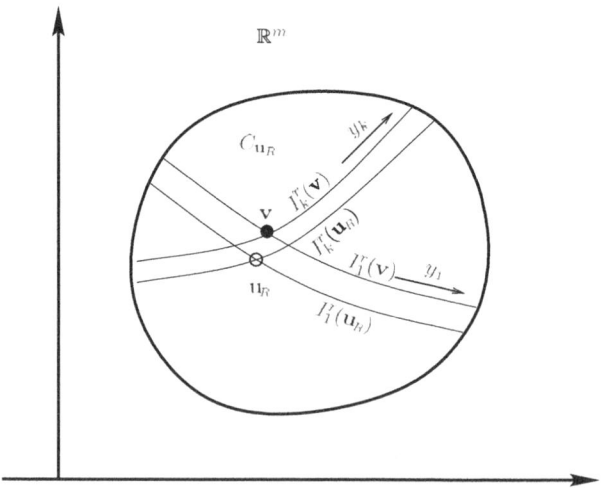

Figure 4.8. A neighborhood of a state $\mathbf{u}_R$ "covered" by interaction curves, and their parametrization.

Along the curve $I_m^r(\mathbf{u_R})$ we use a parameter y_m, so that $y_m = 0$ at $\mathbf{u_R}$. In other words, this is the parameter y of Definition 4.19 (with $\tilde{\mathbf{u}} = \mathbf{u_R}, k = m$) shifted by μ. Next, for every $\mathbf{u}_m \in I_m^r(\mathbf{u_R})$ we use a parameter y_{m-1} along the curve $I_{m-1}^r(\mathbf{u}_m)$, so that $y_{m-1} = 0$ at $\mathbf{u}_m$ (see Figure 4.8). Note that if $\mathbf{u}_m^1, \mathbf{u}_m^2$ are two different points on $I_m^r(\mathbf{u_R})$, the tangent vectors to the curves $I_{m-1}^r(\mathbf{u}_m^1)$ and $I_{m-1}^r(\mathbf{u}_m^2)$ are, respectively, $\mathbf{r}_{m-1}(\mathbf{u}_m^1)$ and $\mathbf{r}_{m-1}(\mathbf{u}_m^2)$ [see (4.25)]. Since both of them are close to $\mathbf{r}_{m-1}(\mathbf{u_R})$, the curves $I_{m-1}^r(\mathbf{u}_m^1)$, $I_{m-1}^r(\mathbf{u}_m^2)$ do not intersect in the small neighborhood $C_{\mathbf{u_R}}$. We continue in this fashion. The last step is to parametrize, for every $\mathbf{u}_2 \in I_2^r(\mathbf{u}_3)$, the curve $I_1^r(\mathbf{u}_2)$ by means of a parameter y_1, which vanishes at $\mathbf{u}_2$. The tangent to this curve at $\mathbf{u}_2$ is $\mathbf{r}_1(\mathbf{u}_2)$ [which is close to $\mathbf{r}_1(\mathbf{u_R})$]. We may summarize this construction by saying that for every $\mathbf{y} = (y_1, \ldots, y_m) \in \mathbb{R}^m$, in a small neighborhood of $\mathbf{y} = \mathbf{0}$, we find a point $\mathbf{u}_1 = \Phi(\mathbf{y}) \in C_{\mathbf{u_R}}$ [going through the sequence in which $\mathbf{u}_m$ corresponds to y_m on $I_m^r(\mathbf{u_R})$, $\mathbf{u}_{m-1}$ to y_{m-1} on $I_{m-1}^r(\mathbf{u}_m)$, . . .]. Clearly, the rows of the Jacobian matrix $D_\mathbf{y}\Phi(\mathbf{y})$ at $\mathbf{y} = \mathbf{0}$ are $\mathbf{r}_1(\mathbf{u_R}), \ldots, \mathbf{r}_m(\mathbf{u_R})$, so that the matrix is nonsingular. The Implicit Function Theorem now implies that a small ball $B_\delta = \{\mathbf{y} \in \mathbb{R}^m, ||\mathbf{y}|| < \delta\}$ is mapped (one-to-one) unto a small neighborhood of $\mathbf{u_R}$, contained in $C_{\mathbf{u_R}}$. This concludes the proof of our theorem, since any point $\mathbf{u}_1 = \Phi(\mathbf{y})$ ($\mathbf{y} \in B_\delta$) is connected to $\mathbf{u_R}$ by a sequence (resulting from the construction) $\mathbf{u}_{m+1} = \mathbf{u_R}, \mathbf{u}_m, \ldots, \mathbf{u}_2$, so that $\mathbf{u}_k$ and $\mathbf{u}_{k+1}$ are, respectively, the left and right states of a k-wave, $\Gamma_k, k = 1, 2, \ldots, m$.   □

Remark 4.28 In analogy with the scalar case [see the discussion following Equation (2.26)] it can be shown that the solution constructed in Theorem 4.27 is a weak solution in the sense of (4.16). Note, however, that it is not a classical solution; in addition to the jump discontinuities, the derivatives (but not the solution itself) are not continuous at the two extreme characteristics of a CRW.

4.2 Euler Equations of Quasi-1-D, Compressible, Inviscid Flow

We consider here the time-dependent flow of a compressible, inviscid fluid moving through a duct of variable cross section. The flow is assumed to be *quasi-one-dimensional* (quasi-1-D); namely, at any given time t the flow is uniform over every cross section of the duct (but, of course, may vary from one cross section to another). Thus, for a duct as depicted in Figure 4.9, we let r be a spatial coordinate along the main axis. Our hypothesis is then that all flow quantities (density, pressure, velocity, etc.) depend only on (r, t).

Certainly, in many cases this is just a simplified model approximating a more realistic two- (or three-) dimensional flow, based on a physical assumption that

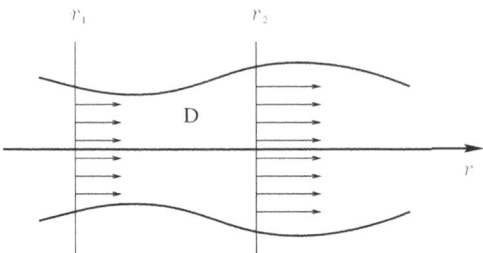

Figure 4.9. Quasi-1-D flow in a duct.

the flow varies primarily along the duct axis. However, there are three cases of substantial physical significance, in which the model is exact:

(A) *Planar flow*, sometimes referred to as "one-dimensional" flow. Here the whole flow is aligned with one direction, say the x axis. All flow quantities depend only on x (which is now the "r coordinate"), in addition to the time t. The velocity vector has only an x component $u(x, t)$.

(B) *Cylindrical flow*. In this case the flow is symmetric about a fixed axis, say the z axis. The coordinate r is now $r = (x^2 + y^2)^{1/2}$ and all flow quantities are functions of (r, t). The velocity $\mathbf{v}(x, y, z, t)$ satisfies $\mathbf{v}(x, y, z, t) = r^{-1}u(r, t)(x\mathbf{i} + y\mathbf{j})$ (where $\mathbf{i}, \mathbf{j}, \mathbf{k}$ are the unit vectors along the x, y, z axes, respectively). Observe that the duct can now be taken as a sector $\lambda_1 x < y < \lambda_2 x$, $0 \leq \lambda_1 < \lambda_2$, $x \geq 0$.

(C) *Spherical flow*. In this case the flow is symmetric about a fixed center O, the origin. The coordinate r is now the distance $(x^2 + y^2 + z^2)^{1/2}$ and the velocity is radial, namely, $\mathbf{v}(x, y, z, t) = r^{-1}u(r, t)(x\mathbf{i} + y\mathbf{j} + z\mathbf{k})$.

The Flow Equations

We now turn to the equations governing our quasi-1-D flow. They express the three basic physical laws governing the flow: conservation of mass, conservation of momentum (or, alternatively, Newton's second law), and conservation of energy. These laws are most easily derived by a "control volume" (or "integral") approach: One considers a fixed mass of fluid and applies the conservation laws to it. As is usually the case, we assume that there are no external forces (such as gravity or electromagnetic) so that the only existing force is due to the hydrodynamic pressure. We shall give here a brief outline of the derivation and refer the reader to fluid dynamics books, such as those by Chorin and Marsden [28], Courant and Friedrichs [30], or Landau and Lifshitz [74], for a detailed derivation of the equations. The reader is also advised to consult these monographs for closely related topics, such as the isotropic character of the hydrodynamic pressure (which makes it a scalar function) or basic thermodynamic facts concerning entropy, internal energy, and their dependence on pressure and density ("equation of state"). In the course of our presentation of the equations, we shall only refer briefly to these facts.

Let $A(r)$ be the cross-sectional area of the duct at r. Clearly, $A(r) \equiv 1$ for planar flow, $A(r) = r$ for cylindrical flow, and $A(r) = r^2$ for spherical flow. There are three unknown functions. The velocity $u(r, t)$, which is the (scalar) component along the r axis, the density (mass per unit volume) $\rho(r, t)$, and the total specific energy (per unit mass) $E(r, t)$. The energy E consists of the kinetic energy (per unit mass) $\frac{1}{2}u^2$ and the internal (thermodynamic) energy e, so $E = \frac{1}{2}u^2 + e$. A basic thermodynamic postulate in the derivation of the equations is that the hydrodynamic pressure p is a function $p = p(e, \rho)$. We refer to this function as the *equation of state* of the fluid. The resultant force on a given volume D of the fluid is then $-\int_{\partial D} p\mathbf{n}\,d\sigma = -\int_{D} \nabla p\,d\tau$, where ∂D is the boundary of D, $d\sigma$ is the surface element, $d\tau$ is the volume element, and $\mathbf{n}$ is the outward (unit) normal to ∂D. The preceding equality follows by Stokes' theorem. Similarly, the work done by the pressure on the fluid (per unit time) is $-\int_{D} \nabla \cdot (p\mathbf{v})\,d\tau$, where $\mathbf{v}$ is the velocity vector. Note also that the outflux (per unit time) across ∂D, for any quantity ψ (in our case $\psi = \rho$ or $\psi = \rho E$), is $\int_{\partial D} \psi(\mathbf{v} \cdot \mathbf{n})\,d\sigma$. We can now take D as the segment $r_1 \leq r \leq r_2$ of the duct (see Figure 4.9). Incorporating these considerations into the balance equations for mass, momentum, and energy, and letting $r_2 - r_1$ go to zero, we obtain the conservation equations in differential form:

(i) $\frac{\partial}{\partial t}\rho + A^{-1}\frac{\partial}{\partial r}[A\rho u] = 0$ (conservation of mass),

(ii) $\frac{\partial}{\partial t}(\rho u) + A^{-1}\frac{\partial}{\partial r}[A\rho u^2] + \frac{\partial p}{\partial r} = 0$ (conservation of momentum),   (4.44)

(iii) $\frac{\partial}{\partial t}(\rho E) + A^{-1}\frac{\partial}{\partial r}[A(\rho E + p)u] = 0$ (conservation of energy),

where the equation of state $p = p(e, \rho)$ is given.

Let us inspect the structure of Equations (4.44) in light of the general framework given by (4.1). In our case $m = 3$ and we can write the system (4.44) as

$$\frac{\partial}{\partial t}\mathbf{U} + A^{-1}\frac{\partial}{\partial r}\left[A\mathbf{F}(\mathbf{U})\right] + \frac{\partial}{\partial r}\mathbf{G}(\mathbf{U}) = 0, \qquad (4.45)$$

$$\mathbf{U} = \begin{bmatrix} \rho \\ \rho u \\ \rho E \end{bmatrix}, \qquad \mathbf{F}(\mathbf{U}) = \begin{bmatrix} \rho u \\ \rho u^2 \\ (\rho E + p)u \end{bmatrix}, \qquad \mathbf{G}(\mathbf{U}) = \begin{bmatrix} 0 \\ p \\ 0 \end{bmatrix}.$$

Writing

$$U = \begin{bmatrix} u_1 \\ u_2 \\ u_3 \end{bmatrix}$$

we have readily

$$F(U) = \begin{bmatrix} u_2 \\ \frac{u_2^2}{u_1} \\ \frac{u_2 u_3}{u_1} + \frac{u_2}{u_1} p \end{bmatrix}, \qquad G(U) = \begin{bmatrix} 0 \\ p \\ 0 \end{bmatrix}, \tag{4.46}$$

where

$$p = p(e, \rho) = p\left(\frac{2u_1 u_3 - u_2^2}{2u_1^2}, u_1 \right),$$

so that F and G are indeed functions of U, the vector of unknown functions. However, the presence of the function $A(r)$ means that the system (4.45) cannot be cast in the form (4.1) [i.e., $A^{-1}(AF(U))_r$ cannot, in general, be written as $(\tilde{F}(U))_r$ for some $\tilde{F}$]. However, this is possible in the planar case $A(r) \equiv 1$. We record it for future use (using again "x" for the spatial coordinate):

$$\frac{\partial}{\partial t} U + \frac{\partial}{\partial x} [F(U) + G(U)] = 0. \tag{4.47}$$

We conclude that the study of solutions of (4.45) must be carried out in its special context.[5] However, we shall see that the basic observations and results developed for the general case (4.1) can be adapted here with minor modifications. From a "physical" point of view, this is possible because $A(r)$ varies "slowly" near any point $r = r_0$, since it is continuously differentiable (an assumption that we retain throughout this section). Thus, the physical features near that point (such as the wave pattern, characteristics, discontinuities, etc.) are "close" to those of (4.47).

Eigenvalues and Characteristic Equations

Differentiation of the middle term in (4.45) yields

$$\frac{\partial}{\partial t} U + \frac{\partial}{\partial r} [F(U) + G(U)] + A^{-1} A' F(U) = 0. \tag{4.48}$$

[5] Note that in the cylindrical and spherical cases the coordinate r is restricted to nonnegative values, and suitable boundary conditions must be imposed at the "singularity" $r = 0$.

Using the representation (4.46) we could now proceed in parallel to the derivation of (4.6), finding the eigenvalues [of the Jacobian of $\mathbf{F}(\mathbf{U}) + \mathbf{G}(\mathbf{U})$ with respect to $\mathbf{U}$] and the associated characteristic relations. However, it turns out to be easier, in our case, to replace the three unknown functions (components of $\mathbf{U}$) by the three functions ρ, u, S, where $S = S(e, \rho)$ is the entropy.[6] We recall that the function S may be defined (see Courant and Friedrichs [30, Chapter I]) by the first law of thermodynamics; it asserts that in an "infinitesimal" reversible process, the change in internal energy of a fixed element of an ideal fluid is equal to the heat absorbed by the element and the work done on it by the pressure force. Thus, with $\tau = 1/\rho$,

$$de = T\, dS - p\, d\tau, \tag{4.49}$$

where T is the temperature and $T\, dS = dQ$ is the absorbed heat.[7] When Equation (4.49) is incorporated into our flow setup, we must keep in mind that it applies to *fixed* mass elements. Thus, time derivatives must be taken along "particle paths" $\frac{dr}{dt} = u$. If we denote such derivatives as $\frac{D}{Dt} = \frac{\partial}{\partial t} + u\frac{\partial}{\partial r}$ (later on we shall see the important role played by such derivatives in the Lagrangian framework), we can write (4.49) as

$$\frac{De}{Dt} = T\frac{DS}{Dt} - p\frac{D\tau}{Dt}. \tag{4.50}$$

Note that with this notation Equation (4.44)(iii) can be written as

$$\frac{D}{Dt}(\rho E) + \frac{\partial}{\partial r}(pu) + \rho E\frac{\partial u}{\partial r} + A^{-1}A'(\rho E + p)u = 0. \tag{4.51}$$

Equation (4.44)(i) can be written as

$$\frac{D\rho}{Dt} + \rho\frac{\partial u}{\partial r} + A^{-1}A'\rho u = 0, \tag{4.52}$$

so that, multiplying it by E and subtracting the result from (4.51) we have

$$\rho\frac{D}{Dt}E + \frac{\partial}{\partial r}(pu) + A^{-1}A'pu = 0. \tag{4.53}$$

---

[6] It is known (Evans [36, Chapter 11]) that a transformation (even smooth) of the unknown variables is in general not admissible, as it leads to wrong jump conditions across discontinuities. It is allowed, however, in domains where the solution is smooth, as we assume here.

[7] Mathematically speaking, we can consider $\frac{1}{T}$ as an "integrating factor" for the form $de + p\, d\tau$, so that $\frac{1}{T}(de + p\, d(\frac{1}{\rho}))$, with $p = p(e, \rho)$, is a total differential of a function of (e, ρ).

Similarly, writing (4.44)(ii) as

$$\frac{D}{Dt}(\rho u) + \rho u \frac{\partial u}{\partial r} + \frac{\partial p}{\partial r} + A^{-1}A'\rho u^2 = 0 \tag{4.54}$$

and subtracting from it (4.52) (multiplied by u) we get the following form of the momentum equation:

$$\rho \frac{Du}{Dt} + \frac{\partial p}{\partial r} = 0. \tag{4.55}$$

Using $E = e + \frac{1}{2}u^2$ and (4.55) in (4.53) we get

$$\rho \frac{De}{Dt} + p \frac{\partial u}{\partial r} + A^{-1}A'pu = 0. \tag{4.56}$$

However, from (4.52), with $\tau = 1/\rho$, we obtain

$$\rho \frac{D\tau}{Dt} = -\frac{p}{\rho^2}\frac{D\rho}{Dt} = \frac{p}{\rho}\left(\frac{\partial u}{\partial r} + A^{-1}A'u\right), \tag{4.57}$$

and so the energy equation (4.56) takes the form

$$\rho \frac{De}{Dt} + \rho p \frac{D\tau}{Dt} = 0, \qquad \tau = \frac{1}{\rho}. \tag{4.58}$$

Comparing this equation with (4.50) we conclude finally that the energy equation (4.44)(iii) is equivalent to[8]

$$\frac{D}{Dt}S = 0. \tag{4.59}$$

In summary, Equations (4.52), (4.55), and (4.59) form an equivalent system to (4.44), which expresses the basic conservation laws. We should keep in mind, however, that this equivalence holds only in domains of smooth flow. The new set of unknown functions is now (ρ, u, S). Hence in the momentum equation (4.55) the pressure p is interpreted as a function of the two thermodynamic variables ρ, S and

$$\frac{\partial p}{\partial r} = \left(\frac{\partial p}{\partial \rho}\right)_S \frac{\partial \rho}{\partial r} + \left(\frac{\partial p}{\partial S}\right)_\rho \frac{\partial S}{\partial r}. \tag{4.60}$$

---

[8] This means that the flow is "adiabatic," namely, that every mass element retains its entropy as long as it undergoes only a slow, reversible process, and, in particular, as long as the flow is smooth.

The term $\left(\frac{\partial p}{\partial \rho}\right)_S$, expressing the rate of change of pressure with respect to density, at fixed entropy, is of prime significance in the study of compressible fluid flow. A basic hypothesis is that it is positive. (Physically, it means simply that compressing a fluid element, without allowing for heat exchange across its boundary, leads to increased pressure.) Thus, we can define a thermodynamic function $c = c(\rho, S) > 0$ by

$$c^2 = \left(\frac{\partial p}{\partial \rho}\right)_S.$$

(4.61)

The function c is the "speed of sound" [at (ρ, S)], a term that will be clarified in the following. Grouping together Equations (4.52), (4.55), and (4.59), and noting (4.60) and (4.61) in the second one, we can write the system (in domains of smooth flow) as

$$\frac{\partial}{\partial t}\begin{bmatrix} \rho \\ u \\ S \end{bmatrix} + \begin{bmatrix} u & \rho & 0 \\ \frac{c^2}{\rho} & u & \frac{1}{\rho}\left(\frac{\partial p}{\partial S}\right)_\rho \\ 0 & 0 & u \end{bmatrix} \cdot \frac{\partial}{\partial r}\begin{bmatrix} \rho \\ u \\ S \end{bmatrix} + A^{-1}A'\begin{bmatrix} \rho u \\ 0 \\ 0 \end{bmatrix} = \begin{bmatrix} 0 \\ 0 \\ 0 \end{bmatrix}.$$

(4.62)

We can now proceed as in the derivation of (4.4). First, it is easily seen that the eigenvalues of the matrix

$$\mathbf{A}(\rho, u, S) = \begin{bmatrix} u & \rho & 0 \\ \frac{c^2}{\rho} & u & \frac{1}{\rho}\left(\frac{\partial p}{\partial S}\right)_\rho \\ 0 & 0 & u \end{bmatrix}.$$

(4.63)

are given by

$$\lambda_1(\rho, u, S) = u - c, \qquad \lambda_2(\rho, u, S) = u, \qquad \lambda_3(\rho, u, S) = u + c, \quad (4.64)$$

with corresponding (left) eigenvectors

$$\mathbf{l}_1^T = \left[\frac{c}{\rho}, -1, \frac{1}{\rho c}\left(\frac{\partial p}{\partial S}\right)_\rho\right], \quad \mathbf{l}_2^T = [0, 0, 1], \quad \mathbf{l}_3^T = \left[\frac{c}{\rho}, 1, \frac{1}{\rho c}\left(\frac{\partial p}{\partial S}\right)_\rho\right].$$

(4.65)

The characteristic curves corresponding to the slopes λ_1, λ_3 are labeled, respectively, as C_-, C_+. Those corresponding to λ_2 are labeled as C_0. Note that the C_0 curves are trajectories traced out by fluid particles. They are referred to as "particle paths." Taking the product of each of the eigenvectors with (4.62) we get the characteristic equations as in (4.6). Using again the notation of total time differentiation [see (2.7)] we can write these equations as follows:

(i) $\frac{c}{\rho}\frac{d\rho}{dt} - \frac{du}{dt} + \frac{1}{\rho c}\left(\frac{\partial p}{\partial S}\right)_\rho \frac{dS}{dt} = -A^{-1}A'uc$ along $\quad C_- : \dfrac{dr}{dt} = u - c,$

(ii) $\qquad\qquad\qquad \frac{DS}{Dt} = 0 \qquad\qquad$ along $\quad C_0 : \dfrac{dr}{dt} = u,$ $\qquad$ (4.66)

(iii) $\frac{c}{\rho}\frac{d\rho}{dt} + \frac{du}{dt} + \frac{1}{\rho c}\left(\frac{\partial p}{\partial S}\right)_\rho \frac{dS}{dt} = -A^{-1}A'uc$ along $\quad C_+ : \dfrac{dr}{dt} = u + c.$

It is common to write the characteristic equations in "total differential" form (see, e.g., Courant and Friedrichs [30, Section 34]). In doing so we note that in view of (4.61), we can write

$$\frac{c}{\rho}\frac{d\rho}{dt} + \frac{1}{\rho c}\left(\frac{\partial p}{\partial S}\right)_\rho \frac{dS}{dt} = \frac{1}{\rho c}\frac{dp}{dt},$$

so that (4.66) yields

$\qquad$ (i) $\frac{1}{\rho c}dp \pm du = -A^{-1}A'uc\,dt$ along $\quad C_\pm : \dfrac{dr}{dt} = u \pm c,$

$\qquad$ (ii) $\qquad dS = 0 \qquad$ along $\quad C_0 : \dfrac{dr}{dt} = u.$ $\qquad$ (4.67)

Remark 4.29 (Adiabatic character of the flow) As pointed out in the footnote to (4.59), the entropy remains invariant along particle paths (i.e., $\frac{dr}{dt} = u$), as long as the flow is smooth. Hence, along these lines the pressure $p = p(\rho, S)$ varies only with the density, and invoking (4.61) we get

$$dp = c^2\,d\rho \qquad \text{along} \qquad \frac{dr}{dt} = u. \qquad (4.68)$$

This equation is equivalent to (4.67)(ii).

Finally, we examine the hyperbolic character of our system, in light of the general Definition 4.6.

Claim 4.30 *The system (4.44) [or, equivalently, (4.62)] is strictly hyperbolic. The λ_1, λ_3 characteristic families are genuinely nonlinear (Remark 4.9) whereas the λ_2 family is linearly degenerate (Definition 4.16).*

Proof The assumption $c > 0$ means that the eigenvalues satisfy $\lambda_1 < \lambda_2 < \lambda_3$ [see (4.64)], in compliance with (4.10). The right eigenvectors of $\mathbf{A}$

[Equation (4.63)] are given by

$$
\mathbf{r}_1 = \begin{bmatrix} \frac{\rho}{c} \\ -1 \\ 0 \end{bmatrix}, \qquad
\mathbf{r}_2 = \begin{bmatrix} \frac{1}{c^2}\left(\frac{\partial p}{\partial S}\right)_\rho \\ 0 \\ 1 \end{bmatrix}, \qquad
\mathbf{r}_3 = \begin{bmatrix} \frac{\rho}{c} \\ 1 \\ 0 \end{bmatrix}.
\tag{4.69}
$$

We compute

$$
\nabla_{\rho,u,S}\lambda_1 \cdot \mathbf{r}_1 = \left[-\left(\frac{\partial c}{\partial \rho}\right)_S, 1, -\left(\frac{\partial c}{\partial S}\right)_\rho \right] \cdot \mathbf{r}_1 = -\frac{\rho}{c}\left(\frac{\partial c}{\partial \rho}\right)_S - 1 \le -1,
$$

where we have imposed the plausible thermodynamic assumption that the speed of sound increases with density, at a fixed entropy [compare to the analogous assumption on the pressure, prior to (4.61)]. Thus, we can normalize $\mathbf{r}_1$ so that it satisfies (4.13). A similar argument applies to λ_3, $\mathbf{r}_3$. For $\mathbf{r}_2$ we have

$$
\nabla_{\rho,u,S}\lambda_2 \cdot \mathbf{r}_2 = [0, 1, 0] \cdot \mathbf{r}_2 \equiv 0,
$$

so it satisfies (4.34) and hence is linearly degenerate.   □

Remark 4.31 Note that except for the planar case ($A' \equiv 0$) the system is not in strict conservation form [see the discussion following (4.46)]. The concepts of hyperbolicity in the claim refer to the Jacobian matrix $\mathbf{F}'(\mathbf{U}) + \mathbf{G}'(\mathbf{U})$, or, equivalently, to the matrix $\mathbf{A}(\rho, u, S)$ [see (4.63)]. This matrix determines (locally) the main features of the flow. In other words, given initial values of the flow at $t = 0$, the flow evolves, near a point $r = r_0$ and for a short time, approximately as it would in the planar case (4.47). In particular, this matrix determines the interaction curves (Definition 4.19) of the system and thus the local, short-time wave pattern (shocks, centered rarefaction waves, and contact discontinuities). It is only at later times that the "geometric" factor (associated with $A' \ne 0$) comes into play, and the interaction of waves deviates from the planar case.

We note that in all cases the characteristic families split into two groups. The linearly degenerate one (C_0) consists of "particle paths," namely, trajectories in the (r, t) plane satisfying $\frac{dr}{dt} = u$. The genuinely nonlinear families $C_\pm$ consist of trajectories satisfying $\frac{dr}{dt} = u \pm c$ and hence the waves are propagating at speeds $\pm c$ relative to the particle paths. Physically, these waves can be viewed as "traveling ripples" of pressure, caused by the compressive effect of particles moving at different velocities, as expressed by (4.67)(i). In other words, these are the characteristics of the system obtained by linearizing (4.62) that represent the propagation of "infinitesimal pressure waves." It justifies the terminology "sound waves" that is often used in this context and the role of c as the "speed of sound" (Landau and Lifshitz [74]).

Isentropic Flow

A special case of considerable interest is a flow in which the entropy is uniform (in space) and constant (in time), that is, $S(r, t) \equiv S_0$. We refer to this flow as an "isentropic flow." Note that, in general, the flow is adiabatic as long as it is smooth [see (4.59)], so that S is conserved along particle paths. Thus, assuming that the entropy is initially uniform, it remains constant as long as the flow is smooth.

In the case of an isentropic flow the third equation in (4.62) is automatically satisfied and can be left out. The system then reduces to

$$\frac{\partial}{\partial t}\begin{bmatrix} \rho \\ u \end{bmatrix} + \begin{bmatrix} u & \rho \\ \frac{c^2}{\rho} & u \end{bmatrix} \cdot \frac{\partial}{\partial r}\begin{bmatrix} \rho \\ u \end{bmatrix} + A^{-1}A'\begin{bmatrix} \rho u \\ 0 \end{bmatrix} = \begin{bmatrix} 0 \\ 0 \end{bmatrix}. \qquad (4.70)$$

The eigenvalues are $u \pm c$ and the characteristic relations (4.66)(i),(iii) can be written as [see also (4.67)(i), where $dp = c^2 d\rho$]

$$\frac{c}{\rho}d\rho \pm du = -A^{-1}A'\, uc\, dt \qquad \text{along} \qquad C_\pm : \frac{dr}{dt} = u \pm c. \qquad (4.71)$$

Note that $c^2 = \left(\frac{\partial p}{\partial \rho}\right)_S$ is now a function of ρ alone. In the special case of planar flow ($A' = 0$) the relations (4.71) can be integrated to yield

$$R_\pm(\rho, u) = \int_{\rho_0}^{\rho} \frac{c(\beta)}{\beta}d\beta \pm u = constant \qquad \text{along} \qquad C_\pm : \frac{dr}{dt} = u \pm c. \qquad (4.72)$$

The functions $R_\pm(\rho, u)$ are called the "Riemann invariants" of the flow. They serve as very useful tools in solving initial value problems of isentropic flow (Courant and Friedrichs [30, Sections 37 and 38]).

Weak Solutions and Jump Conditions

The definition of a weak solution to the system (4.45) follows the procedure of Definition 4.12. If we take a test function $\phi(r, t) \in C_0^1$, a weak solution $\mathbf{U}(r, t)$ to (4.45) must satisfy

$$\int_{\mathbb{R}} \int_0^{\infty} \left\{ A(r)\left[\mathbf{U}\phi_t + (\mathbf{F}(\mathbf{U}) + \mathbf{G}(\mathbf{U}))\phi_r\right] + A'(r)\mathbf{G}(\mathbf{U})\phi \right\}\, dt\, dr$$

$$+ \int_{\mathbb{R}} A(r)\phi(r, 0)\mathbf{U}_0(r)\, dr = \mathbf{0}, \qquad (4.73)$$

where $U_0(r) = U(r, 0)$. This definition is formally obtained by multiplying (4.45) by $A\phi$ and integrating by parts. Note that if the coordinate r is restricted (say, $r \geq 0$ as in the cylindrical or spherical cases), then the integration in r is accordingly restricted, and $\phi(r, t)$ vanishes near the endpoints of the r-domain.

We can now proceed to study jump discontinuities of the solution $U(r, t)$. If Ω is a domain in the (r, t) plane and $\Gamma = \partial\Omega$ (as in Figure 2.2) we obtain, as in Claim 2.6 and (4.17),

$$\oint_\Gamma A(r)\left[-U(r, t)\,dr + (F(U) + G(U))\,dt\right] - \iint_\Omega A'(r)G(U)\,dr\,dt = 0.$$

$$(4.74)$$

Note that the fact that the system (4.45) is not in conservation form is reflected in the presence of the term $A'(r)G(U)$ in (4.73), (4.74).[9] Using the setup of Figure 2.2 we see that

$$\iint_\Omega A'(r)G(U)\,dr\,dt = \iint_{\Omega_1} + \iint_{\Omega_2} A'(r)G(U)\,dr\,dt$$

(although U is discontinuous across C). Letting $C : r = r(t)$, we set

$$U_\pm(r(t), t) = \lim_{r \to r(t)_\pm} U(r(t), t),$$

$$[U](t) = U_+(r(t), t) - U_-(r(t), t),$$

$$[F(U)](t) = F(U_+(r(t), t)) - F(U_-(r(t), t)),$$

$$[G(U)](t) = G(U_+(r(t), t)) - G(U_-(r(t), t)).$$

Following the derivation of (2.20) we get

$$\int_C A(r(t))\left\{[U](t)r'(t) - [F(U) + G(U)](t)\right\}dt = 0. \qquad (4.75)$$

We conclude, as in Corollary 2.8, that the speed $\sigma(t) = r'(t)$ of the discontinuity satisfies the Rankine–Hugoniot jump condition,

$$[F(U) + G(U)](t) = \sigma(t)[U](t). \qquad (4.76)$$

---

[9] This term appears only in the momentum equation and has the form $A'(r)p$. It constitutes a "geometric source" term for the motion of the fluid. Assuming $p > 0$, this term is positive (resp. negative) in a diverging (resp. converging) duct.

Thus, we obtain a jump condition that is identical to (4.18). In other words, because it "ignores" the variable cross section, the resulting speed σ is equal to the one obtained in the planar case (4.47) (note that the flux here is $\mathbf{F} + \mathbf{G}$). Heuristically, this can be explained by the fact that the jump condition (4.76) concerns a narrow domain, bounded by the cross sections at $r(t) \pm \epsilon$ (small $\epsilon > 0$). The variation of the area $A(r)$ in this domain can be taken as small, converging to the planar case as $\epsilon \to 0$.

In view of Claim 4.30 and the general Definition 4.23, the discontinuity associated with the eigenvalue $\lambda_2 = u$ [see (4.64)] is a *contact discontinuity*, which coincides with the particle path C_0. When it separates two constant states $\mathbf{U}_\pm$, we infer from (4.40) that the velocity u is uniform on the two sides $u_\pm = \sigma$. Inserting this in the Rankine–Hugoniot condition (4.76) and using the explicit form (4.45) of $\mathbf{F}, \mathbf{G}$, we get

$$
\begin{aligned}
\sigma(\rho_+ - \rho_-) &= \sigma(\rho_+ - \rho_-), \\
\sigma^2(\rho_+ - \rho_-) &= \sigma^2(\rho_+ - \rho_-) + p_+ - p_-, \\
\sigma(\rho_+ E_+ - \rho_- E_-) &= \sigma \left[(\rho_+ E_+ + p_+) - (\rho_- E_- + p_-)\right].
\end{aligned}
\tag{4.77}
$$

We conclude that $p_+ = p_-$, whereas $\rho_+ - \rho_-$ is arbitrary. Thus, the velocity and pressure are continuous across C_0, whereas the jump in density is arbitrary. These facts justify the term "material interface," which is used sometimes to describe this discontinuity. The interface travels at the "particle" (or "fluid") speed and separates two regions that carry equal pressures but differ in densities. No mass is crossing over from one side to the other, so they can be regarded as two separate materials.

The discontinuities associated with the eigenvalues $\lambda_{1,3} = u \pm c$ are shock waves, since the families are genuinely nonlinear (Claim 4.30). They are designated, following Definition 4.18, as 1-shocks and 3-shocks. The Lax entropy condition for a 3-shock implies, by (4.36) [see Figure 4.5(b)],

$$
u_- + c_- > \sigma_3 > u_+ + c_+.
\tag{4.78}
$$

Because c is the speed of sound relative to the fluid, $u + c$ is the speed of acoustic waves in the (r, t)-frame. The inequalities (4.78) express the very fundamental fact concerning admissible shocks, namely, that *they are supersonic with respect to the "front" (preshock) state and subsonic with respect to the "back" (postshock) state.* (See the shock σ_3 in Figure 4.10.) Clearly, the same conclusion applies to the case of 1-shocks. However, the inequality $u_+ - c_+ < \sigma_1 < u_- - c_-$ (see the shock σ_1 in Figure 4.10) now means $\sigma_1 - u_- < -c_-, \sigma_1 - u_+ > -c_+$, so that the acoustic wave moving to the left at speed (relative to the fluid) $-c_-$ (resp. $-c_+$) is slower (resp. faster) than the shock. Thus, the shock is supersonic with respect to the state on its *left*, which is now regarded as the "front"

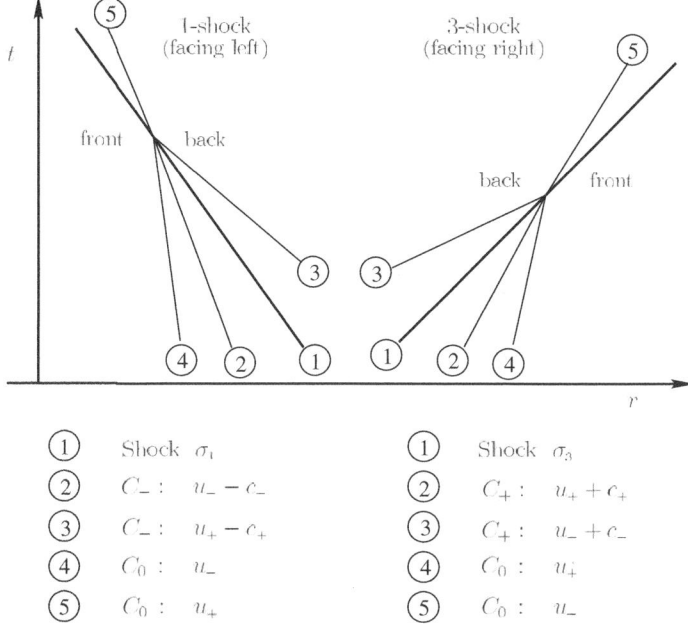

Figure 4.10. Right-facing (σ_3) and left-facing (σ_1) shocks.

(preshock) state. This is expressed by saying that 1-shocks "face left" whereas 3-shocks "face right." We note also that the fluid crosses the shock trajectory from "front" to "back." Indeed, we have $\sigma_3 > u_+$, whereas by the entropy condition (4.38) $\sigma_3 > u_-$ (see Figure 4.10). The same conclusion holds in the case of 1-shocks. Note that these conclusions hold *regardless* of the signs of the speeds σ_1, σ_3. Both can be positive or negative (or the speeds can even be zero, in which case the shocks are stationary).

The jump conditions (4.76) for 1- and 3-shocks can be further developed, so as to give more specific algebraic expressions for the Hugoniot curves (i.e., the "shock" part of the interaction curve; see Definition 4.19). Such expressions are basic in the study of compressible fluid dynamics (and the GRP method). We shall present them later in the chapter, following the introduction of the Lagrangian framework.

Lagrangian Coordinates

The Lagrangian representation of the flow is based on the flow map, that is, on tracking particles as they move about. This means that each particle is assigned a fixed set of parameters that uniquely characterize it at any time. These parameters are called the "Lagrangian coordinates" of the particle. They can serve as a substitute of the standard underlying coordinate system (the "Eulerian" system) as long as the flow is sufficiently well "organized" (e.g., no two particles collide). In many cases, a natural choice for the Lagrangian coordinates of a particle is the set of its (Eulerian) coordinates at $t = 0$. However, in our case of quasi-1-D flow there exists an alternative definition of the Lagrangian coordinate, which will prove to be instrumental in the solution of generalized Riemann problems.

Definition 4.32 The Lagrangian coordinate ξ of the particle initially located ($t = 0$) at r is given by

$$\xi = \int_{r_0}^{r} A(s)\rho(s, 0)\, ds, \tag{4.79}$$

where r_0 is a fixed point in the initial domain.

The coordinate ξ is therefore (see Figure 4.9) the total mass initially enclosed in the duct section $r_0 \leq s \leq r$ (or $r \leq s \leq r_0$ with a negative sign, if $r < r_0$). Since by assumption particle paths issuing from two different points $r_1 \neq r_2$ do not intersect each other, the transformation $r \to \xi(r, t)$ and its inverse $\xi(r, t) \to r$ are well defined at all later times. The trajectory $t \to r(\xi, t)$ is the particle path starting at $r(\xi, 0)$ [which is the r related to ξ by (4.79)]. Thus,

$$\frac{\partial}{\partial t} r(\xi, t) = u(r(\xi, t), t). \tag{4.80}$$

Also, the mass enclosed between any two particle paths $r(\xi, t)$ and $r(\xi + \Delta\xi, t)$ is constant in time and equal to its initial value $\Delta\xi$. Thus,

$$\Delta\xi = \int_{r(\xi,t)}^{r(\xi+\Delta\xi,t)} A(s)\rho(s, t)\, ds.$$

Dividing by $\Delta\xi$ and letting $\Delta\xi \to 0$ we obtain

$$1 = A(r(\xi, t)) \cdot \rho(r(\xi, t), t) \cdot \frac{\partial}{\partial \xi} r(\xi, t),$$

namely,

$$\frac{\partial}{\partial \xi} r(\xi, t) = [A(r(\xi, t))\rho(r(\xi, t), t)]^{-1}. \tag{4.81}$$

Let Q denote any flow variable (such as density or pressure) that is a function $Q = Q(r, t)$. By changing $r = r(\xi, t)$ it becomes a function $Q = Q(r(\xi, t), t)$ of the Lagrangian coordinate and time. For simplicity we shall persist with the notation $Q = Q(\xi, t)$, while indicating clearly (especially in differentiations) which variables are being used.

By (4.80), (4.81) the Lagrangian time derivative satisfies

$$\frac{\partial}{\partial t} Q(\xi, t) = \frac{\partial}{\partial t} Q(r, t)\Big|_{r=r(\xi,t)} + u(\xi, t) \frac{\partial}{\partial r} Q(r, t)\Big|_{r=r(\xi,t)} \tag{4.82}$$

and is identical to the "total time derivative" $\frac{D}{Dt}$ introduced earlier [see the paragraph preceding Equation (4.50)]. Thus, we can write the flow equations in terms of the basic variables $\tau = \frac{1}{\rho}$, u, and E by simply recalling Equations (4.52), (4.53), and (4.55) and using (4.81), (4.82). We obtain the system

$$\frac{\partial}{\partial t} \mathbf{V} + \frac{\partial}{\partial \xi} (A\mathbf{\Phi}(\mathbf{V})) + A \frac{\partial}{\partial \xi} \mathbf{\Psi}(\mathbf{V}) = 0,$$

$$\tag{4.83}$$

$$\mathbf{V} = \begin{bmatrix} \tau \\ u \\ E \end{bmatrix}, \quad \mathbf{\Phi}(\mathbf{V}) = \begin{bmatrix} -u \\ 0 \\ pu \end{bmatrix}, \quad \mathbf{\Psi}(\mathbf{V}) = \begin{bmatrix} 0 \\ p \\ 0 \end{bmatrix},$$

where it is understood that all flow variables, as well as the coordinate r, are functions of ξ, t. In particular, $A = A(r(\xi, t))$. This system is equivalent to the basic system (4.45). As in the Eulerian case, we can replace the energy equation (in regions of smooth flow) by $\frac{\partial}{\partial t} S(\xi, t) = 0$ [see (4.59)]. Doing so, and carrying out the differentiations in (4.83), we get, parallel to (4.62),

$$\frac{\partial}{\partial t} \begin{bmatrix} \tau \\ u \\ S \end{bmatrix} + A \begin{bmatrix} 0 & -1 & 0 \\ -c^2\tau^{-2} & 0 & \left(\frac{\partial p}{\partial S}\right)_\rho \\ 0 & 0 & 0 \end{bmatrix} \cdot \frac{\partial}{\partial \xi} \begin{bmatrix} \tau \\ u \\ S \end{bmatrix} + \frac{\partial A}{\partial \xi} \begin{bmatrix} -u \\ 0 \\ 0 \end{bmatrix} = \begin{bmatrix} 0 \\ 0 \\ 0 \end{bmatrix}.$$

$$\tag{4.84}$$

The eigenvalues of the system are now $\mu_1 = -A\rho c$, $\mu_2 = 0$, $\mu_3 = A\rho c$. As in Claim 4.30, the μ_1, μ_3 families are genuinely nonlinear whereas the μ_2 family is linearly degenerate. In fact, the acoustic "impedance" $g = \rho c$ is the speed of sound relative to the fluid in the planar Lagrangian frame ($A \equiv 1$). The μ_1, μ_3 characteristic curves are, respectively, the sound waves propagating in the $\pm\xi$ directions. The μ_2 characteristics ($\xi = constant$) are the particle paths. There is no need to derive the characteristic relations anew from (4.84), as they are already given in total differential form by Equations (4.67). Retaining the notation $A' = \frac{d}{dr} A(r)\big|_{r=r(\xi,t)}$, we obtain

$$dp \pm g\, du = -A^{-1} A' g u c\, dt \quad \text{along} \quad \frac{d\xi}{dt} = \pm Ag, \quad g = \rho c. \quad (4.85)$$

Remark 4.33 The definition of a weak solution can be based on the system (4.83), as a substitute to (4.45). However, it is not at all evident that the resulting discontinuous waves (shocks or contact discontinuities) are those obtained from the Eulerian frame by the transformation $\xi = \xi(r, t)$. In fact, as the case of $v_t + v v_x = 0$ and $(v^2)_t + \frac{2}{3}(v^3)_x = 0$ (obtained by multiplying the first by v) shows, algebraic manipulation of a nonlinear equation may lead to different waves and jump conditions. However, the Lagrangian and Eulerian frames are indeed equivalent, as is shown in Wagner [116].

Shock Waves – Detailed Study of the Jump Condition

The considerations following the jump condition (4.76) led us to some important conclusions concerning the nature of flow discontinuities. The density undergoes a jump across a contact discontinuity, whereas pressure and velocity remain continuous. In particular, there is no mass flux across such a discontinuity. However, mass is moving from front to back across 1- and 3-shocks. The entropy condition implies that the shock is supersonic (resp. subsonic) with respect to the front (resp. back) state. In the following paragraphs we shall study the jump conditions, as well as the characteristic relations throughout centered rarefaction waves, in more detail. The formulas derived here will serve as the basis for the solution of the Riemann and generalized Riemann problems.

Note first that the jump condition (4.76) is independent of the variable cross section $A(r)$. Using the basic form (4.45) it can be written in terms of the three algebraic relations

> (i) $\sigma(\rho_+ - \rho_-) = \rho_+ u_+ - \rho_- u_-,$
>
> (ii) $\sigma(\rho_+ u_+ - \rho_- u_-) = \rho_+ u_+^2 + p_+ - (\rho_- u_-^2 + p_-),$ $\qquad$ (4.86)
>
> (iii) $\sigma(\rho_+ E_+ - \rho_- E_-) = (\rho_+ E_+ + p_+)u_+ - (\rho_- E_- + p_-)u_-.$

To fix the ideas, we assume that "+" and "−" denote, respectively, the front (preshock) and back (postshock) states (corresponding to a 3-shock as in Figure 4.10, with $\sigma = \sigma_3$). Let $v_\pm = u_\pm - \sigma$ be the velocities of the fluid relative to the shock. It is then easy to see that Equations (4.86) take the form

> (i) $\rho_+ v_+ = \rho_- v_-$
>
> (ii) $\rho_+ v_+^2 + p_+ = \rho_- v_-^2 + p_-,$ $\qquad$ (4.87)
>
> (iii) $\left[\rho_+\left(e_+ + \frac{1}{2}v_+^2\right) + p_+\right]v_+ = \left[\rho_-\left(e_- + \frac{1}{2}v_-^2\right) + p_-\right]v_-,$

where $e = E - \frac{1}{2}u^2$ is the specific internal energy. These equations have a clear physical meaning: They express the balance of mass, momentum, and energy in a reference frame moving with the discontinuity. Since $v_+ = u_+ - \sigma < 0$, the quantity $M = -\rho_+ v_+ > 0$ is the mass flux across the shock front (from the preshock to the postshock state) per unit time and unit area.

Remark 4.34 The jump condition can be written in Lagrangian coordinates, using the notion of a weak solution to Equation (4.83). If $\sigma_L = \xi'(t)$ is the shock speed, the first condition yields

$$\sigma_L(\tau_+ - \tau_-) = -A(u_+ - u_-) = -A(v_+ - v_-)$$

$(A = A(r(\xi(t), t)))$. By the first equation in (4.87) this reduces to

$$\sigma_L = AM. \tag{4.88}$$

In view of Definition (4.79) of the coordinate ξ, the equality (4.88) is easily understood; the shock speed is simply the total amount of mass crossing the shock front in unit time.

We now proceed as in Proposition 4.17, with regard to 3-shocks (associated with C_+ characteristics). Given a state (ρ_+, p_+, u_+), viewed as a preshock state, we search algebraic expressions for the locus of all possible left states (ρ_-, p_-, u_-) that can be connected by an admissible 3-shock. To relate it to the general theory presented in Section 4.1, we might say that we look for explicit expressions for $\mathbf{u}_k(y)$, as in Proposition 4.17(ii). It turns out that in our case, the parameter y is replaced by a thermodynamic quantity, most commonly the density ρ.

It is more convenient to use the form (4.87) for the jump conditions. Also, we shall use ρ and $\tau = 1/\rho$ interchangeably. Combining (4.87)(i),(ii) we get, by $M\tau_\pm = -v_\pm$,

$$\text{(i)} \quad M = \frac{p_+ - p_-}{v_+ - v_-},$$

$$\text{(ii)} \quad -M^2 = \frac{p_+ - p_-}{\tau_+ - \tau_-}, \tag{4.89}$$

$$\text{(iii)} \quad (p_+ - p_-)(\tau_+ - \tau_-) = -(v_+ - v_-)^2$$

and

$$(p_+ - p_-)(\tau_+ + \tau_-) = -(v_+^2 - v_-^2). \tag{4.90}$$

Using (4.87)(iii) in (4.90) we obtain

$$\frac{p_+ - p_-}{2}(\tau_+ + \tau_-) = i_+ - i_-,$$

where $i = e + p\tau$ is the enthalpy, so that

$$\frac{p_+ + p_-}{2}(\tau_- - \tau_+) = e_+ - e_-. \tag{4.91}$$

Note the similarity between (4.91) and (4.49) by setting $dS = 0$ in the latter. The shock transition is thus formally equivalent to an adiabatic process between (τ_+, e_+) and (τ_-, e_-), at an average pressure $(p_+ + p_-)/2$.

From Equations (4.89) [either (ii) or (iii)] we infer that either $p_- > p_+$ and $\rho_- > \rho_+$ or $p_- < p_+$ and $\rho_- < \rho_+$. We intend to show, on the basis of the entropy condition, that the second case (i.e., that for which the postshock state has lower pressure and density) must be excluded. To this end we make the physically plausible hypothesis that along the Hugoniot curve the speed of sound $c = c(\rho, p)$ increases as ρ (or, equivalently p) increases.

Claim 4.35 *The entropy condition (4.78) is satisfied only if $\rho_- > \rho_+$, $p_- > p_+$.*

Proof Suppose to the contrary that $\rho_- < \rho_+$, $p_- < p_+$. Then by hypothesis $c_- = c(\rho_-, p_-) < c_+ = c(\rho_+, p_+)$. Since $M > 0$ we have $u_- < u_+$ by (4.89)(i). Thus, $u_- + c_- < u_+ + c_+$, contradicting (4.78). □

From the general discussion in Proposition 4.17 we know that only "one-half" of the curve representing all states (ρ_-, p_-, u_-) satisfying (4.89), (4.91) [with a fixed right state (ρ_+, p_+, u_+)] actually qualifies as the "Hugoniot curve" of admissible states. It corresponds to P_3^+ in Definition 4.19. In view of Claim 4.35 the Hugoniot curve consists of states for which $\rho_- > \rho_+$, $p_- > p_+$. These states satisfy (4.91), where $e_- = e(\tau_-, p_-)$ is obtained by solving (for e) the equation of state $p = p(e, \rho)$. Under very general thermodynamical assumptions the relation (4.91) actually defines a smooth decreasing and convex curve in the (τ, p) plane. We refer the reader to Courant and Friedrichs [30] for a comprehensive treatment of this topic. It should be noted that (4.91) is purely thermodynamic. Strictly speaking, it represents the "projection" of the Hugoniot curve (which includes velocities) on the thermodynamic plane (τ, p). We summarize as follows:

Summary 4.36 *For a fixed state (τ_+, p_+) the equation*

$$H(\tau, p) = \frac{p_+ + p}{2}(\tau - \tau_+) + e(\tau, p) - e(\tau_+, p_+) = 0 \tag{4.92}$$

defines a smooth, monotonically decreasing, convex curve in the (τ, p) plane. It represents all thermodynamic states (τ, p) satisfying the Rankine–Hugoniot

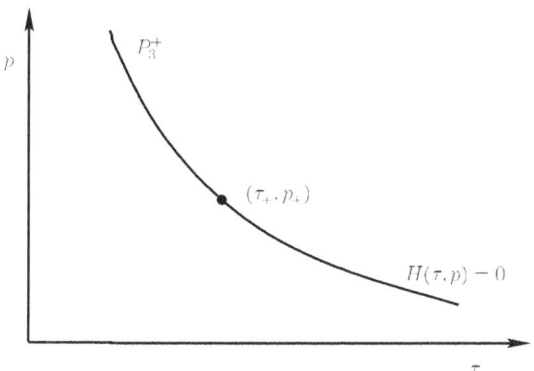

Figure 4.11. The thermodynamic Hugoniot curve P_3^+, left states connected to (τ_+, p_+) by a 3-shock.

jump conditions relative to (τ_+, p_+). In particular, if (τ_+, p_+) is the right state, the part $P_3^+ \{\tau < \tau_+, p > p_+\}$ is the Hugoniot curve of all admissible left (postshock) states [connected to (τ_+, p_+) by a 3-shock; see Figure 4.11].

Remark 4.37 The lower part of the graph $H = 0$ in Figure 4.11 is also mean-ingful in this context. If we designate (τ_+, p_+) as a *left state*, then it comprises all *right states* (τ, p) connected to (τ_+, p_+) by a 3-shock (see Remark 4.20). Since (4.92) is purely thermodynamic, it applies also to 1-shocks. In other words, renaming (τ_+, p_+) as (τ_-, p_-) the part P_3^+ is identical to P_1^+, the locus of all *right states* (τ, p) connected to the *left state* (τ_-, p_-) by an admissible 1-shock.

The entropy S increases along P_3^+, as τ decreases (and p increases). In fact, this is another (physical) aspect of the "entropy condition," which asserts that the entropy of admissible postshock states is higher than that of the preshock state.

Proposition 4.38 Let $p = p(\tau)$ represent the curve P_3^+. Then the entropy $S(\tau) = S(p(\tau), \tau)$ satisfies

$$(i) \qquad \frac{dS}{d\tau} < 0, \qquad \tau < \tau_+,$$

$$(ii) \qquad \frac{dS}{d\tau}\bigg|_{\tau=\tau_+} = \frac{d^2 S}{d\tau^2}\bigg|_{\tau=\tau_+} = 0. \tag{4.93}$$

Proof In terms of differentials along P_3^+, the relation (4.92) can be written as

$$2\,de + (\tau - \tau_+)\,dp + (p + p_+)\,d\tau = 0,$$

and from the general relation (4.49) we get

$$2T\,dS + (p_+ - p)\,d\tau + (\tau - \tau_+)\,dp = 0.$$

Equation (4.93)(i) now follows by $\frac{dp}{d\tau} < 0$. We also get $S'(\tau) = 0$ at $\tau = \tau_+$. Differentiating once more we have

$$2T\,S''(\tau) + 2T'(\tau)S'(\tau) + (\tau - \tau_+)p''(\tau) = 0;$$

hence $S''(\tau)_{\tau=\tau_+} = 0$. $\square$

The admissible part P_3^+ of $H = 0$ is, strictly speaking, the projection of the full Hugoniot curve (Definition 4.19) on the subspace (τ, p) of the thermodynamic variables. We shall still need the "velocity" part of the curve. Since either τ or p can serve as parameters, we shall display it in the (u, p) plane. As before, we consider the case of a 3-shock, $\sigma = \sigma_3$. Since $M > 0$, we obtain, from (4.89)(i) and $p_- > p_+$,

$$u_- > u_+ \tag{4.94}$$

(recall that $v_\pm = u_\pm - \sigma$). We now use (4.89)(iii) to get

$$u_- = u_+ + [(p_- - p_+)(\tau_+ - \tau_-)]^{1/2}. \tag{4.95}$$

Inserting τ_- as a function of p_- [using (4.92)] we obtain a curve $u = u_+ + \phi_3(p)$ (Figure 4.12), which represents the velocity behind an admissible 3-shock as a function of the postshock pressure [given a right preshock state (ρ_+, p_+, u_+)]. Combining this curve with P_3^+ of Figure 4.11 we obtain the "full" Hugoniot curve for the 3-shock.

Remark 4.39 (Hugoniot for 1-shocks) Unlike the thermodynamic part of the Hugoniot curve (see Remark 4.37), the u–p Hugoniot depends on direction. In other words, for a 1-shock, the postshock relative velocity $v_+ = u_+ - \sigma$ is positive [see Figure 4.10 and the discussion following Equation (4.78)], so $M = -\rho_+ v_+ < 0$ (which simply means that the fluid crossing the shock is moving from left to right). The equations (4.89)–(4.91) are still valid, with the roles of "$\pm$" states now reversed; the "$+$" is the postshock state, while the "$-$" is the preshock state. For admissible 1-shocks we therefore have $p_+ > p_-$,

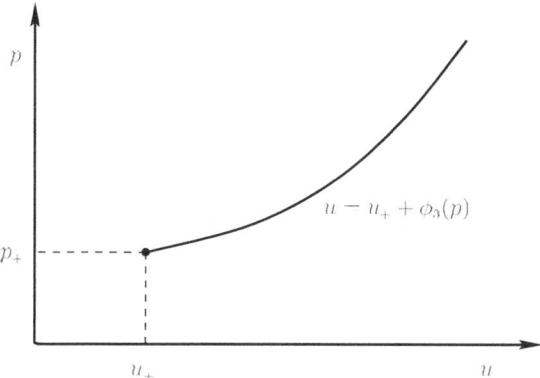

Figure 4.12. The 3-Hugoniot curve in the (u, p) plane. (u_+, p_+) is the preshock state.

$p_+ > p_-$, and $u_+ < u_-$ [by (4.89)(i)]. In particular, if we fix (τ_+, p_+) (now as a right postshock state) the *lower* part of the curve $H = 0$ (Figure 4.11) is the locus of all states (τ, p) connected (as preshock states) to (τ_+, p_+) by an admissible 1-shock. As was noted in Remark 4.37, if (τ_+, p_+) is renamed (τ_-, p_-) and taken as a left (preshock) state, the *upper* part of $H = 0$ (denoted as P_3^+ in Figure 4.11) serves again as the locus of admissible postshock states connected to it by a 1-shock. However, the expression for u_+ [instead of (4.95)], where "+" is still the right state, is now

$$u_+ = u_- - [(p_- - p_+)(\tau_+ - \tau_-)]^{1/2}, \qquad (4.96)$$

and expressing τ_+ in terms of p_+, we obtain the curve $u = u_- - \phi_1(p)$ (Figure 4.13) representing the velocity behind an admissible 1-shock as a function of pressure.

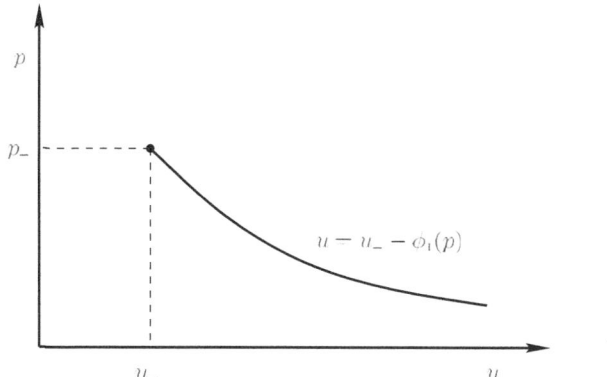

Figure 4.13. The 1-Hugoniot curve in the (u, p) plane. (u_-, p_-) is the preshock state.

Remark 4.40 It is worth observing (see Figures 4.12 and 4.13) that in all cases (1- or 3-shocks) we have $u_- > u_+$ for admissible shocks.

We know from the general theory (Summary 4.14 and Proposition 4.17) that admissible postshock states form one-parameter families. In general, this is true only for "small" jumps. However, in the case of compressible fluid flow, this is true for arbitrary jumps, provided only that the thermodynamic assumptions leading to the existence of the curve $H = 0$ [see Equation (4.92)] are valid. As we shall see later, this is true in particular for the important class of perfect gases. Furthermore, under the assumptions leading to Summary 4.36 the Hugoniot curves can be parametrized by any one of the flow variables, including the shock speed. This is stated as follows:

Proposition 4.41 *Given a preshock state* (ρ_0, p_0, u_0), *any admissible postshock state* (ρ_1, p_1, u_1) *is fully determined either by one of the variables* ρ_1, p_1, u_1 *or by the corresponding shock speed* σ.

Proof We may suppose that the shock is a 3-shock, in which case $(\rho_0, p_0, u_0) = (\rho_+, p_+, u_+)$. From Summary 4.36 and Equation (4.95) we see that indeed any of the values ρ_-, p_-, u_- fully determines the other two. Furthermore, we have by (4.86)(i)

$$\sigma_3 = \frac{\rho_+ u_+ - \rho_- u_-}{\rho_+ - \rho_-}.$$

Assume now that σ_3 is given. Then we know $v_+ = u_+ - \sigma_3$ and hence $M = -\rho_+ v_+$. According to (4.89)(ii), $-M^2$ is the slope of the segment connecting (τ_+, p_+) to the point $(\tau_-, p_-) \in P_3^+$ (see Figure 4.11). However, the convexity of P_3^+ implies that the slope uniquely determines the point (τ_-, p_-) and hence also the velocity u_-.   □

Centered Rarefaction Waves

The Rankine–Hugoniot jump condition (4.76) turned out to be independent of the variation in cross-sectional area [see the discussion following (4.76)]. This however is not the case with centered rarefaction waves; the cross-sectional area $A(r)$ plays a significant role in determining their structure. Indeed, unlike the case of a shock, where the speed is determined by the limiting values of the flow variables, the CRW represents a "full structure" of characteristic curves fanning out of a common center. In the strict conservation form, as discussed in Section 4.1, the characteristic curves are straight lines, each carrying constant values of all components of the solution (see Definition 4.5). In our context of quasi-1-D flow, this applies only to planar flow (see Remark 4.31). In the more general case $[A'(r) \neq 0]$ the planar case serves as a limiting ("asymptotic") part of the solution, for infinitesimally short time. In other words, the instantaneous values of the flow variables at the center are determined by the planar approximation. The characteristic curves emanating from the center then bend, modified by the cross-sectional area variation. The flow variables evolve along each curve, subject to the characteristic relations (4.66), (4.67).

We note that the treatment of a CRW, as given in Definition 4.5, was based on the fact that it connects two *constant* states, $\mathbf{u}_r$ and $\mathbf{u}_l$. If, however, the "head" or "tail" characteristic curve (the extreme curves of the CRW fan) propagates into a nonuniform region, its slope varies according to Equation (4.64). In particular, all characteristic curves in the CRW become curvilinear and the solution is no longer self-similar. This observation is valid even in the planar case. As we shall see in Chapter 5, the treatment of the "non-self-similar" CRW is fundamental in the solution of the generalized Riemann problem.

We shall study only the planar case here [Equation (4.47)], deferring the quasi-1-D case to Section 5.1. Thus, we may invoke the general theory of Section 4.1 and study in detail the features of a 1-CRW. Recall that in view of Claim 4.30 both λ_1 and λ_3 characteristic families are genuinely nonlinear, thus enabling CRW solutions.

As in the shock case, we use "+" and "−" to indicate, respectively, right and left states. They are now separated by the CRW fan, which consists of straight C_- characteristics (see Figure 4.14). As in Definition 4.7, we let y be the slope of the line $C_-^y, u_- - c_- \leq y \leq u_+ - c_+$. All flow variables are constant

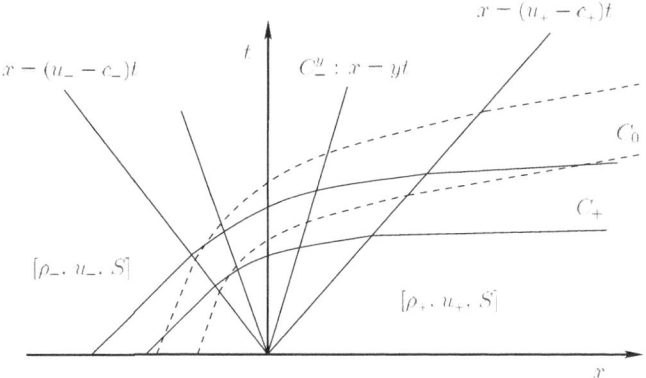

Figure 4.14. A planar 1-CRW. It is necessarily isentropic.

along C_-^y, and we denote them as $u(y)$, $\rho(y)\ldots$. In particular, $u(u_\pm - c_\pm) = u_\pm$, $c(u_\pm - c_\pm) = c_\pm$, and in general

$$y = u(y) - c(y), \qquad u_- - c_- \leq y \leq u_+ - c_+. \tag{4.97}$$

The particle paths C_0 are also depicted in Figure 4.14. They intersect the C_-^y curves transversally, since their slopes at the point of intersection are $u(y) > u(y) - c(y)$. The flow within the CRW (outside of the singularity at 0) is smooth, so that by (4.67)(ii) the particle paths are adiabatic, carrying constant values of the entropy S. Since, by assumption, the entropy is uniform along the extreme characteristic lines $y = u_\pm - c_\pm$ (being adjacent to constant states), it is uniform throughout the CRW. We have therefore established the following:

Claim 4.42 *A planar CRW (connecting two constant states) is isentropic. In particular, the entropies of the two constant states are equal.*

This observation enables us to fully resolve the CRW. Let S be the constant entropy. Then $c^2 = \left(\frac{\partial p}{\partial \rho}\right)_S$ is a function of ρ. Any characteristic curve C_+ traverses the CRW and can be parametrized by y, designating its point of intersection with C_-^y. The density $\rho(y)$ along C_+ determines $c(y)$. Implementing Equation (4.72) we obtain

$$R_+\left(\rho(y), u(y)\right) \equiv \int\limits_{\rho_-}^{\rho(y)} \frac{c(\rho)}{\rho}\, d\rho + u(y) = k, \qquad u_- - c_- \leq y \leq u_+ - c_+,$$

$$\tag{4.98}$$

where k is a constant. It can be determined by using either endpoint $y = u_\pm - c_\pm$,

$$k = u_- = u_+ + \int\limits_{\rho_-}^{\rho_+} \frac{c(\rho)}{\rho}\, d\rho \qquad \text{(1-CRW)}. \tag{4.99}$$

If $u_+ < u_-$ we must have $\rho_+ > \rho_-$, but then $c_+ = c(\rho_+) > c_- = c(\rho_-)$ and $u_+ - c_+ < u_- - c_-$, which contradicts the assumed structure of the CRW (see Figure 4.14). Furthermore, we may apply the same considerations to any sector of the CRW, corresponding to $y_1 \leq y \leq y_2$, which is also a CRW. We obtain therefore, $u(y_2) \geq u(y_1)$, whereas $p(y_2) \leq p(y_1)$ and $\rho(y_2) \leq \rho(y_1)$. Thus, we have the following corollary.

Corollary 4.43 *Across the CRW (in the direction of particle paths) the velocity increases whereas density and pressure decrease.*

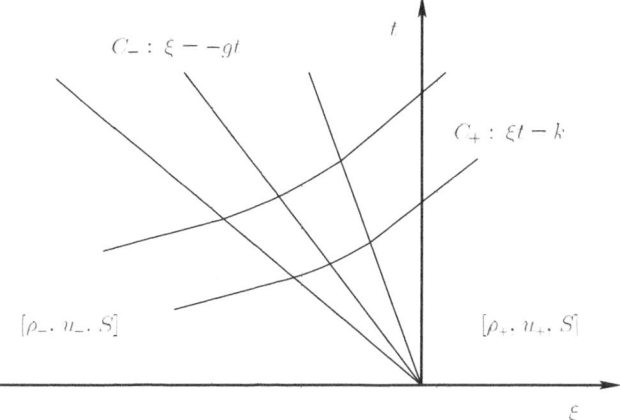

Figure 4.15. A 1-CRW in Lagrangian representation. The C_+ curves are hyperbolas within the CRW.

This corollary justifies the terminology "rarefaction wave," inasmuch as the fluid "rarefies" as it moves across the wave. (Observe that the "$\pm$" in (4.72) refer to the two different Riemann invariants and should not be confused with the "$\pm$" used here to designate the two constant states. Here we are using the Riemann invariant R_+, which is constant along C_+.) Of course, a more explicit relation between $u(y)$ and $\rho(y)$ can be obtained only when we have a more explicit equation $c = c(\rho)$. This will be done in the following for the case of a perfect (polytropic) gas.

Finally, we note that once $u(y)$, $\rho(y)$, and $c(y)$ are determined by (4.98), they are the constant values of the solution along the line $x = yt$.

We conclude the treatment of a planar CRW by studying its representation in Lagrangian coordinates. Specializing to a 1-CRW, we recall [see Equation (4.84)] that the slopes of the C_- characteristic lines are $-g = -\rho c$ [since $A(r) \equiv 1$]. In particular, all slopes are negative (see Figure 4.15). By definition, the particle paths are vertical lines $\xi = constant$, whereas the C_+ characteristic curves satisfy $\frac{d\xi}{dt} = g$. Here the slopes $-g$ play a role analogous to that of y in the preceding case. In particular, $-g = \xi/t$ throughout the CRW, and the equation for the C_+ curves becomes $\frac{d\xi}{dt} = -\frac{\xi}{t}$; hence $\xi t = constant$. We summarize in the following claim.

Claim 4.44 *Consider a 1-CRW for planar flow, in the Lagrangian framework. Then the C_- characteristic lines are given by*

$$\xi = -gt, \qquad g_+ \le g \le g_-$$

(necessarily $g_- = \rho_- c_- \geq g_+ = \rho_+ c_+$). The C_+ curves are the hyperbolas

$$C_+ : \xi t = k, \quad k < 0,$$

where $k = -g_-^{-1}\xi_-^2$ if the curve intersects $\xi = -g_- t$ at $(\xi_-, -g_-^{-1}\xi_-)$.

We shall return to the study of non-self-similar centered rarefaction waves in Section 5.1.

The Riemann Problem (RP) for Planar Flows

The system (4.47) of the equations governing planar flow is in conservation form and is thus subject to the general framework discussed in Section 4.1. In particular, one can study the Riemann problem associated with the system (see Definition 4.26).

The role played by the RP in all aspects of fluid dynamics is indeed remarkable. It was initiated by Riemann in his pioneering work on mathematical fluid dynamics (see Courant and Friedrichs [30, Chapter III, Section 80] and historical comments there). Since the early times of computational fluid dynamics it served as a cornerstone in the construction of discrete schemes (Godunov [56], Glimm [49], and references in Godlewski and Raviart [55]). As we shall see in Chapter 6 (and, in fact, have already seen in Chapter 3) the Riemann problem is accepted as a universal tool in measuring the efficiency and accuracy of various numerical schemes. In the field of physical fluid dynamics it is better known as the "shock tube" problem. It is extensively used in the analysis of a wide variety of experiments (see the *Handbook of Shock Waves* [16]).

The theory developed in Section 4.1 allows us to conclude that, given two sufficiently close states U_R, U_L there exists a unique, self-similar solution to the IVP (4.47) [or, equivalently, to (4.62), with $A \equiv 1$], subject to the initial data

$$U(x, 0) = \begin{cases} U_L, & x < 0, \\ U_R, & x > 0. \end{cases} \tag{4.100}$$

We denote this solution by $R(\frac{x}{t}; U_L, U_R)$. It consists in general of three waves $\Gamma_1, \Gamma_2, \Gamma_3$. In view of Claim 4.30, Γ_1 and Γ_3 can be either admissible shocks or centered rarefaction waves. Γ_2 is necessarily a contact discontinuity. The flow between any two adjacent waves is uniform. For suitably chosen initial data, one or two waves may be absent.

It is our intention to develop in more detail the solution to the RP in the context of planar flow. As a rule, the RP can now be solved for quite general initial data

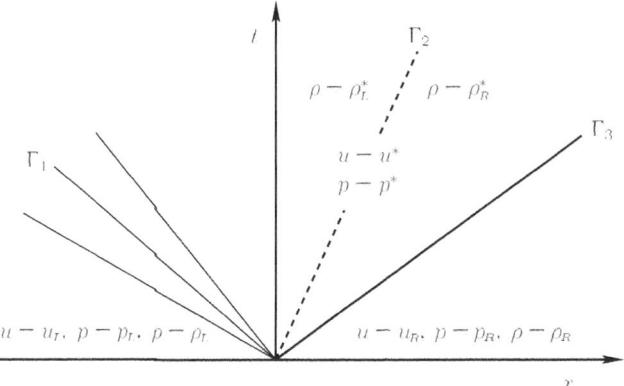

Figure 4.16. A schematic solution to the RP for planar flow.

(not necessarily close) as long as the equation of state is not "pathological."
By this we mean that, in addition to the thermodynamical assumptions underlying Summary 4.36, the "Interaction Curves", as given in Summary 4.45
below, remain convex for a large range of pressures and densities. However,
a comprehensive discussion of this topic is beyond the scope of this monograph, and we refer the reader to Godlewski and Raviart [55] and references
therein.

As a starting point in our treatment, we recall that the variables p, u are
continuous across a contact discontinuity [see (4.77)]. It follows that $p =
p^*, u = u^*$ are constant in the whole sector between the waves Γ_1 and Γ_3 (see
Figure 4.16). The basic idea is then to locate u^*, p^* as the common point of the
$u-p$ interaction curves (see Definition 4.19) for the states $\mathbf{U}_R, \mathbf{U}_L$.

Recall that the part P_3^+ of the curve $H(\tau, p) = 0$ (see Summary 4.36) is the
locus of all left states (τ, p) connected to the right state (τ_+, p_+) by an admissible
3-shock. The lower part of that curve $(\tau > \tau_+)$ is not the lower part of the 3-
interaction curve (see Remark 4.37). Indeed, the lower part of the interaction
curve (Definition 4.19) consists of the left states (τ, p) connected to (τ_+, p_+)
by a 3-CRW. Thus, in view of Claim 4.42, it is simply the *isentropic curve*
$S(p, \tau) = S(p_+, \tau_+)$ (restricted to $\tau \geq \tau_+$) through the point (τ_+, p_+) in the
(τ, p) thermodynamic plane. We can refer to this full curve as the "projection"
of the (right) 3-interaction curve on the (τ, p) plane. We note that in view of the
general theory [see Equation (4.25)] the joint curve has a continuous tangent at
(τ_+, p_+).[10]

---

[10] Observe that in our case, by Proposition 4.38, the curve is actually C^2 at (τ_+, p_+).

Turning now to the representation of the interaction curve in the (u, p) plane, we repeat the same procedure, patching together the "shock" part [projection of the full Hugoniot curve on the (u, p) plane] and the "rarefaction" part. Specializing now to the 1-interaction curve we obtain the shock part from (4.96) (see Figure 4.13) and the rarefaction part from (4.98). For future reference we summarize as follows.

Summary 4.45 *Given a left state* (τ_-, p_-, u_-), *of entropy* $S_- = S(p_-, \tau_-)$, *let* I_1^l *be its (left) 1-interaction curve, consisting of all right states that can be connected to it by an admissible 1-shock or a CRW. Then its projection on the* (u, p) *plane is given by*

$$
u = \begin{cases} u_- - \phi_1(p) = u_- - [(p_- - p)(\tau - \tau_-)]^{1/2}, & p > p_-, \\[2mm] u_- - \int\limits_{\rho_-}^{\rho} \frac{c(\beta)}{\beta}\, d\beta, & p < p_-. \end{cases}
$$

$$(4.101)$$

In the top line τ *is determined from* p *by the implicit relation* $H(\tau, p) = 0$ *[through* (τ_-, p_-); *see (4.92) and Remark 4.37] and in the bottom line* $c = c(\rho)$ *and* $\rho = \rho(p)$ *along the isentropic curve* $S(p, \tau) = S_-$.

We obtain in the same way the curve I_3^r. It is the locus of all left states connected to the right state (τ_+, p_+, u_+) by a 3-wave (admissible shock or CRW). Combining (4.95) and the Riemann invariant $R_-(\rho, u)$ [see Equation (4.72)] gives its projection on the (u, p) plane [in analogy with (4.101)]:

$$
u = \begin{cases} u_+ + \phi_3(p) = u_+ + [(p - p_+)(\tau_+ - \tau)]^{1/2}, & p > p_+, \\[2mm] u_+ + \int\limits_{\rho_+}^{\rho} \frac{c(\beta)}{\beta}\, d\beta, & p < p_+. \end{cases}
$$

$$(4.102)$$

Note that by (4.95), (4.96) the curves are symmetric with respect to $u = u_\pm$.[11] We refer to Figure 4.17 where both curves are schematically shown.

In view of the foregoing discussion, the solution to the RP is a straightforward matter.

---

[11] Basically, this follows from the Galilean invariance of the flow equations. By translation we may assume $u_\pm = 0$, and then the reflection $x \to -x$, $u \to -u$ transforms a 1-wave to a 3-wave and vice versa.

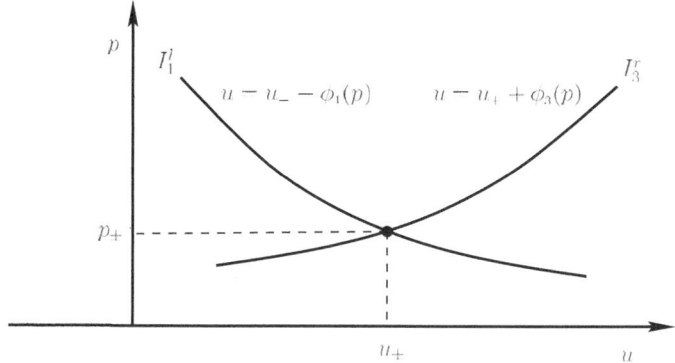

Figure 4.17. $u - p$ representation of I_1^l, I_3^r interaction curves. The parts $p > p_\pm$ represent shocks.

Construction 4.46 (Solution to the RP–planar flow) *Let the right and left states $\mathbf{U}_R$ and $\mathbf{U}_L$, respectively, be given. In the (u, p) plane construct the curves*

I_3^r *with respect to the right state $\mathbf{U}_R$,*
I_1^l *with respect to the left state $\mathbf{U}_L$.*

Let (u^*, p^*) be their point of intersection in the (u, p) plane. Then the solution to the RP with initial data $\mathbf{U}(x, 0)$ as in (4.100) is depicted in Figure 4.16, where

Γ_2 *is a contact discontinuity moving at speed u^*,*
Γ_1 *is an admissible 1-shock (resp. 1-CRW) if $p^* > p_L$ (resp. $p^* < p_L$),*
Γ_3 *is an admissible 3-shock (resp. 3-CRW) if $p^* > p_R$ (resp. $p^* < p_R$),*

The pressure $p = p^*$ is uniform in the sectorial region between Γ_1 and Γ_3. The density ρ_L^* is uniform in the sectorial region $[\Gamma_1, \Gamma_2]$, and is determined by the respective interaction curve in the (τ, p) plane from the knowledge of $\mathbf{U}_L$ and p^*. If Γ_1 is a shock, use Proposition 4.41; if Γ_1 is a CRW, use the fact that (ρ_L, p_L) and (ρ_L^*, p^*) are isentropic. Similarly, ρ_R^* is uniform in the sectorial region $[\Gamma_2, \Gamma_3]$, and is likewise determined from the state $\mathbf{U}_R$ and the wave Γ_3.

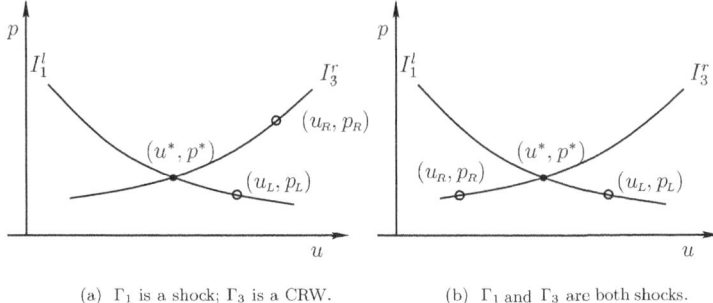

(a) Γ_1 is a shock; Γ_3 is a CRW.   (b) Γ_1 and Γ_3 are both shocks.

Figure 4.18. Solutions to the Riemann problem.

In Figure 4.18 we present two possible solutions of Riemann problems based on the preceding analysis.

Remark 4.47 The analysis of the RP presumes the intersection of I_1^l and I_3^r [in the (u, p) plane]. As has been observed earlier, the general theory does not necessarily guarantee such an intersection, if the states U_R and U_L are not sufficiently close. We shall see shortly that this question can be fully resolved in the case of a perfect gas.

Remark 4.48 (The RP in Lagrangian coordinates) The planar RP in Lagrangian coordinates is the IVP for the system (4.83), where $A \equiv 1$ and

$$V(\xi, 0) = \begin{cases} V_L, & \xi < 0, \\ V_R, & \xi > 0. \end{cases} \tag{4.103}$$

In view of Definition 4.32 [and Equation (4.79) there, with $r_0 = 0$], $\xi = \rho_R r$ (resp. $\xi = \rho_L r$) for $r > 0$ (resp. $r < 0$). The features of the solution, including its self-similar character, are then readily obtained from Construction 4.46. It is further simplified by the fact that the contact discontinuity (the wave Γ_2) coincides with the line $\xi = 0$ and carries the constant (and equal on both sides) values $u = u^*$, $p = p^*$. Thus, the wave Γ_1 necessarily propagates into the domain $\{\xi < 0\}$, whereas Γ_3 propagates into the domain $\{\xi > 0\}$. Note further that in this case the CRW has a simple form as in Claim 4.44.

Perfect (γ-Law) Gas

The most common equation of state for gases is that of a *perfect gas* (see Landau and Lifshitz [74]), sometimes also referred to as a *polytropic gas* (see Courant and Friedrichs [30]). In addition to serving as a common model in physics, it is used almost exclusively in the investigation of advanced numerical methods for fluid dynamics. In terms of the specific internal energy e and the density ρ, the pressure is given by

$$p = (\gamma - 1)\rho e, \tag{4.104}$$

where $\gamma > 1$ is a constant that depends on the gas.[12] Furthermore, in such gases one has $e = c_v T$, where T is the temperature, so that the first law of thermodynamics (4.49) can be explicitly integrated. Regarding now p as a function of (ρ, S) [see (4.60)] we can write

$$p = A(S)\rho^\gamma, \tag{4.105}$$

where $A(S)$ is a function of entropy given by

$$A(S) = \exp\left(\frac{S}{c_v}\right). \tag{4.106}$$

We refer to Courant and Friedrichs [30, Chapter I, Sections 3 and 4] for a detailed discussion of this topic.

Owing to the form (4.105) of the isentropic curves, the term "γ-law gas" is also used for perfect gases.

It follows from the definition (4.61) that the speed of sound is given explicitly by

$$c^2 = \frac{\gamma p}{\rho}, \tag{4.107}$$

and the Riemann invariants for isentropic flow [see (4.72)] take the form

$$R_\pm(\rho, u) = \frac{2}{\gamma - 1} c \pm u. \tag{4.108}$$

Next, we can find explicit expressions for the Hugoniot curve. To do this it is convenient to introduce a constant μ such that

$$\mu^2 = \frac{\gamma - 1}{\gamma + 1}. \tag{4.109}$$

Inserting (4.104) in (4.92) we find

$$2\mu^2 H(\tau, p) = (\tau - \mu^2 \tau_+)p - (\tau_+ - \mu^2 \tau)p_+. \tag{4.110}$$

Recall (Summary 4.36) that when (τ_+, p_+) is a preshock state, all possible postshock states (τ, p) are given by $H(\tau, p) = 0$, with $p > p_+$, $\tau < \tau_+$. Inserting

---

[12] More specifically, $\gamma = 1 + \frac{R}{\lambda c_v}$, where R is the universal constant of gases, λ is the molecular weight of the gas, and c_v is its specific heat at constant volume. Typically, $\gamma = 5/3$ for a monatomic gas, and $\gamma = 7/5$ for a diatomic gas. (See Fermi [46] for a detailed account.)

this in (4.95), we find the explicit form for the projection of the Hugoniot curve of a 3-shock on the (u, p) plane,

$$u = u_+ + \phi_3(p) = u_+ + (p - p_+)\sqrt{\frac{(1 - \mu^2)\tau_+}{p + \mu^2 p_+}}. \tag{4.111}$$

Using (4.107) this can also be written as

$$u = u_+ + \frac{c_+}{\gamma p_+} \cdot (p - p_+)\left[1 + \left(\frac{\gamma + 1}{2\gamma}\right)\frac{p - p_+}{p_+}\right]^{-1/2} \tag{4.112}$$

and similarly, by (4.96) for a 1-shock,

$$u = u_- - \phi_1(p) = u_- - (p - p_-)\sqrt{\frac{(1 - \mu^2)\tau_-}{p + \mu^2 p_-}}, \tag{4.113}$$

which, in analogy with (4.112), can be written as

$$u = u_- - \frac{c_-}{\gamma p_-} \cdot (p - p_-)\left[1 + \left(\frac{\gamma + 1}{2\gamma}\right)\frac{p - p_-}{p_-}\right]^{-1/2}, \tag{4.114}$$

where now (τ_-, p_-, u_-) is the preshock state, and $p > p_-, \tau < \tau_-$.

Note that if ρ_+, p_+ and ρ, p are isentropic then the corresponding speeds of sound satisfy

$$\frac{c}{c_+} = \left[\frac{p}{p_+}\right]^{\frac{\gamma - 1}{2\gamma}}.$$

The general expressions (4.101), (4.102) for the (u, p) interaction curves can now be rewritten in explicit algebraic form, by combining the expressions (4.108) for the Riemann invariants and Equations (4.111)–(4.114). For later reference, we summarize as follows:

Summary 4.49 *Given a left (resp. right) state* $\mathbf{U}_L = (\tau_L, p_L, u_L)$ *[resp.* $\mathbf{U}_R = (\tau_R, p_R, u_R)$]*, with entropy* S_L *and speed of sound* c_L *(resp.* S_R, c_R*), let* $I_1^!$ *and* I_3^r *be the corresponding left and right interaction curves, as in Summary 4.45. Then their projections on the* (u, p) *plane are given by*

$$u = \begin{cases} u_L - \frac{c_L}{\gamma p_L} \cdot (p - p_L)\left[1 + \left(\frac{\gamma+1}{2\gamma}\right)\frac{p - p_L}{p_L}\right]^{-1/2}, & p \geq p_L, \\[2ex] u_L - \left(\frac{2}{\gamma-1}\right)c_L \cdot \left[\left(\frac{p}{p_L}\right)^{\frac{\gamma-1}{2\gamma}} - 1\right], & p \leq p_L, \end{cases} \tag{4.115}$$

for the I_1^l interaction curve and by

$$u = \begin{cases} u_R + \frac{c_R}{\gamma p_R} \cdot (p - p_R)\left[1 + \left(\frac{\gamma+1}{2\gamma}\right)\frac{p-p_R}{p_R}\right]^{-1/2}, & p \geq p_R, \\[4mm] u_R + \left(\frac{2}{\gamma-1}\right)c_R \cdot \left[\left(\frac{p}{p_R}\right)^{\frac{\gamma-1}{2\gamma}} - 1\right], & p \leq p_R, \end{cases}$$

(4.116)

for the I_3^r interaction curve.
The projections of the interaction curves on the (τ, p) plane are given by

$$\tau = \begin{cases} \tau_L \frac{p_L + \mu^2 p}{p + \mu^2 p_L}, & p \geq p_L, \\[4mm] \tau_L \left(\frac{p_L}{p}\right)^{\frac{1}{\gamma}}, & p \leq p_L, \end{cases}$$

(4.117)

for the I_1^l interaction curve and by

$$\tau = \begin{cases} \tau_R \frac{p_R + \mu^2 p}{p + \mu^2 p_R}, & p \geq p_R, \\[4mm] \tau_R \left(\frac{p_R}{p}\right)^{\frac{1}{\gamma}}, & p \leq p_R, \end{cases}$$

(4.118)

for the I_3^r interaction curve.

The interaction curves I_1^l, I_3^r intersect the zero pressure line at the points $u_l = u_L + \left(\frac{2}{\gamma-1}\right)c_L$ and $u_r = u_R - \left(\frac{2}{\gamma-1}\right)c_R$, respectively. These points have a clear meaning. For example, when a 1-CRW propagates into a state (τ_L, p_L, u_L), the velocity u_l is the maximum velocity attained by the rarefied gas, when the tail pressure (at the back state) vanishes. Inspection of the expressions for the interaction curve shows clearly that they do not intersect (at a nonnegative pressure) if and only if $u_r > u_l$. In view of Construction 4.46 we conclude that this is also the only case in which the Riemann problem is not solvable. We can summarize this as follows.

Corollary 4.50 *Given left and right states, U_L and U_R, respectively, as in Summary 4.49, the planar Riemann problem (4.47), (4.100) for a γ-law gas is solvable in all cases, except for the case where $u_r > u_l$.*

Remark 4.51 The solution to the RP in this case consists of solving a pair of nonlinear algebraic equations. Note that if the variable p is replaced by $\zeta = p^{\frac{\gamma-1}{2\gamma}}$, the rarefaction parts of the interaction curves are transformed into linear relations between u and ζ. In Appendix C we describe in detail an efficient

"Riemann solver," where this fact is exploited. Because of the basic role played by the Riemann problem in many modern numerical methods, the literature concerning efficient Riemann solvers (for perfect as well as general gases) is very rich. We refer the reader to the papers by Alcrudo and Garcia-Navarro [1], Chorin [25], Dai and Woodward [35], Garcia-Navarro, Hubbard, and Priestley [47], Saurel, Larini, and Loraud [99], Schulz-Rinne [100], Teng [109], and Yang and Przekwas [122], as well as to Toro's book [111] and references therein.

Another interesting phenomenon in connection with the Hugoniot curve of a perfect gas is related to its "high end," namely, when the pressure increases to infinity. From Equation (4.110) we see that if, along the curve $H(\tau, p) = 0$, we let $\frac{p}{p_+}$ go to infinity, the ratio $\frac{\tau}{\tau_+}$ tends to μ^2. We therefore have the following conclusion.

Corollary 4.52 (An "infinite shock") *As the pressure increases to infinity along the Hugoniot curve of a perfect gas, the "compression ratio" $\frac{\rho}{\rho_+}$ approaches the limiting value $\frac{\gamma+1}{\gamma-1}$.*

5

The Generalized Riemann Problem (GRP) for Compressible Fluid Dynamics

This chapter is concerned with the main topic of the monograph, namely, the solution of the GRP for quasi-1-D, inviscid, compressible, nonisentropic, time-dependent flow. In Section 5.1 we formulate the problem and study its solution in the Lagrangian and Eulerian frames. In particular, we state and prove the main ingredient in the GRP method, Theorem 5.7. A weaker form of this theorem leads to the "acoustic approximation" (Proposition 5.9). Summary 5.24 gives a step-by-step description of the GRP analysis. In Section 5.2 we present the GRP methodology for the construction of second-order, high-resolution finite-difference (or finite-volume) schemes. Starting out from the (first-order) Godunov scheme, we present the basic (E_1) GRP scheme. It is based on the acoustic approximation and constitutes the simplest second-order extension of Godunov's scheme. This is followed by a presentation of the full array of GRP schemes (as well as MUSCL). Generally speaking, the presentation in this chapter follows closely the GRP papers [7] and [10].

5.1 The GRP for Quasi-1-D, Compressible, Inviscid Flow

In Section 4.2 we studied the Euler equations (4.45) governing the quasi-1-D flow in a duct of variable cross section. We emphasized in particular the role of the Riemann problem ("shock tube problem"), namely, the IVP subject to initial data (4.100). As we shall see in this chapter, the solution to the Riemann problem is a basic ingredient in the numerical resolution of the flow. It was observed that in the planar case (4.47) of uniform cross-sectional area this solution is "self-similar" and is readily obtained by solving a pair of algebraic equations (see Construction 4.46 and Figure 4.16). In the important special case of a perfect gas this procedure is further simplified, as seen in Appendix C. The simplicity of the solution is conducive to the selection of a numerical scheme based on the Riemann problem solution.[1] However, as already noted in Section 3.2, such a scheme ("Godunov's scheme") yields poor results in the resolution of singularities, even in the scalar case. It was further demonstrated there that the resolution is greatly improved when switching to *piecewise-linear* data, thus leading to the *generalized Riemann problem*. Turning back to the quasi-1-D system (4.45), we noted in Remark 4.31 that it is not in strict conservation form. In particular, one cannot expect here a self-similar solution to the Riemann problem, implying that it cannot be reduced to an algebraic problem, as in the planar case. Thus, in dealing with the general case, we have to resort to further analysis of the problem, even in the case of initial data as in (4.100) (see Glimm, Marshall, and Plohr [53] in this case). Replacing the piecewise-constant by piecewise-linear initial data brings about a dramatic improvement in the numerical results. This has been observed in Section 3.2 in the scalar case and was first established for the fluid-dynamical case in the pioneering work of van Leer (see van Leer [112]). We shall therefore concentrate in this section on the IVP for the system (4.45), with initial data that are linear on the two sides of the singularity, across which both the functions and their slopes may experience a jump. The term "generalized Riemann problem" (GRP) has been attached to this problem (see Section 3.1 and Ben-Artzi and Falcovitz [7]) and we shall employ it henceforth.

The analytic treatment of the GRP given here relies on some basic theoretical results, which we now proceed to describe. We refer the reader to Harabetian [60] and to Godlewski and Raviart [55] and references therein for detailed analysis.

Structure of the Solution to the GRP

Let $U(r, t)$ be the solution to the GRP, namely, the IVP for the system (4.45), subject to the initial data

$$U(r, 0) = U_0(r) = \begin{cases} U_L + rU'_L, & r < 0, \\ U_R + rU'_R, & r > 0, \end{cases} \tag{5.1}$$

where U_R, U'_R, U_L, U'_L are constant vectors.

Remark 5.1 The location of the initial discontinuity at $r = 0$ was selected for notational convenience. Clearly, in the case of cylindrical or spherical coordinates, where $r = 0$ is a geometric singularity, this location must be shifted to $r \neq 0$, with obvious modifications.

---

[1] We refer to Godlewski and Raviart [55] for further simplifications based on approximate solutions to the Riemann problem, such as the "Roe scheme."

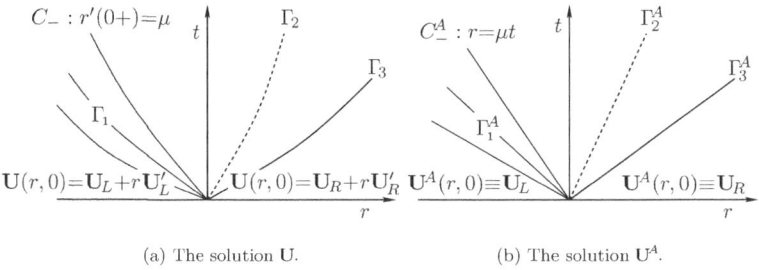

(a) The solution $\mathbf{U}$. (b) The solution $\mathbf{U}^A$.

Figure 5.1. The solutions to the GRP (a), and its associated RP (b).

The initial structure of the solution $\mathbf{U}(r, t)$ is determined by the limiting values (at $r = 0\pm$) $\mathbf{U}_R$, $\mathbf{U}_L$. We therefore associate with the GRP its "limiting planar problem."

Definition 5.2 The "associated Riemann problem" [to the GRP with initial data (5.1)] is the Riemann problem [for the *planar* system (4.47)] subject to the piecewise-constant initial data

$$\mathbf{U}_0^A(r) = \begin{cases} \mathbf{U}_L, & r < 0, \\ \mathbf{U}_R, & r > 0. \end{cases} \tag{5.2}$$

In accordance with the notation introduced following (4.100) we denote by $\mathbf{U}^A(r, t) = \mathbf{R}^A\left(\frac{r}{t}; \mathbf{U}_L, \mathbf{U}_R\right)$ the solution to the associated problem, and we retain the notation "r" for the spatial coordinate (instead of "x" used in Section 4.2 for the planar case). Observe that $\mathbf{U}^A$ is self-similar, depending only on the direction $\frac{r}{t}$.

A schematic description of $\mathbf{U}(r, t)$ and $\mathbf{U}^A(r, t)$ is given in Figures 5.1(a) and 5.1(b) respectively. Note that Figure 5.1(b) is identical to Figure 4.16, except for the superscript "A". It should be emphasized that the solutions are shown only for a short time t, following the "disintegration" of the initial discontinuity at $r = 0$. The waves (in terms of type and initial strength) emanating from that discontinuity are completely determined by the limiting values $\mathbf{U}_R$, $\mathbf{U}_L$ and the planar solution $\mathbf{U}^A$. This is due to the fundamental property of "finite propagation speed"; the solution at a point (r, t) depends only on a finite interval of the initial data.[2] As (r, t) approaches $(0, 0)$, this interval shrinks to the point $r = 0$, and $\mathbf{U}(r, t)$ "approaches" $\mathbf{U}^A(r, t)$. In general, the solution $\mathbf{U}(r, t)$ consists of three waves Γ_1, Γ_2, Γ_3,[3] in tandem with the three planar waves Γ_1^A, Γ_2^A, Γ_3^A (Figure 5.1). Whereas Γ_2 and Γ_2^A are always contact discontinuities, the waves

---

[2] This interval, referred to as the "domain of dependence" of (r, t), consists of all points that can be attained by tracing waves from (r, t) "backward" to the initial line $t = 0$.

[3] In special cases one or two waves are missing and can be referred to as waves of "zero strength."

Γ_i, and $\Gamma_i^A (i = 1, 3)$ are either shocks or centered rarefaction waves. The type (shock or CRW) of Γ_i is identical to that of Γ_i^A. Furthermore, in the shock case (Γ_3, Γ_3^A in Figure 5.1) the initial speeds (slopes) and initial jumps of all flow variables are identical. In the case of a CRW the head and tail characteristics (of Γ_1, Γ_1^A in Figure 5.1) emanate from $r = 0$ with equal slopes. For any characteristic C_- within the fan Γ_1, approaching the origin with limiting slope μ (see Figure 5.1), there is a matching characteristic C_-^A in Γ_1^A of (constant) slope μ. Recall that the flow variables along C_- are not constant. However, the limiting values of the flow variables along C_-, as $t \to 0$, are equal to the corresponding (constant) values of these variables along C_-^A. Finally, the solution $\mathbf{U}(r, t)$ in the regions between the three waves is smooth and approaches, along any direction $r = \mu t$, the corresponding value of $\mathbf{U}^A$, which is constant along the full ray $r = \mu t$, $t > 0$. For future reference we record this fact in the following equation:

$$\lim_{t \to 0+} \mathbf{U}(\mu t, t) = \mathbf{U}^A(\mu t, t) = \mathbf{R}^A(\mu; \mathbf{U}_L, \mathbf{U}_R), \qquad -\infty < \mu < \infty. \quad (5.3)$$

The smoothness of $\mathbf{U}(r, t)$ (between waves) implies in particular that the wave trajectories (discontinuities, characteristics) are smooth curves.

The solution $\mathbf{U}(r, t)$ can be represented by an asymptotic expansion in terms of r, t. However, for the purpose of this monograph we shall need only the *first-order* terms of this expansion.[4] More explicitly, we need the following definition.

Definition 5.3 (The linear GRP) Given the initial data (5.1), let $\mathbf{U}(r, t)$ be the solution to the GRP. The *linear GRP* is the following: Evaluate the limiting value

$$\left(\frac{\partial}{\partial t} \mathbf{U} \right)_0 = \lim_{t \to 0+} \frac{\partial}{\partial t} \mathbf{U}(0, t). \quad (5.4)$$

The rest of this section will be devoted to a detailed discussion of the solution to the linear GRP.

Remark 5.4

(a) As we shall see in the course of the solution, our method is equally applicable to the evaluation of the directional derivative

$$\left(\frac{\partial}{\partial t} \mathbf{U} \right)_\alpha = \lim_{t \to 0+} \frac{d}{dt} \mathbf{U}(r = \alpha t, t), \qquad -\infty < \alpha < \infty. \quad (5.5)$$

---

[4] Note that the "zero-order" term is given by Equation (5.3).

As indicated here, this means that we obtain the full first-order perturbation built into $\mathbf{U}(r, t)$, with respect to the associated $\mathbf{U}^A(r, t)$. This will be important in the application of the GRP numerical method to general "moving grids."

(b) The evaluation of the directional derivatives (5.5) given in this section can be extended to higher order derivatives, using the same methodology. This is not done here since the entire numerical treatment in this monograph is based solely on the linear GRP, combining the simplicity of the algorithm with its high-resolution capability.

The Linear GRP in Lagrangian Coordinates – Setup and Statement of the Main Theorem

Defining the Lagrangian coordinate ξ as in (4.79), we replace the system (4.47) by (4.83). In particular, the three unknown flow variables $\mathbf{V} = (\tau, u, E)$ now replace those of the Eulerian representation $\mathbf{U} = (\rho, \rho u, \rho E)$, where $\tau = 1/\rho$. Recall [see the discussion following (4.84)] that the Lagrangian formulation leads to certain simplifications in the structure of the solution to the GRP. The most significant is the fact that the contact discontinuity carries a constant value of ξ. Thus, taking $r_0 = 0$ so that $\xi = 0$ at the initial discontinuity ensures that the contact discontinuity Γ_2 stays along $\xi = 0$ for $t > 0$. The wave pattern for the solution to the GRP in (ξ, t) coordinates [analogous to the one depicted in Figure 5.1(a)] is schematically given in Figure 5.2.

Observe that the limiting values $\mathbf{V}_L$, $\mathbf{V}_R$ are related to $\mathbf{U}_L$, $\mathbf{U}_R$, respectively, as indicated previously. The linear initial data (5.1) are replaced by the linear

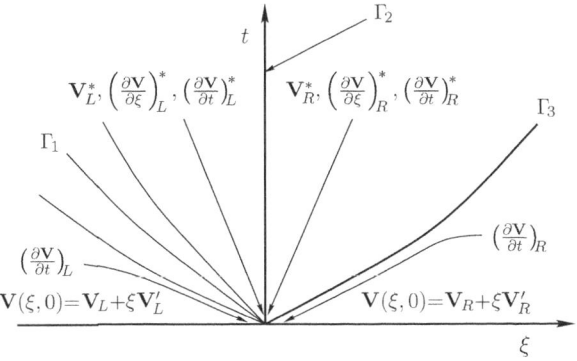

Figure 5.2. Structure of the solution to the GRP in Lagrangian coordinates.

(in ξ) initial data

$$V(\xi, 0) = V_0(\xi) = \begin{cases} V_L + \xi\, V_L', & \xi < 0, \\ V_R + \xi\, V_R', & \xi > 0. \end{cases} \tag{5.6}$$

This is justified as follows. Let Q be any flow variable (say, $Q = \rho$ or $Q = u$). The initial value $Q(r, 0)$ is linear (say, for $r > 0$) and using (4.81) we get

$$\frac{\partial}{\partial \xi} Q(\xi, 0)\bigg|_{\xi = 0+} = [A(0)\rho_R]^{-1} \frac{\partial}{\partial r} Q(r, 0)\bigg|_{r = 0+}. \tag{5.7}$$

In this equation we are using $\frac{\partial}{\partial \xi}\big|_{\xi = 0+}$, $\frac{\partial}{\partial r}\big|_{r = 0+}$ to denote the one-sided (from the right) derivatives and $\rho_R = \rho(0+, 0)$ to denote the value of the density in U_R [i.e., the limiting value of $\rho(r, 0)$ as $r \to 0+$]. Clearly, if $Q(r, 0)$ is linear, $Q(\xi, 0)$ is generally not linear (in ξ). However, the solution to the *linear GRP*, as we shall see, depends *only on the limiting slopes* $\frac{\partial}{\partial \xi} Q(\xi, 0)\big|_{\xi = 0\pm}$. Thus, we are justified in assuming that $V_0(\xi)$ is given by (5.6), where the relation of U_R' to V_R' is obtained from (5.7). Similarly, the left-hand-side derivatives are given by

$$\frac{\partial}{\partial \xi} Q(\xi, 0)\bigg|_{\xi = 0-} = [A(0)\rho_L]^{-1} \frac{\partial}{\partial r} Q(r, 0)\bigg|_{r = 0-}, \tag{5.8}$$

for any flow variable Q. Note that the cross-sectional area $A(r)$ is assumed to be continuous (and even continuously differentiable) at all points, including $r = 0$.

We denote by $V(\xi, t)$ the solution to the GRP (in Lagrangian coordinates) subject to the initial data (5.6). The associated Riemann solution (Definition 5.2) $V^A(\xi, t) = R^A\left(\frac{\xi}{t}; V_L, V_R\right)$ is obtained as in Remark 4.48. As in the Eulerian case, the solution V^A is the "limit of V" as $t \to 0$ [see (5.3)]. In particular, it will be useful to denote by V_L^*, V_R^* the (constant) values of the solution along the two sides of the contact discontinuity (compare Figure 4.16)

$$V_L^* = V^A(0-, t), \qquad V_R^* = V^A(0+, t). \tag{5.9}$$

Clearly, in the case of the pressure and the velocity $p_L^* = p_R^* = p^*$ and $u_L^* = u_R^* = u^*$.

The linear GRP in this framework is transformed into the problem of evaluating the instantaneous time derivatives

$$\left(\frac{\partial}{\partial t} V\right)^* = \lim_{t \to 0+} \frac{\partial}{\partial t} V(0, t) \tag{5.10}$$

Table 5.5. *Notation for the linear GRP in* Lagrangian coordinates

Symbol	Definition
$\mathbf{V}_L$, $\mathbf{V}_R$	$\lim \mathbf{V}(\xi, 0)$ as $\xi \to 0-, 0+$
$\mathbf{V}'_L$, $\mathbf{V}'_R$	$\dfrac{\partial}{\partial \xi} \mathbf{V}(\xi, 0)$ for $\xi < 0, \xi > 0$
$\mathbf{R}^A \left(\dfrac{\xi}{t}; \mathbf{V}_L, \mathbf{V}_R \right)$	Lagrangian solution of the associated RP
$\mathbf{V}^*$, $\mathbf{V}^*_L$, $\mathbf{V}^*_R$	$\mathbf{R}^A (0; \mathbf{V}_L, \mathbf{V}_R)$
	$(\rho^*_L, \rho^*_R, \dots$ for discontinuous variables)
$\left(\dfrac{\partial}{\partial t} \mathbf{V} \right)^*$	$\lim\limits_{t \to 0+} \dfrac{\partial}{\partial t} \mathbf{V}(0, t)$
	$\left(\text{for discontinuous variables: } \left(\dfrac{\partial}{\partial t} \rho \right)^*_L , \left(\dfrac{\partial}{\partial t} \rho \right)^*_R \dots \right)$
$\left(\dfrac{\partial}{\partial \xi} \mathbf{V} \right)^*_R$	$\lim\limits_{t \to 0+} \lim\limits_{\xi \to 0+} \dfrac{\partial}{\partial \xi} \mathbf{V}(\xi, t)$
$\left(\dfrac{\partial}{\partial \xi} \mathbf{V} \right)^*_L$	$\lim\limits_{t \to 0+} \lim\limits_{\xi \to 0-} \dfrac{\partial}{\partial \xi} \mathbf{V}(\xi, t)$
$\left(\dfrac{\partial}{\partial t} \mathbf{V} \right)_R$	$\lim\limits_{\xi \to 0+} \lim\limits_{t \to 0+} \dfrac{\partial}{\partial t} \mathbf{V}(\xi, t)$
$\left(\dfrac{\partial}{\partial t} \mathbf{V} \right)_L$	$\lim\limits_{\xi \to 0-} \lim\limits_{t \to 0+} \dfrac{\partial}{\partial t} \mathbf{V}(\xi, t)$
λ	$\dfrac{A'(0)}{A(0)} = \dfrac{1}{A(0)} \dfrac{d}{dr} A(r) \Big|_{r=0}$
	= rate of change of cross-sectional area

along the contact discontinuity $\xi = 0$. As already noted in Remark 5.4, the determination of all other directional derivatives [and in particular $\left(\frac{\partial}{\partial t} \mathbf{U} \right)_0$ along $r = 0$, as in (5.4)] will then be an easy matter. Clearly, for discontinuous variables (such as density) we must distinguish sides, using the obvious notation

$$\left(\frac{\partial}{\partial t} \rho \right)^*_L = \lim_{t \to 0+} \frac{\partial}{\partial t} \rho(0-, t), \qquad \left(\frac{\partial}{\partial t} \rho \right)^*_R = \lim_{t \to 0+} \frac{\partial}{\partial t} \rho(0+, t). \qquad (5.11)$$

In Table 5.5 we summarize the notation needed in the solution of the linear GRP. This notation is also indicated in Figure 5.2. The Riemann solution $\mathbf{R}^A \left(\frac{\xi}{t}; \mathbf{V}_L, \mathbf{V}_R \right)$ refers to the Lagrangian solution. Observe also that although $\mathbf{V} = (\tau, u, E)$, the notation applies equally to all other flow variables, for example $\left(\frac{\partial}{\partial t} p \right)^*$, etc.

Remark 5.6 The order in which the limits in t and ξ are taken is important. Thus, $\left(\frac{\partial}{\partial t}\mathbf{V}\right)_R$ is obtained by evaluating first the time derivative at $t = 0$ (and $\xi > 0$) and then letting $\xi \to 0$. In particular, $\left(\frac{\partial}{\partial t}\mathbf{V}\right)_R$ can be evaluated directly from the initial data, using the system (4.83). However, $\left(\frac{\partial}{\partial t}\mathbf{V}\right)^*$ can only be evaluated in the context of the full solution of the linear GRP. Similarly, $\left(\frac{\partial}{\partial \xi}\mathbf{V}\right)^*_R$ is obtained by taking the ξ derivative on the right side of the contact discontinuity (at $t > 0$) and then letting $t \to 0$.

The main ingredient in the solution of the linear GRP and, indeed, the fundamental building block of the GRP method, is the following theorem.

Theorem 5.7 (Main theorem of linear GRP) *Let $\left(\frac{\partial u}{\partial t}\right)^*$ and $\left(\frac{\partial p}{\partial t}\right)^*$ be the time derivatives of the velocity and pressure along the contact discontinuity, evaluated at $t = 0+$. These derivatives are determined by a pair of linear equations*

$$a_L\left(\frac{\partial u}{\partial t}\right)^* + b_L\left(\frac{\partial p}{\partial t}\right)^* = d_L, \tag{5.12$_L$}$$

$$a_R\left(\frac{\partial u}{\partial t}\right)^* + b_R\left(\frac{\partial p}{\partial t}\right)^* = d_R. \tag{5.12$_R$}$$

The coefficients depend on the equation of state,[5] and in addition

a_L, b_L, d_L	*depend on*	$\lambda, \mathbf{V}^*_L, \mathbf{V}_L, \mathbf{V}'_L$	*(see Claim 5.17) and*
a_R, b_R, d_R	*depend on*	$\lambda, \mathbf{V}^*_R, \mathbf{V}_R, \mathbf{V}'_R$	*(see Claim 5.18).*

All six coefficients can be explicitly evaluated from the indicated data, as will be seen in the course of the proof. (See Corollary 5.15 and Corollary 5.20 in the case of a γ-law gas.)

Because the proof of the theorem is rather long, we break it up into several steps, terminating in Claims 5.17 and 5.18. For the reader's convenience we include in Summary 5.24 a general layout of the various steps.

Remark 5.8

(a) Note that equations (5.12)$_{L,R}$ are coupled only through the dependence of all six coefficients on the associated Riemann solution $\mathbf{V}^*$. Apart from that, the "left" (resp. "right") coefficients a_L, b_L, d_L (resp. a_R, b_R, d_R) depend only on the "left-side" (resp. "right-side") initial data.

---

[5] More specifically, on the isentropic curve (5.28) and the Hugoniot curve (5.50).

(b) The cross-sectional area $A(r)$ and hence also the Lagrangian coordinate ξ are arbitrary up to a constant factor, which means that V_L', V_R' are only determined up to this factor. In the formulas this is balanced by suitable multiplication by A. Mathematically speaking, it means that we could use the normalization $A(0) = 1$. However, we refrain from doing that for the following two reasons: (i) to retain the "dimensional correctness" of formulas such as those appearing in Proposition 5.9 below and (ii) to facilitate the translation of the formulas to the actual GRP numerical algorithm as presented in Section 5.2 (where obviously the cross-sectional area varies from one node to the other).

The proof of Theorem 5.7 requires some preparations, most notably the treatment of a CRW in the GRP setting. This will be done in the next subsections. However, before dealing with the general case, let us specialize to the "acoustic case."

The Acoustic Case

Assume that the initial flow variables are all continuous at $\xi = 0$ so that $V_L = V_R$, but we allow jumps in their slopes $V_L' \neq V_R'$. Clearly, the associated Riemann solution is now constant:

$$V^A(\xi, t) \equiv V_L = V_R;$$

hence, according to (5.3) the GRP solution $V(\xi, t)$ is continuous at $\xi = t = 0$. It follows that the initial wave pattern of $V(\xi, t)$ does not contain a jump discontinuity (shock or contact), nor does it contain a CRW. The "waves" emanating from the origin are therefore just characteristic curves C_-, C_+ as in Figure 5.3 (the curve C_0 coincides with the particle path $\xi = 0$). These curves

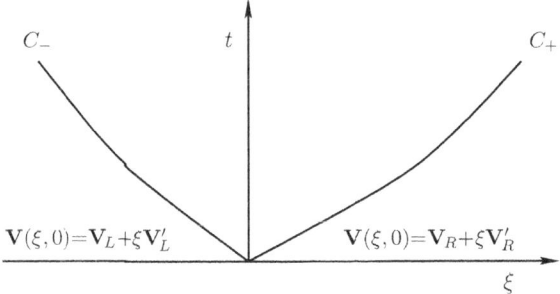

Figure 5.3. The "acoustic case" $V_L = V_R$, $V_L' \neq V_R'$.

are characterized as "sound waves" (see Remark 4.31), justifying the terminology "acoustic case" used here. In view of (4.85) their slopes are $-A(0)g_L = -A(0)\rho_L c_L$ (for C_-) and $A(0)g_R = A(0)\rho_R c_R$ (for C_+), where c_L, c_R are the initial speeds of sound (in the Eulerian frame), as $r \to 0-$, $0+$, respectively. When viewed from the side of the contact discontinuity, these slopes are, respectively, $-A(0)g_L^*$ and $A(0)g_R^*$. Of course, these values are equal to the previous ones, since $\mathbf{V}^* = \mathbf{V}_L = \mathbf{V}_R$. However, we retain the two-sided notation (such as g_L, g_L^*, g_R, g_R^*, which are all equal) in the formulas. This will enable us to use them in the numerical application based on the acoustic case (see Construction 5.38) where $\mathbf{V}_L \neq \mathbf{V}_R$, but their difference is sufficiently "small" (see Remark 5.11). We shall see later (Remark 5.21) that the coefficients depend continuously on $\mathbf{V}_L$, $\mathbf{V}_R$.

Proposition 5.9 (The acoustic case–Lagrangian framework) *Assume* $\mathbf{V}_L = \mathbf{V}_R$, $\mathbf{V}_L' \neq \mathbf{V}_R'$. *Then the coefficients in Equations* (5.12)$_{L,R}$ *are given by*

$$\begin{aligned}
a_L &= 1, b_L = (g_L^*)^{-1} = (\rho_L^* c_L^*)^{-1}, \\
d_L &= -(g_L^*)^{-1} g_L \left\{ A(0)[g_L u_L' + p_L'] + \lambda u_L c_L \right\}, \\
a_R &= -1, b_R = (g_R^*)^{-1} = (\rho_R^* c_R^*)^{-1}, \\
d_R &= -(g_R^*)^{-1} g_R \left\{ A(0)[g_R u_R' - p_R'] + \lambda u_R c_R \right\}.
\end{aligned} \tag{5.13}$$

In particular, the coefficients in (5.12)$_{L,R}$ *depend only on the initial data (including the speed of sound).*

Note that the derivatives in d_L, d_R are the ξ derivatives (conforming with the notation $\mathbf{V}_L'$, $\mathbf{V}_R'$). Using (5.8) we get, for example,

$$u_L' = A(0)^{-1} \rho_L^{-1} \frac{\partial}{\partial r} u(r, 0) \bigg|_{r=0-}.$$

Proof (of Proposition 5.9) Since u, p are continuous across the line $\xi = 0$, the same holds true for their time derivatives $\frac{\partial u}{\partial t}$, $\frac{\partial p}{\partial t}$. They are therefore continuous in the full domain between C_- and C_+ (see Figure 5.3) and approach, respectively, $\left(\frac{\partial u}{\partial t}\right)^*$, $\left(\frac{\partial p}{\partial t}\right)^*$ as $(\xi, t) \to (0, 0)$ in this domain. In view of the second equation in (4.83) the derivative $\frac{\partial p}{\partial \xi}$ is also continuous in this domain and approaches the value $A(0)\left(\frac{\partial p}{\partial \xi}\right)^* = -\left(\frac{\partial u}{\partial t}\right)^*$ as $(\xi, t) \to (0, 0)$. The pressure is also continuous across C_-; hence the same is true for its derivative $\frac{dp}{dt}$ (along C_-). Using the chain rule, we can express this derivative in two ways, approaching

C_- from either side. For $t \to 0+$, we record separately the two limiting values of the slope as $-A(0)g_L$ and $-A(0)g_L^*$, obtaining

$$\left(\frac{\partial p}{\partial t}\right)^* - A(0)g_L^* \left(\frac{\partial p}{\partial \xi}\right)^* = \left(\frac{\partial p}{\partial t}\right)_L - A(0)g_L p_L'. \tag{5.14}$$

The flow is isentropic along $\xi = constant$ [see (4.59)] so that [in view of (4.61)]

$$\left(\frac{\partial p}{\partial t}\right)_L = c_L^2 \left(\frac{\partial \rho}{\partial t}\right)_L = -g_L^2 \left(\frac{\partial \tau}{\partial t}\right)_L = -g_L^2 (Au)_L', \tag{5.15}$$

where in the last step we have used the first equation in (4.83). We now observe that, by (4.81),

$$(Au)_L' = \frac{d}{d\xi} [A(r(\xi, 0))u(\xi, 0)]_{\xi=0-} = \lambda \rho_L^{-1} u_L + A(0)u_L', \tag{5.16}$$

and inserting this in (5.15) yields

$$\left(\frac{\partial p}{\partial t}\right)_L = -\lambda g_L c_L u_L - A(0)g_L^2 u_L'. \tag{5.17}$$

Using (5.17) and $A(0)\left(\frac{\partial p}{\partial \xi}\right)^* = -\left(\frac{\partial u}{\partial t}\right)^*$ in (5.14) we get

$$g_L^* \left(\frac{\partial u}{\partial t}\right)^* + \left(\frac{\partial p}{\partial t}\right)^* = -A(0)g_L^2 u_L' - A(0)g_L p_L' - \lambda g_L c_L u_L,$$

which is identical to $(5.12)_L$ with a_L, b_L, d_L as in (5.13).

The values of a_R, b_R, d_R in (5.13) are obtained in exactly the same way (or by using the previous argument for the transformed setting $r \to -r$, $\xi \to -\xi$, $u \to -u$, and $p \to p$). $\qquad \square$

Remark 5.10 Once the time derivative $\left(\frac{\partial p}{\partial t}\right)^*$ is known, the time derivatives for the density are given by

$$\left(\frac{\partial \rho}{\partial t}\right)_{L,R}^* = (c^*)_{L,R}^{-2} \left(\frac{\partial p}{\partial t}\right)^*. \tag{5.18}$$

The density is evidently continuous across $\xi = 0$ when $V_L = V_R$. However, Equation (5.18) will be used in the more general setting as discussed in the paragraph preceding Proposition 5.9.

Remark 5.11 As we shall see in Section 5.2, the acoustic case [i.e., the coefficients as given in (5.13)] is fully adequate for the numerical simulation of practically all compressible flow problems. The versions of the GRP method based on this observation (labeled E_1 and L_1; see Remarks 5.38 and 5.40) combine simplicity with accuracy in resolving sharp flow discontinuities ("high-resolution" property).

Resolution of a CRW in the Lagrangian Framework

We now turn back to the general setting of the linear GRP (5.10) and assume that the wave pattern is as in Figure 5.2. In particular, the Γ_1 wave is a CRW. Whereas the associated Riemann solution $\mathbf{R}^A\left(\frac{\xi}{t}; \mathbf{V}_L, \mathbf{V}_R\right)$ is self-similar (so that all characteristic curves in the CRW are straight lines carrying constant values of $\mathbf{V}$), the CRW in the GRP setting is curvilinear [compare Figures 5.1(a) and (b)]. The values of $\mathbf{V}$ vary along each characteristic curve. However, in view of (5.3), they converge at the singularity to the corresponding values of the associated solution.

Our first objective is to study the (nonzero) directional derivatives of the flow variables inside the CRW at the singularity. As will be seen later, this constitutes the main technical step in the proof of Theorem 5.7. We start out by "mapping" the CRW in terms of "characteristic coordinates." This is done as shown in Figure 5.4. The CRW consists of C_- characteristic curves fanning out of the origin and C_+ transversal curves. Their slopes (and characteristic relations) are given by Equation (4.85). The structure of the associated CRW is described in Claim 4.44. In particular, the slopes of corresponding C_- characteristic curves coincide at the singularity; these limiting slopes range from $-A(0)g_L = -A(0)\rho_L c_L$ at the leading (head) curve to $-A(0)g_L^* = -A(0)\rho_L^* c_L^*$

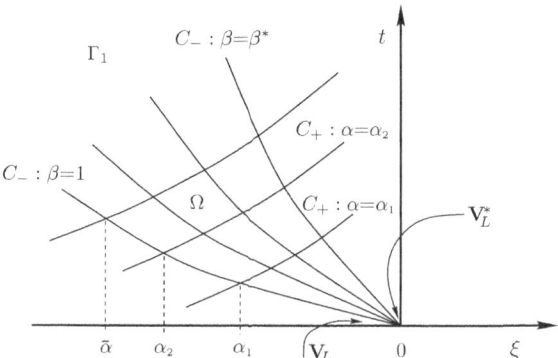

Figure 5.4. Characteristic coordinates in a 1-CRW.

at the tail curve.[6] We parametrize the C_- curves by the normalized slope $\beta = \frac{g}{g_L}$, so that $\beta = 1$ along the leading curve whereas $\beta = \beta^* = \frac{g_L^*}{g_L} \leq 1$ along the tail curve. The transversal family of characteristic curves C_+ is parametrized by assigning to each curve a value α that is the ξ coordinate of its point of intersection with the leading C_- curve ($\beta = 1$). In particular, the C_+ curve at the singularity degenerates to a single point and carries the value $\alpha = 0$. In analogy with "polar coordinates" at the center, this degenerate curve still carries all values of $\beta \in [\beta^*, 1]$ (the center in polar coordinates carries all the angular values in $[0, 2\pi)$). Fixing a small negative value $\bar{\alpha} < 0$, we conclude that the "triangular" sector Ω of the CRW shown in Figure 5.4 is "mapped" onto the rectangle [in the (α, β) plane]

$$D = \left\{ (\alpha, \beta), \;\; \bar{\alpha} \leq \alpha \leq 0, \;\; 0 < \beta^* \leq \beta \leq 1 \right\}. \tag{5.19}$$

The coordinates (ξ, t) in Ω can be expressed in terms of α and β,

$$\xi = \xi(\alpha, \beta), \qquad t = t(\alpha, \beta), \qquad (\alpha, \beta) \in D. \tag{5.20}$$

These functions are smooth, but they are not one-to-one, of course, since $\xi(0, \beta) = t(0, \beta) = 0$ for all $\beta^* \leq \beta \leq 1$. In the case of the associated Riemann problem, explicit expressions can easily be deduced from Claim 4.44. Indeed, the C_- curves are straight lines satisfying $\xi = -A(0)g_L\beta t$, whereas the C_+ curves are hyperbolas satisfying $\xi t = -A(0)^{-1}g_L^{-1}\alpha^2$, so that

$$\xi = \alpha\beta^{1/2}, \qquad t = -A(0)^{-1}g_L^{-1}\alpha\beta^{-1/2}, \qquad (\alpha, \beta) \in D. \tag{5.21}$$

In the general case, the expressions (5.21) are the leading terms (in powers of α) in the transformations (5.20), which can therefore be rewritten as

$$\begin{aligned} \xi(\alpha, \beta) &= \alpha\beta^{1/2} + \epsilon(\alpha, \beta)\alpha^2, & (\alpha, \beta) \in D, \\ t(\alpha, \beta) &= -A(0)^{-1}g_L^{-1}\alpha\beta^{-1/2} + \eta(\alpha, \beta)\alpha^2, & (\alpha, \beta) \in D, \end{aligned} \tag{5.22}$$

where $\epsilon(\alpha, \beta)$ and $\eta(\alpha, \beta)$ are smooth functions in D.

These arguments may now be extended to all flow variables defined in the CRW. Thus, if $Q = Q(\xi, t)$ stands for such a variable (say, $Q = \rho$ or $Q = u$), we can substitute for ξ, t the expressions (5.22), so that $Q = Q(\alpha, \beta)$ becomes a smooth function of $(\alpha, \beta) \in D$. It should be emphasized that in general $Q(\xi, t)$ is not even continuous at $(\xi, t) = (0, 0)$, as it may approach different limiting

---

[6] Recall that the values at the singularity are determined by $\mathbf{R}^A\left(\frac{\xi}{t}; \mathbf{V}_L, \mathbf{V}_R\right)$ and range from $\mathbf{V}_L$ at the head to $\mathbf{V}_L^*$ at the tail. We shall always assume that $g_L^* > 0$.

values along different C_- curves. However, this is resolved in the (α, β) representation, where two different C_- curves correspond to two different $\beta : \beta_1$ and β_2. As $\alpha \rightarrow 0$, they approach different points $(0, \beta_1)$, $(0, \beta_2)$, carrying different limiting values $Q(0, \beta_1)$, $Q(0, \beta_2)$ of any flow variable Q. Note finally that the relation (5.3) between the solution to the GRP and the associated Riemann solution implies that the limiting values $Q(0, \beta)$, $\beta^* \leq \beta \leq 1$, are identical in both solutions for any flow variable Q.

In the associated solution, all flow variables are uniform along C_- curves, namely, $Q(\alpha, \beta) \equiv Q(0, \beta)$, $\bar{\alpha} \leq \alpha \leq 0$, and in particular $\frac{\partial}{\partial \alpha} Q(\alpha, \beta)|_{\alpha=0} = 0$. This (directional) derivative expresses the initial variation of $Q(\alpha, \beta)$ in the direction of C_- (corresponding to the normalized slope β) as it emanates from the singularity. In the GRP case it does not generally vanish. In fact, the evaluation of $\frac{\partial}{\partial \alpha} Q(\alpha, \beta)|_{\alpha=0}$ is essential to our treatment of the linear GRP and the proof of Theorem 5.7, as was already indicated.

To simplify notation we shall write $\frac{\partial}{\partial \alpha} Q(0, \beta)$ for $\frac{\partial}{\partial \alpha} Q(\alpha, \beta)|_{\alpha=0}$ when there is no risk of confusion.

It is instructive at this point to see how Equations (5.22) can be formally justified. In fact, at every point in the CRW the cross-sectional area A can be expressed as a function $\widetilde{A}(\alpha, \beta)$,

$$\widetilde{A}(\alpha, \beta) = A(r(\xi, t)) = A(r(\xi(\alpha, \beta), t(\alpha, \beta))).$$

At $\xi = 0$, $\widetilde{A}(0, \beta) \equiv A(0)$. The equations (4.85) for the characteristic directions can be written as

$$
\text{(i)} \quad \frac{\partial \xi}{\partial \alpha} = -g\widetilde{A}\frac{\partial t}{\partial \alpha},
$$

$$
\text{(ii)} \quad \frac{\partial \xi}{\partial \beta} = g\widetilde{A}\frac{\partial t}{\partial \beta}.
\tag{5.23}
$$

Differentiating the first equation with respect to β and the second with respect to α, and noting that at $\alpha = 0$, $\frac{\partial t}{\partial \beta}(0, \beta) = \frac{\partial \widetilde{A}}{\partial \beta}(0, \beta) = 0$ we obtain

$$2g(0, \beta) \cdot \frac{\partial}{\partial \beta}\left(\frac{\partial t}{\partial \alpha}(0, \beta)\right) + \frac{\partial g}{\partial \beta}(0, \beta) \cdot \frac{\partial t}{\partial \alpha}(0, \beta) = 0.$$

By definition $g(0, \beta) = g_L\beta$ and $\frac{\partial t}{\partial \alpha}(0, 1) = -A(0)^{-1}g_L^{-1}$. Hence $\frac{\partial t}{\partial \alpha}(0, \beta) = -A(0)^{-1}g_L^{-1}\beta^{-1/2}$, so that $\frac{\partial \xi}{\partial \alpha}(0, \beta) = \beta^{1/2}$, by (5.23)(i). This establishes (5.22) by Taylor's theorem.

The key result in the GRP treatment of the CRW is that, if we single out the velocity $u(\alpha, \beta)$, its derivative $\frac{\partial u}{\partial \alpha}(0, \beta)$ can be readily determined. This is stated in the following proposition. We refer to Table 5.5 for some of the notation used here.

Proposition 5.12 *Consider the CRW as in Figure 5.4, parametrized by* $(\alpha, \beta) \in$ *D [see (5.19)]. Let* $a(\beta) = \frac{\partial u}{\partial \alpha}(0, \beta)$, $\beta^* \leq \beta \leq 1$. *Then there exists a function* $H(\beta)$ *such that*

$$\frac{d}{d\beta}a(\beta) + H(\beta) = -\frac{1}{2}\lambda A(0)^{-1}g_L^{-1}\beta^{-1/2}\frac{d}{d\beta}[u(0, \beta)c(0, \beta)] . \quad (5.24)$$

The function $H(\beta)$ *depends only on* V_L *and on the associated Riemann solution. Furthermore, if the state ahead (in our case to the left) of the CRW is isentropic, in the sense that* $S'_L = 0$ *(where S is the entropy) then* $H(\beta) \equiv 0$ *for* $\beta^* \leq \beta \leq 1$. *The relation (5.24) is supplemented by the initial condition*

$$a(1) = u'_L + g_L^{-1}p'_L \quad (5.25)$$

to yield a unique solution $a(\beta)$. *(See Corollary 5.15 for explicit expressions in the case of a* γ-*law gas.)*

The proof given here follows Ben-Artzi and Falcovitz [10] and is rather technical. The reader may skip it on first reading.

Proof As in (5.23), we rewrite the characteristic relations (4.85) in terms of (α, β), so that [note that $A' = \frac{d}{dr}A(r)$, and then substitute $r = r(\xi(\alpha, \beta), t(\alpha, \beta))$ to get $\widetilde{A}'(\alpha, \beta)))$]

$$\text{(i)} \quad \frac{\partial p}{\partial \alpha} - g\frac{\partial u}{\partial \alpha} + \left(\widetilde{A}\right)^{-1}\widetilde{A}'guc\frac{\partial t}{\partial \alpha} = 0,$$

$$\text{(ii)} \quad \frac{\partial p}{\partial \beta} + g\frac{\partial u}{\partial \beta} + \left(\widetilde{A}\right)^{-1}\widetilde{A}'guc\frac{\partial t}{\partial \beta} = 0. \quad (5.26)$$

Differentiate the first equation with respect to β and the second with respect to α and then evaluate the difference at $\alpha = 0$ (where of course $\frac{\partial t}{\partial \beta} \equiv 0$). This leads to

$$2g(0, \beta)a'(\beta) + \frac{\partial g}{\partial \alpha}(0, \beta)\frac{\partial u}{\partial \beta}(0, \beta) + a(\beta)\frac{\partial g}{\partial \beta}(0, \beta)$$

$$- \lambda\frac{\partial t}{\partial \alpha}(0, \beta)\frac{\partial}{\partial \beta}[u(0, \beta)c(0, \beta)g(0, \beta)] = 0 \quad (5.27)$$

[note that $\frac{\partial}{\partial \beta}[\left(\widetilde{A}\right)^{-1}\widetilde{A}'] \equiv 0$ at $\alpha = 0$ since $\widetilde{A}(0, \beta)^{-1} \cdot \widetilde{A}'(0, \beta) \equiv A(0)^{-1}$ $A'(0)$]. To establish (5.24) we need to express $\frac{\partial g}{\partial \alpha}(0, \beta)$ in terms of $a(\beta)$ [and flow variables of the associated solution at $(0, \beta)$]. Observe that $\frac{\partial p}{\partial \alpha}(0, \beta)$ is readily available in these terms, in view of (5.26)(i). However, the CRW is not isentropic, so g is not only a function of p but also of the entropy S. We therefore proceed as follows.

Let (ρ_I, p_I) be any state given by its density and pressure. The thermodynamic states isentropic to this state can be parametrized by p. In particular, the variable $g = \rho c$ can be expressed as a function

$$g = G(p; \rho_I, p_I) \tag{5.28}$$

along this curve. The function G is a smooth function of p, ρ_I, p_I.

Given any point (α, β) in the CRW, the entropy is invariant along the particle path $\xi = \xi(\alpha, \beta)$. We can therefore take the initial value

$$\rho_I = \rho_0(\xi(\alpha, \beta)) = \rho(\xi(\alpha, \beta), 0), \qquad p_I = p_0(\xi(\alpha, \beta)) = p(\xi(\alpha, \beta), 0)$$

to obtain, at time $t = t(\alpha, \beta)$,

$$g(\alpha, \beta) = G(p(\alpha, \beta); \rho_0(\xi(\alpha, \beta)), p_0(\xi(\alpha, \beta))). \tag{5.29}$$

In view of (5.22) (and the notation in Table 5.5) we have

$$\frac{\partial}{\partial \alpha} \rho_0(\xi(\alpha, \beta)) \bigg|_{\alpha=0} = \rho'_L \cdot \beta^{1/2},$$

and similarly for p_0. Differentiating (5.29) with respect to α and setting $\alpha = 0$ yields

$$\frac{\partial g}{\partial \alpha}(0, \beta) = G_p\left(p(0, \beta); \rho_L, p_L\right) \cdot \frac{\partial p}{\partial \alpha}(0, \beta) + I(\beta) \cdot \beta^{1/2},$$
$$I(\beta) = G_{\rho_I}\left(p(0, \beta); \rho_L, p_L\right) \cdot \rho'_L + G_{p_I}\left(p(0, \beta); \rho_L, p_L\right) \cdot p'_L. \tag{5.30}$$

Clearly $I(\beta)$ depends only on the equation of state (more specifically on G), the values $p(0, \beta)$ obtained from the associated Riemann solution, and the initial values $\rho_L, \rho'_L, p_L, p'_L$ ahead of the CRW.

Using (5.26)(i) we can rewrite (5.30) as

$$\frac{\partial g}{\partial \alpha}(0, \beta) = G_p\left(p(0, \beta); \rho_L, p_L\right)\Bigg[g(0, \beta)a(\beta)$$

$$- \lambda \frac{\partial t}{\partial \alpha}(0, \beta)g(0, \beta)u(0, \beta)c(0, \beta)\Bigg] + I(\beta)\beta^{1/2}, \quad (5.31)$$

and inserting this in (5.27) yields

$$2g(0, \beta)a'(\beta) + F(\beta)a(\beta) + \tilde{H}(\beta)$$

$$= \lambda \frac{\partial t}{\partial \alpha}(0, \beta) \left\{ \frac{\partial}{\partial \beta} [g(0, \beta)u(0, \beta)c(0, \beta)] + G_p(p(0, \beta); \rho_{\mathrm{L}}, p_{\mathrm{L}}) \right.$$

$$\left. \times \frac{\partial u}{\partial \beta}(0, \beta)g(0, \beta)u(0, \beta)c(0, \beta) \right\}, \tag{5.32}$$

where

$$F(\beta) = \frac{\partial g}{\partial \beta}(0, \beta) + G_p(p(0, \beta); \rho_{\mathrm{L}}, p_{\mathrm{L}})g(0, \beta)\frac{\partial u}{\partial \beta}(0, \beta),$$

$$\tilde{H}(\beta) = I(\beta)\frac{\partial u}{\partial \beta}(0, \beta)\beta^{1/2}. \tag{5.33}$$

We claim that $F(\beta) \equiv 0$. Indeed, by (5.26)(ii), taken at $\alpha = 0$, we have

$$\frac{\partial p}{\partial \beta}(0, \beta) = -g(0, \beta)\frac{\partial u}{\partial \beta}(0, \beta).$$

However, the CRW in the associated Riemann solution is isentropic; hence $g(0, \beta) = G(p(0, \beta); \rho_{\mathrm{L}}, p_{\mathrm{L}})$ and we get

$$F(\beta) = \frac{\partial g}{\partial \beta}(0, \beta) - G_p(p(0, \beta); \rho_{\mathrm{L}}, p_{\mathrm{L}})\frac{\partial p}{\partial \beta}(0, \beta) \equiv 0. \tag{5.34}$$

Using the same consideration and noting that $\frac{\partial t}{\partial \alpha}(0, \beta) = -A(0)^{-1}g_{\mathrm{L}}^{-1}\beta^{-1/2}$, the right-hand side of (5.32) becomes

$$-\lambda A(0)^{-1}g_{\mathrm{L}}^{-1}\beta^{-1/2} \left\{ \frac{\partial}{\partial \beta}[u(0, \beta)c(0, \beta)g(0, \beta)] - \frac{\partial g}{\partial \beta}(0, \beta)u(0, \beta)c(0, \beta) \right\}$$

$$= -\lambda A(0)^{-1}\beta^{1/2}\frac{\partial}{\partial \beta}[u(0, \beta)c(0, \beta)], \tag{5.35}$$

where we have used $g(0, \beta) = g_{\mathrm{L}}\beta$. Incorporating (5.34), (5.35) in (5.32) we obtain (5.24), with

$$H(\beta) = (2g_{\mathrm{L}}\beta)^{-1}\tilde{H}(\beta) = \frac{1}{2}g_{\mathrm{L}}^{-1}\beta^{-1/2}I(\beta)\frac{\partial u}{\partial \beta}(0, \beta). \tag{5.36}$$

Now, if $S'_L = 0$, the initial entropy S_0 satisfies $S_0(\xi) - S_{\mathrm{L}} = O(\xi^2)$; hence the isentropic curves through $(\rho_0(\xi), p_0(\xi))$ and $(\rho_{\mathrm{L}}, p_{\mathrm{L}})$ differ by $O(\xi^2)$. In particular, for all values p close to p_{L},

$$G(p; \rho_{\mathrm{L}}, p_{\mathrm{L}}) - G(p; \rho_0(\xi), p_0(\xi)) = O(\xi^2),$$

and taking $\xi = \xi(\alpha, \beta)$, $p = p(0, \beta)$, we have

$$G\left(p(0, \beta); \rho_L, p_L\right) - G\left(p(0, \beta); \rho_0(\xi(\alpha, \beta)), p_0(\xi(\alpha, \beta))\right) = O(\alpha^2),$$

$$(5.37)$$

where we have used (5.22) to replace $\xi = O(\alpha)$. Taking the α derivative of (5.37) at $\alpha = 0$ we get

$$I(\beta) = G_{\rho_I}\left(p(0, \beta); \rho_L, p_L\right) \rho'_L + G_{p_I}\left(p(0, \beta); \rho_L, p_L\right) p'_L \equiv 0$$

$$\left(\text{when } S'_L = 0\right); \quad (5.38)$$

hence $H(\beta) \equiv 0$ by (5.36).

Finally, we obtain (5.25) by the chain rule, (4.83), and (5.22),

$$a(1) = \frac{\partial u}{\partial \alpha}(0, 1) = u'_L \cdot \frac{\partial \xi}{\partial \alpha}(0, 1) + \left(\frac{\partial u}{\partial t}\right)_L \cdot \frac{\partial t}{\partial \alpha}(0, 1)$$

$$= u'_L + \left(-A(0)\frac{\partial p}{\partial \xi}\right)_L \cdot \left(-A(0)^{-1} g_L^{-1}\right) = u'_L + g_L^{-1} p'_L. \qquad \square$$

Remark 5.13 Notice the role played here by the entropy S. The CRW in the associated solution is isentropic, so $g(0, \beta) = G\left(p(0, \beta); \rho_L, p_L\right)$, and $Q(\alpha, \beta) = Q(0, \beta)$ for every flow variable Q. However, in our case, if $S'_L = 0$, we have $H(\beta) \equiv 0$. Nevertheless, if $a(1) \neq 0$ then still, in view of (5.24) (even in the planar case $\lambda = 0$) $a(\beta) \neq 0$ and in particular, by (5.26)(i), $\frac{\partial p}{\partial \alpha}(0, \beta) \neq 0$.

Remark 5.14 Equation (5.24) expresses the "decoupled" dependence of $a(\beta)$ on the "thermodynamic" data $H(\beta)$ (vanishing if $S'_L = 0$) and the "geometric" right-hand side (vanishing in the planar case or more generally if $\lambda = A'(0)A(0)^{-1} = 0$).

Explicit Formulas for the (Lagrangian) GRP in the γ-Law Case

In the case of a perfect (γ-law) gas [see (4.104), (4.105)] we can obtain fully explicit formulas for the CRW.

In view of (4.105)–(4.107), the function G of (5.28) takes the form

$$g = g_I \left(\frac{p}{p_I}\right)^{\frac{\gamma+1}{2\gamma}} = (\gamma p_I \rho_I)^{1/2} \left(\frac{p}{p_I}\right)^{\frac{\gamma+1}{2\gamma}}. \qquad (5.39)$$

Since $g(0, \beta) = g_L \beta$, we easily obtain

$$p(0, \beta) = p_L \beta^{\frac{2\gamma}{\gamma+1}}, \qquad \rho(0, \beta) = \rho_L \beta^{\frac{2}{\gamma+1}}, \qquad c(0, \beta) = c_L \beta^{\frac{\gamma-1}{\gamma+1}}. \quad (5.40)$$

The Riemann invariant R_+ [see (4.108)] is constant throughout the (isentropic) CRW in the associated solution, so that

$$u(0, \beta) = u_L + \frac{2}{\gamma - 1}c_L - \frac{2}{\gamma - 1}c(0, \beta) = u_L + \frac{2c_L}{\gamma - 1}\left(1 - \beta^{\frac{\gamma-1}{\gamma+1}}\right).$$

(5.41)

From the definition of $I(\beta)$ in (5.30) and (5.39) it follows that

$$I(\beta) = \left[\frac{1}{2}c_L\rho_L' - \frac{1}{2c_L}p_L'\right] \cdot \beta;$$

(5.42)

hence, by (5.36) and (5.41),

$$H(\beta) = \frac{1}{2\rho_L(\gamma + 1)}\left[\frac{1}{c_L}p_L' - c_L\rho_L'\right] \cdot \beta^{\frac{\gamma-3}{2(\gamma+1)}}.$$

(5.43)

The right-hand side of (5.24) can be evaluated explicitly using (5.40), (5.41):

$$-\frac{1}{2}\lambda A(0)^{-1}g_L^{-1}\beta^{-1/2}\frac{d}{d\beta}[u(0, \beta)c(0, \beta)]$$

$$= -\frac{1}{2}\frac{\lambda A(0)^{-1}}{(\gamma + 1)\rho_L}\left\{[(\gamma - 1)u_L + 2c_L]\beta^{-\frac{\gamma+5}{2(\gamma+1)}} - 4c_L\beta^{\frac{\gamma-7}{2(\gamma+1)}}\right\}.$$

(5.44)

Incorporating (5.43) and (5.44) in (5.24) we obtain the following explicit formula for $a(\beta)$.

Corollary 5.15 *In the case of a γ-law gas, the function $a(\beta) = \frac{\partial u}{\partial \alpha}(0, \beta)$ of Proposition 5.12 is given by*

$$a(\beta) = a(1) + \frac{1}{g_L(3\gamma - 1)}\left[c_L^2\rho_L' - p_L'\right]\left(\beta^{\frac{3\gamma-1}{2(\gamma+1)}} - 1\right)$$

$$- \frac{\lambda A(0)^{-1}}{\rho_L(\gamma - 3)}[(\gamma - 1)u_L + 2c_L]\left(\beta^{\frac{\gamma-3}{2(\gamma+1)}} - 1\right)$$

$$+ \frac{4\lambda A(0)^{-1}c_L}{\rho_L(3\gamma - 5)} \cdot \left(\beta^{\frac{3\gamma-5}{2(\gamma+1)}} - 1\right), \quad \gamma \neq \frac{5}{3}, 3.$$

(5.45)

We refer to Ben-Artzi and Falcovitz [10, Equations (3.21) and (3.22)] for the exceptional cases $\gamma = \frac{5}{3}, 3$.

Concluding the Treatment of the CRW

In the preceding subsections we established the general expression (5.24) for the directional derivative $a(\beta) = \frac{\partial u}{\partial \alpha}(0, \beta)$ and then derived its explicit form for a γ-law gas. Once $a(\beta)$ is known, all other directional derivatives are readily available. In the following corollary we collect these facts. We use $\rho = J(p; \rho_I, p_I)$ for the function representing all states (ρ, p) isentropic to a given state (ρ_I, p_I). This is related to the analogous expression (5.28) for g by $G = \rho c = J \cdot \left(\frac{\partial J}{\partial p}\right)^{-1/2}$.

Corollary 5.16 (Directional derivatives in a CRW) *Using notation as in Proposition 5.12, we have*

$$\frac{\partial p}{\partial \alpha}(0, \beta) = g_L \beta \left[a(\beta) + \lambda A(0)^{-1} g_L^{-1} u(0, \beta) c(0, \beta) \beta^{-1/2} \right],$$

$$\frac{\partial \rho}{\partial \alpha}(0, \beta) = \left[J_{\rho_I}(p(0, \beta); \rho_L, p_L) \cdot \rho_L' + J_{p_I}(p(0, \beta); \rho_L, p_L) \cdot p_L' \right] \beta^{1/2}$$

$$+ c(0, \beta)^{-2} \frac{\partial p}{\partial \alpha}(0, \beta). \tag{5.46}$$

Proof The expression for $\frac{\partial p}{\partial \alpha}$ follows by combining (5.26)(i) with (5.22). The one for $\frac{\partial \rho}{\partial \alpha}$ is obtained from $\rho(\alpha, \beta) = J(p(\alpha, \beta); \rho_0(\xi(\alpha, \beta)), p_0(\xi(\alpha, \beta)))$ as in (5.30). □

In the case of a γ-law gas, we have

$$\rho = J(p; \rho_I, p_I) = \rho_I \left(\frac{p}{p_I}\right)^{\frac{1}{\gamma}}, \tag{5.47}$$

and explicit expressions can be derived for $\frac{\partial p}{\partial \alpha}$ and $\frac{\partial \rho}{\partial \alpha}$ by using (5.40), (5.41), and (5.45).

Time Derivatives of p, u on the Interface – Proof of the Main Theorem

We are now in a position to prove the main theorem of this section, Theorem 5.7. We assume that the solution is as shown in Figure 5.2 and treat separately the coefficients in (5.12)$_L$ and (5.12)$_R$. The notation employed here is that used already and in Table 5.5. This applies in particular to the function $a(\beta)$ as in Proposition 5.12.

Claim 5.17 (The coefficients a_L, b_L, d_L) *Equation (5.12)$_L$ holds with*

$$a_L = 1, \qquad b_L = \left(g_L^*\right)^{-1}, \qquad d_L = -\left(g_L g_L^*\right)^{1/2} A(0) a(\beta^*) - \lambda u^* c_L'. \tag{5.48}$$

Proof The flow in the region $\xi(\alpha, \beta^*) \leq \xi \leq 0$ (i.e., between the tail charac-teristic of the rarefaction wave and the contact discontinuity) is smooth. Using the chain rule at (α, β^*) and then letting $\alpha \to 0$ we obtain

$$\frac{\partial p}{\partial \alpha}(0, \beta^*) = \left(\frac{\partial p}{\partial t}\right)^* \frac{\partial t}{\partial \alpha}(0, \beta^*) + \left(\frac{\partial p}{\partial \xi}\right)^*_L \frac{\partial \xi}{\partial \alpha}(0, \beta^*). \tag{5.49}$$

By (4.83) we have $A(0)\left(\frac{\partial p}{\partial \xi}\right)^*_L = -\left(\frac{\partial u}{\partial t}\right)^*$ and by (5.22)

$$\frac{\partial \xi}{\partial \alpha}(0, \beta^*) = (\beta^*)^{1/2}, \qquad \frac{\partial t}{\partial \alpha}(0, \beta^*) = -A(0)^{-1} g_L^{-1} (\beta^*)^{-1/2}.$$

Also, by (5.46),

$$\frac{\partial p}{\partial \alpha}(0, \beta^*) = g_L \beta^* \left[a(\beta^*) + \lambda A(0)^{-1} g_L^{-1} u^* c_L^* (\beta^*)^{-1/2}\right].$$

Incorporating these equations in (5.49) and noting $\beta^* = g_L^{-1} g_L^*$ we obtain (5.48). $\qquad\qquad \square$

We now turn to the treatment of the coefficients in $(5.12)_R$. Two difficulties are encountered here. First, all flow variables are discontinuous across the shock. Second, the state behind the shock is not uniform, since the initial data are not uniform. The flow in the two regions separated by the shock is smooth, and so is the shock trajectory. For clarity, we shall denote by $Q_+(\xi, t)$ the value of the flow variable Q ($Q = \rho, u$, etc.) in the region ahead of the shock (i.e., the region between Γ_3 and the positive ξ axis in Figure 5.2) while retaining the notation $Q(\xi, t)$ for the values behind the shock (i.e., between Γ_3 and the contact discontinuity $\xi = 0$). We parametrize the shock trajectory Γ_3 as $\xi(\theta)$, $t(\theta)$, with $(\xi(0), t(0)) = (0, 0)$. Along this curve we can simplify the notation of the flow variables on the two sides in an obvious way,

$$Q(\theta) = Q(\xi(\theta), t(\theta)), \qquad Q_+(\theta) = Q_+(\xi(\theta), t(\theta)).$$

The shock relation (4.102) can be written here as

$$u(\theta) = u_+(\theta) + \Phi(p(\theta); \rho_+(\theta), p_+(\theta)) \tag{5.50}$$

[since $p(\theta) \geq p_+(\theta)$, $\Phi = \phi_3$ of (4.102)]. According to (4.88), (4.89) the shock speed $\sigma = \xi'(t) = \frac{\xi'(\theta)}{t'(\theta)}$ can be expressed as

$$\sigma(\theta) = A(\theta) \frac{p(\theta) - p_+(\theta)}{u(\theta) - u_+(\theta)}, \qquad A(\theta) = A(r(\xi(\theta), t(\theta))), \tag{5.51}$$

and the derivative $\frac{df}{d\theta}$ can be expressed as

$$\frac{df}{d\theta} = \left(\sigma(\theta)\frac{\partial f}{\partial \xi} + \frac{\partial f}{\partial t}\right) \cdot t'(\theta) \tag{5.52}$$

for any function $f = f(\xi, t)$. Clearly, if $f(\xi, t) = Q(\xi, t)$ the derivatives $\frac{\partial Q}{\partial \xi}$, $\frac{\partial Q}{\partial t}$ in (5.52) are evaluated behind the shock [i.e., at $\xi = \xi(\theta)-$, $t = t(\theta)+$] and similarly for $Q_+(\xi, t)$.

The assertion of the next claim is that the coefficients a_R, b_R, d_R of $(5.12)_R$ are obtained by differentiating (5.50).

Claim 5.18 (The coefficients a_R, b_R, d_R) *Equation $(5.12)_R$ holds true with coefficients a_R, b_R, d_R, which depend only on the following: the initial data V_R, V'_R to the right of the discontinuity, the function Φ of (5.50), and the values p^*, u^*, ρ^*_R obtained in the solution of the associated Riemann problem.*

Detailed expressions for a_R, b_R, d_R are given in the following [Equation (5.58)] for the general case, as well as Equation (5.62) for the special case of a γ-law gas].

Proof We consider the function Φ as a function of three variables, $\Phi = \Phi$ $(p; \rho_l, p_l)$ [compare (5.28)]. Differentiating (5.50) with respect to θ and using (5.52) we get

$$\frac{\partial u}{\partial t} + \sigma(\theta)\frac{\partial u}{\partial \xi} = \frac{\partial u_+}{\partial t} + \sigma(\theta)\frac{\partial u_+}{\partial \xi}$$

$$+ \Phi_p\left(p(\theta); \rho_+(\theta), p_+(\theta)\right) \cdot \left[\frac{\partial p}{\partial t} + \sigma(\theta)\frac{\partial p}{\partial \xi}\right]$$

$$+ \Phi_{\rho_l}\left(p(\theta); \rho_+(\theta), p_+(\theta)\right) \cdot \left[\frac{\partial \rho_+}{\partial t} + \sigma(\theta)\frac{\partial \rho_+}{\partial \xi}\right]$$

$$+ \Phi_{p_l}\left(p(\theta); \rho_+(\theta), p_+(\theta)\right) \cdot \left[\frac{\partial p_+}{\partial t} + \sigma(\theta)\frac{\partial p_+}{\partial \xi}\right]. \tag{5.53}$$

Inspecting Figure 5.2 and using the notation in Table 5.5 we see that as $\theta \to 0$ we have the following limits:

$$\frac{\partial p}{\partial t} \to \left(\frac{\partial p}{\partial t}\right)^*, \qquad \frac{\partial u}{\partial t} \to \left(\frac{\partial u}{\partial t}\right)^*,$$

$$\frac{\partial p}{\partial \xi} \to \left(\frac{\partial p}{\partial \xi}\right)^*, \qquad \frac{\partial u}{\partial \xi} \to \left(\frac{\partial u}{\partial \xi}\right)^*_R, \tag{5.54}$$

and, for $Q_+ = p_+, u_+, \rho_+$,

$$\frac{\partial Q_+}{\partial t} \to \left(\frac{\partial Q}{\partial t}\right)_R, \qquad \frac{\partial Q_+}{\partial \xi} \to Q'_R. \tag{5.55}$$

Furthermore, using Equations (4.83) and (5.18), we have

$$A(0)\left(\frac{\partial p}{\partial \xi}\right)^* = -\left(\frac{\partial u}{\partial t}\right)^*,$$

$$A(0)\left(\frac{\partial u}{\partial \xi}\right)^*_R = \left(\frac{\partial \tau}{\partial t}\right)^*_R - u^*\left(\frac{\partial}{\partial \xi}A(r(\xi,0))\right)^*_R \tag{5.56}$$

$$= -\left(g^*_R\right)^{-2}\left(\frac{\partial p}{\partial t}\right)^* - \lambda\left(\rho^*_R\right)^{-1}u^*.$$

Note that we have used, as in Remark 5.6,

$$\left(\frac{\partial}{\partial \xi}A(r(\xi,0))\right)^*_R = \lim_{t\to 0+}\frac{A'(r(0,t))}{A(r(0,t))}\cdot \rho(0+,t)^{-1} = \lambda\left(\rho^*_R\right)^{-1}.$$

However,

$$\left(\frac{\partial \rho}{\partial t}\right)_R = -\rho^2_R\left(\frac{\partial \tau}{\partial t}\right)_R = -\rho^2_R\frac{\partial}{\partial \xi}[Au]_{\xi=0+}$$

$$= -\rho^2_R\left[\lambda\rho^{-1}_R u_R + A(0)u'_R\right] \qquad \text{[compare (5.15)]},$$

$$\left(\frac{\partial p}{\partial t}\right)_R = c^2_R\left(\frac{\partial \rho}{\partial t}\right)_R \tag{5.57}$$

$$= -g^2_R\left[\lambda\rho^{-1}_R u_R + A(0)u'_R\right] \qquad \text{[compare (5.17)]},$$

$$\left(\frac{\partial u}{\partial t}\right)_R = -A(0)p'_R.$$

We can now let $\theta \to 0$ in (5.53) and use these relations to get

$$[1 + A(0)^{-1}\sigma(0)\Phi_p\left(p^*;\rho_R,p_R\right)]\left(\frac{\partial u}{\partial t}\right)^*$$

$$+ \left[-A(0)^{-1}\sigma(0)\left(g^*_R\right)^{-2} - \Phi_p\left(p^*;\rho_R,p_R\right)\right]\left(\frac{\partial p}{\partial t}\right)^*$$

$$= \left[\sigma(0) - \rho^2_R A(0)\Phi_{\rho_I}\left(p^*;\rho_R,p_R\right) - g^2_R A(0)\Phi_{p_I}\left(p^*;\rho_R,p_R\right)\right]u'_R$$

$$+ \left[-A(0) + \sigma(0)\Phi_{p_I}\left(p^*;\rho_R,p_R\right)\right]p'_R + \sigma(0)\Phi_{\rho_I}\left(p^*;\rho_R,p_R\right)\rho'_R$$

$$+ \lambda\left\{A(0)^{-1}\sigma(0)\left(\rho^*_R\right)^{-1}u^* - \rho_R u_R\Phi_{\rho_I}\left(p^*;\rho_R,p_R\right)\right.$$

$$\left. - g^2_R\rho^{-1}_R u_R\Phi_{p_I}\left(p^*;\rho_R,p_R\right)\right\}. \tag{5.58}$$

The limiting speed $\sigma(0)$ is identical to that of the associated solution, and by (5.51) we have

$$\sigma(0) = A(0)\frac{p^* - p_R}{u^* - u_R}. \tag{5.59}$$

This concludes the proof of the claim, and indeed, the proof of Theorem 5.7. □

Remark 5.19 Observe that in view of (5.58) d_R is linear in u'_R, p'_R, ρ'_R and $\lambda = A'(0)A(0)^{-1}$, and can be written as

$$d_R = L_u u'_R + L_p p'_R + L_\rho \rho'_R + L_\lambda \cdot \lambda, \tag{5.60}$$

where L_u, L_p, L_ρ, L_λ depend on $\mathbf{V}_R$, $\mathbf{V}_R^*$ and the equation of state (in terms of Φ). Specializing to a γ-law gas we have, by (4.111),

$$\Phi(p; \rho_I, p_I) = (p - p_I)\sqrt{\frac{1 - \mu^2}{\rho_I(p + \mu^2 p_I)}}, \qquad \mu^2 = \frac{\gamma - 1}{\gamma + 1}, \tag{5.61}$$

and carrying out the differentiations in (5.58) we get the following explicit expressions.

Corollary 5.20 (The coefficients a_R, b_R, d_R for a γ-law gas) *In the γ-law case we have from (5.58)–(5.61), with $\sigma(0)$ as in (5.59) and d_R as in (5.60),*

$$a_R = 2 - \frac{1}{2}\frac{p^* - p_R}{p^* + \mu^2 p_R},$$

$$b_R = -A(0)^{-1}\sigma(0)(g_R^*)^{-2} - A(0)\sigma(0)^{-1}(a_R - 1),$$

$$L_u = \sigma(0) + \frac{A(0)}{2}\rho_R(u^* - u_R) + (A(0)g_R^*)^2\sigma(0)^{-1}\left[1 + \frac{\mu^2(p^* - p_R)}{2(p^* + \mu^2 p_R)}\right],$$

$$L_p = -\left(2 + \frac{\mu^2}{2}\frac{p^* - p_R}{p^* + \mu^2 p_R}\right)A(0), \tag{5.62}$$

$$L_\rho = -\left(\frac{p^* - p_R}{2\rho_R}\right)A(0),$$

$$L_\lambda = A(0)^{-1}\sigma(0)u^*(\rho_R^*)^{-1} + u_R\rho_R^{-1}(u^* - u_R)$$

$$\times\left[g_R^2\left(\frac{1}{p^* - p_R} + \frac{\mu^2}{2}\cdot\frac{1}{p^* + \mu^2 p_R}\right) + \frac{\rho_R}{2}\right].$$

Remark 5.21 (The "acoustic limit" $\mathbf{V}_L = \mathbf{V}_R$) When $\mathbf{V}_L = \mathbf{V}_R$ the solution of the associated Riemann problem is constant and $\mathbf{V}^* = \mathbf{V}_L = \mathbf{V}_R$. The shock speed $\sigma(0)$ is replaced by the characteristic slope $A(0)g_R$. Invoking these facts in (5.62) we obtain the coefficients a_R, b_R, d_R of the acoustic case (5.13) (multiplied by a factor of -2). Similarly, in this case $\beta^* = 1$ and $a(\beta^*) = a(1)$, as in (5.25). The coefficients a_L, b_L, d_L of (5.48) are then identical to those of (5.13).

Conclusion of the Linear GRP in the Lagrangian Case

The linear GRP in the Lagrangian coordinates, as posed in (5.10), is now fully solved. Indeed, the derivatives $\left(\frac{\partial p}{\partial t}\right)^*$, $\left(\frac{\partial u}{\partial t}\right)^*$ are determined by Theorem 5.7. The time derivatives of the density on the two sides of the contact discontinuity are then obtained by

$$\left(\frac{\partial \rho}{\partial t}\right)^*_R = (c^*_R)^{-2}\left(\frac{\partial p}{\partial t}\right)^*, \qquad \left(\frac{\partial \rho}{\partial t}\right)^*_L = (c^*_L)^{-2}\left(\frac{\partial p}{\partial t}\right)^*. \qquad (5.63)$$

(We refer to Figure 5.2 and Table 5.5 for notation.)

The solution to the linear GRP in Eulerian coordinates (see Proposition 5.26) also requires knowledge of the ξ derivatives at the contact discontinuity.

The ξ derivatives of p, u on the two sides of the discontinuity are evaluated as in (5.56):

$$A(0)\left(\frac{\partial p}{\partial \xi}\right)^* = -\left(\frac{\partial u}{\partial t}\right)^*,$$

$$A(0)\left(\frac{\partial u}{\partial \xi}\right)^*_R = -(g^*_R)^{-2}\left(\frac{\partial p}{\partial t}\right)^* - \lambda(\rho^*_R)^{-1}u^* \qquad (5.64)$$

[and an analogous expression for $\left(\frac{\partial u}{\partial \xi}\right)^*_L$]. The remaining derivatives are $\left(\frac{\partial \rho}{\partial \xi}\right)^*_R$, $\left(\frac{\partial \rho}{\partial \xi}\right)^*_L$. Here we face a difficulty, since these ξ derivatives do not appear (directly) in the basic system of equations (4.83). We therefore resort to the method employed in the proof of Claim 5.18.

Claim 5.22 (ξ derivatives of ρ at the contact discontinuity) *Assume again the wave pattern of Figure 5.2. Using the notation of Table 5.5 we can express the derivatives $\left(\frac{\partial \rho}{\partial \xi}\right)^*_L$, $\left(\frac{\partial \rho}{\partial \xi}\right)^*_R$ as follows:*

$$\left(\frac{\partial \rho}{\partial \xi}\right)^*_L = (\beta^*)^{-1/2}\frac{\partial \rho}{\partial \alpha}(0, \beta^*) + A(0)^{-1}(g^*_L)^{-1}(c^*_L)^{-2}\left(\frac{\partial p}{\partial t}\right)^*, \qquad (5.65)$$

where $\frac{\partial \rho}{\partial \alpha}(0, \beta)$ *is given in* (5.46) *and* $\beta^* = \frac{g_L^*}{g_L}$, *and*

$$
\left(\frac{\partial \rho}{\partial \xi}\right)_R^* = \left[-(\rho_R^*)^2 A(0)^2 \sigma(0)^{-3} - 3(c_R^*)^{-2} \sigma(0)^{-1}\right] \left(\frac{\partial p}{\partial t}\right)^*
$$

$$
+ 3(\rho_R^*)^2 A(0) \sigma(0)^{-2} \left(\frac{\partial u}{\partial t}\right)^* + 3(\rho_R^*)^2 A(0)^2 \sigma(0)^{-2} p_R' + \left(\frac{\rho_R^*}{\rho_R}\right)^2 \rho_R'
$$

$$
+ \left[-g_R^2 A(0)^3 (\rho_R^*)^2 \sigma(0)^{-3} - 3A(0) \sigma(0)^{-1} (\rho_R^*)^2\right] u_R'
$$

$$
+ \rho_R^* \sigma(0)^{-1} \left\{\rho_R^{-1} u_R \rho_R^* \left[-g_R^2 A(0)^2 \sigma(0)^{-2} - 1\right] - 2u^*\right\} \lambda, \quad (5.66)
$$

where $\sigma(0)$ *is given by* (5.59).

Proof By the chain rule [compare (5.49)], we have

$$
\frac{\partial \rho}{\partial \alpha}(0, \beta^*) = \left(\frac{\partial \rho}{\partial t}\right)_L^* \frac{\partial t}{\partial \alpha}(0, \beta^*) + \left(\frac{\partial \rho}{\partial \xi}\right)_L^* \frac{\partial \xi}{\partial \alpha}(0, \beta^*),
$$

from which (5.65) follows by solving for $\left(\frac{\partial \rho}{\partial \xi}\right)_L^*$ and noting (5.22) and (5.63).

To establish (5.66) we recall the parametric notation introduced in the paragraph preceding (5.50) and the jump relation (4.89)(iii). This relation can be written as

$$
(p(\theta) - p_+(\theta))(\tau_+(\theta) - \tau(\theta)) = (u(\theta) - u_+(\theta))^2. \quad (5.67)
$$

Differentiating with respect to θ according to (5.52) and setting $\theta = 0$ we get, in view of (5.54), (5.55),

$$
\left[\sigma(0)\left(\left(\frac{\partial p}{\partial \xi}\right)^* - p_R'\right) + \left(\frac{\partial p}{\partial t}\right)^* - \left(\frac{\partial p}{\partial t}\right)_R\right](\tau_R - \tau_R^*)
$$

$$
+ \left[\sigma(0)\left(\tau_R' - \left(\frac{\partial \tau}{\partial \xi}\right)_R^*\right) + \left(\frac{\partial \tau}{\partial t}\right)_R - \left(\frac{\partial \tau}{\partial t}\right)_R^*\right](p^* - p_R)
$$

$$
= 2(u^* - u_R)\left[\sigma(0)\left(\left(\frac{\partial u}{\partial \xi}\right)_R^* - u_R'\right) + \left(\frac{\partial u}{\partial t}\right)^* - \left(\frac{\partial u}{\partial t}\right)_R\right]. \quad (5.68)
$$

We now make the following substitutions:

$$
\left(\frac{\partial p}{\partial \xi}\right)^*, \qquad \left(\frac{\partial u}{\partial \xi}\right)_R^* \qquad \text{by (5.56)},
$$

$$
\left(\frac{\partial p}{\partial t}\right)_R, \qquad \left(\frac{\partial u}{\partial t}\right)_R, \qquad \left(\frac{\partial \tau}{\partial t}\right)_R \qquad \text{by (5.57)}.
$$

Also, by (5.63), $\left(\frac{\partial \tau}{\partial t}\right)_R^* = -(g_R^*)^{-2}\left(\frac{\partial p}{\partial t}\right)^*$, and by (4.89), (5.59),

$$\sigma(0) = A(0)\frac{p^* - p_R}{u^* - u_R}, \qquad \sigma(0)^2 = A(0)^2\frac{p^* - p_R}{\tau_R - \tau^*}. \qquad (5.69)$$

Inserting these substitutions in (5.68) and solving for $\left(\frac{\partial \tau}{\partial \xi}\right)_R^* = -(\rho_R^*)^{-2}\left(\frac{\partial p}{\partial \xi}\right)_R^*$ we obtain (5.66). □

Equations (5.65), (5.66) are more complicated than (5.64). However, in the *acoustic case* they can be considerably simplified.

Claim 5.23 (ξ derivatives of ρ at the contact discontinuity – acoustic case)
In the acoustic case (see Proposition 5.9) we have, instead of (5.65), (5.66),

$$\left(\frac{\partial \rho}{\partial \xi}\right)_L^* = (g_L^*)^{-1}\left[\rho_L^2 u_L' + g_L\rho_L' + \lambda A(0)^{-1}\rho_L u_L + A(0)^{-1}(c_L^*)^{-2}\left(\frac{\partial p}{\partial t}\right)^*\right],$$

$$\left(\frac{\partial \rho}{\partial \xi}\right)_R^* = (g_R^*)^{-1}\left[-\rho_R^2 u_R' + g_R\rho_R' - \lambda A(0)^{-1}\rho_R u_R - A(0)^{-1}(c_R^*)^{-2}\left(\frac{\partial p}{\partial t}\right)^*\right].$$

$$(5.70)$$

Proof Using the same reasoning as in (5.14) we get

$$\left(\frac{\partial \rho}{\partial t}\right)_L^* - A(0)g_L^*\left(\frac{\partial \rho}{\partial \xi}\right)_L^* = \left(\frac{\partial \rho}{\partial t}\right)_L - A(0)g_L\rho_L'.$$

Since

$$\left(\frac{\partial \rho}{\partial t}\right)_L = c_L^{-2}\left(\frac{\partial p}{\partial t}\right)_L, \qquad \left(\frac{\partial \rho}{\partial t}\right)_L^* = (c_L^*)^{-2}\left(\frac{\partial p}{\partial t}\right)^*,$$

we have by (5.17)

$$(c_L^*)^{-2}\left(\frac{\partial p}{\partial t}\right)^* - A(0)g_L^*\left(\frac{\partial \rho}{\partial \xi}\right)_L^* = -\lambda\rho_L u_L - A(0)\rho_L^2 u_L' - A(0)g_L\rho_L',$$

which yields (5.70) for the left derivative. The right derivative is handled similarly. □

All ξ and t derivatives of flow variables (at the singularity) are now accounted for. As a first step in their evaluation, we solve the associated Riemann problem and determine the values $\mathbf{V}^* = \mathbf{R}^A(0; \mathbf{V}_L, \mathbf{V}_R)$ (including discontinuous values such as ρ_L^*, ρ_R^*). The second step is the solution of the linear GRP [see (5.10)], namely, finding all (instantaneous) time derivatives at the singularity. The third step completes the treatment by providing for all ξ derivatives there. For the reader's convenience we summarize here the procedure involved in the second and third steps. The first step, that of solving the Riemann problem, is given in Construction 4.46 and Appendix C. Recall that notation used below is that introduced in Table 5.5.

Summary 5.24 (Solving the linear GRP in Lagrangian coordinates)

(I) *Find the derivatives* $\left(\frac{\partial u}{\partial t}\right)^*$, $\left(\frac{\partial p}{\partial t}\right)^*$ *as in Theorem 5.7.*
γ-*law gas: Explicit formulas are given in Corollary 5.15 and Corollary 5.20.*

(II) *Evaluate* $\left(\frac{\partial \rho}{\partial t}\right)_L^*$, $\left(\frac{\partial \rho}{\partial t}\right)_R^*$ *by (5.63), and evaluate* $\left(\frac{\partial u}{\partial \xi}\right)_L^*$, $\left(\frac{\partial u}{\partial \xi}\right)_R^*$, $\left(\frac{\partial p}{\partial \xi}\right)^*$ *by (5.64).*

(III) *Determine* $\left(\frac{\partial \rho}{\partial \xi}\right)_L^*$, $\left(\frac{\partial \rho}{\partial \xi}\right)_R^*$ *by (5.65), (5.66) [or (5.70) in the acoustic case].*

(IV) *In the acoustic case the coefficients in Theorem 5.7 are given in (5.13) and do not depend on the equation of state. (See also Remark 5.21.) In this case Equations (5.65), (5.66) are replaced by (5.70).*

Remark 5.25 Our formulas were based on the hypothesis that the wave pattern for the GRP is as in Figure 5.2. All other possibilities can be easily derived from these formulas by simple "reflections." For example, if the wave Γ_3 is a CRW, we obtain the coefficients a_L, b_L, d_L, from Proposition 5.12 and Claim 5.17 by reflecting $\xi \to -\xi, t \to t, u \to -u$ and leaving unchanged the thermodynamic variables (so that $-u_R$ replaces u_L but u_R' comes instead of u_L').

The Linear GRP in the Eulerian Framework

We are now in a position to address the (Eulerian) linear GRP as stated in (5.4). Indeed, having the ξ and t derivatives at our disposal, we can obtain any directional derivative by the chain rule. This applies in particular to the line $r = 0$, when represented as a curve in the (ξ, t) plane, $r(\xi, t) = 0$.

We distinguish three typical cases as depicted in Figure 5.5. All three correspond to the wave pattern assumed in Figure 5.2.

The line $r = 0$ is represented in the (ξ, t) plane by a curve $\xi = \xi(t), \xi(0) = 0$. To find its functional form we simply differentiate the identity $r(\xi(t), t) = 0$ to obtain by (4.80), (4.81)

$$\xi'(t) = -A(0)\rho(\xi, t)u(\xi, t), \qquad \xi(0) = 0. \tag{5.71}$$

We basically distinguish two cases:

(a) The "nonsonic" case is when the line $r = 0$ is not "included" in a CRW. This corresponds to Figures 5.5(a) and (b).

(b) The "sonic" case is when the line $r = 0$ is tangent (at $r = t = 0$) to a characteristic curve within a rarefaction fan, such as Figure 5.5(c).

In the setup of Figure 5.5, the instantaneous values behind the shock Γ_3 (at $t = 0+$) are included in $\mathbf{V}_R^*$ (see Table 5.5), whereas those behind the CRW (Γ_1) are included in $\mathbf{V}_L^*$. According to (5.3) the limiting values of the solution to the GRP along $r = 0$, as $t \to 0$, are given by

$$\mathbf{U}_0 = \lim_{t \to 0+} \mathbf{U}(0, t) = \mathbf{R}^{\mathrm{A}}(0; \mathbf{U}_L, \mathbf{U}_R). \tag{5.72}$$

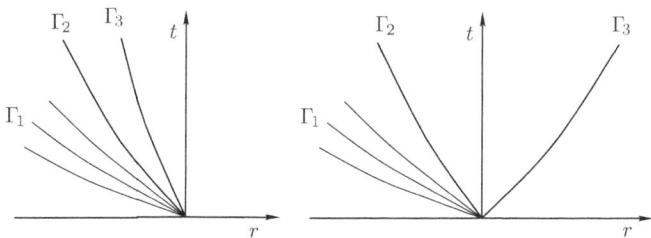

(a) $\Gamma_1, \Gamma_2, \Gamma_3$ propagate to the left.   (b) $r = 0$ separates Γ_2 and Γ_3.

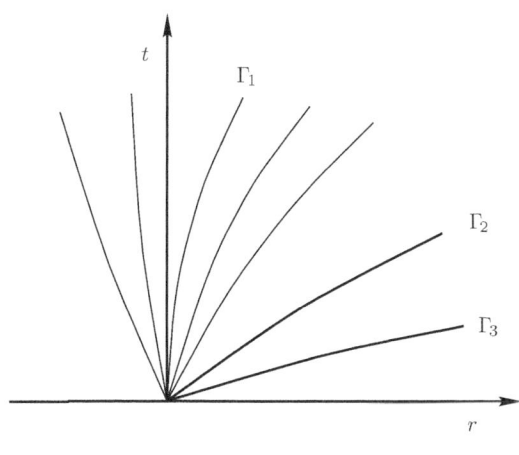

(c) The sonic case.

Figure 5.5. Possible Eulerian GRP solutions for the wave pattern of Figure 5.2.

By the discussion preceding (5.3) we also know that the position of $r = 0$ (near $t = 0$) relative to the waves Γ_1, Γ_2, Γ_3 is fully determined by the solution to the associated Riemann problem. In particular (always assuming the wave pattern of Figure 5.5), the line $r = 0$ is nonsonic if either $u^* - c_L^* < 0$ (the full CRW propagates into the region $r < 0$) or $u_L - c_L > 0$ (it propagates into the region $r > 0$). Since the shock speed σ_E (in Eulerian coordinates) is given by

$$\sigma_E = \frac{\rho_R u_R - \rho_R^* u^*}{\rho_R - \rho_R^*}$$

[see (4.86)], we infer that the line $r = 0$ is situated between the contact discontinuity Γ_2 and the shock Γ_3 [as in Figure 5.5(a)] if $u^* < 0 < \sigma_E$ and so on.

Taking $Q = Q(r, t)$ to be any flow variable (such as p or u) we denote [as in (5.72)]

$$Q_0 = \lim_{t \to 0+} Q(0, t).$$

The initial slope $\xi'(0)$ of the curve $\xi(t)$ is then obtained from (5.71),

$$\xi'(0) = -A(0)\rho_0 u_0. \tag{5.73}$$

We now consider $Q = Q(\xi, t)$ in the Lagrangian framework and take its partial derivative along the line $r = 0$ [which is represented by $\xi = \xi(t)$]. The limiting values of these derivatives at the initial point $(0, 0)$ are determined according to the location of the line $r = 0$. In the nonsonic case, this line lies between two waves, or to the left or right [as in Figure 5.5(a)] of all the three. We denote by $\partial_t^0 Q$, $\partial_\xi^0 Q$ those limiting derivatives:

$$\partial_t^0 Q = \lim_{t \to 0+} \frac{\partial}{\partial t} Q(\xi, t)\bigg|_{\xi = \xi(t)},$$

$$\partial_\xi^0 Q = \lim_{t \to 0+} \frac{\partial}{\partial \xi} Q(\xi, t)\bigg|_{\xi = \xi(t)}. \tag{5.74}$$

For example, in the situation displayed in Figure 5.5(a) ($r = 0$ to the right of Γ_3),

$$\partial_t^0 Q = \left(\frac{\partial}{\partial t} Q\right)_R, \qquad \partial_\xi^0 Q = Q_R',$$

whereas in the case of Figure 5.5(b) ($r = 0$ between Γ_2 and Γ_3),

$$\partial_t^0 Q = \left(\frac{\partial}{\partial t} Q\right)_R^*, \qquad \partial_\xi^0 Q = \left(\frac{\partial}{\partial \xi} Q\right)_R^*$$

(see Table 5.5).

The solution to the linear GRP (5.4) is now straightforward. Since $\mathbf{U}(0, t) = \mathbf{U}(\xi(t), t)$, we have, by the chain rule, the following:

Proposition 5.26 (The linear GRP in Eulerian coordinates – nonsonic case)
The solution $\left(\frac{\partial}{\partial t}\mathbf{U}\right)_0$ *to the linear GRP in the nonsonic case is given by the chain rule,*

$$\left(\frac{\partial}{\partial t}\mathbf{U}\right)_0 = \partial_t^0 \mathbf{U} - A(0)\rho_0 u_0 \, \partial_\xi^0 \mathbf{U}. \tag{5.75}$$

The derivatives in the right-hand side of (5.75) are obtained as in Summary 5.24.

We are now left with the sonic case, as in Figure 5.5(c). Here $u_{\mathrm{L}} - c_{\mathrm{L}} < 0 <$ $u^* - c_{\mathrm{L}}^*$, so that the line $r = 0$ is tangent (at $t = 0$) to a characteristic contained in the rarefaction fan. In the language of the characteristic coordinates introduced for the analysis of the CRW [see the discussion preceding (5.19)] we can identify the line $r = 0$ with a smooth trajectory $\alpha(t)$, $\beta(t)$, so that $(\alpha(0), \beta(0)) = (0, \beta_0)$, $\beta^* \leq \beta_0 \leq 1$. The limiting slope (at $t = 0$) of the characteristic C_- associated with β is $u(0, \beta) - c(0, \beta)$, so that the tangency of $r = 0$ to the β_0 characteristic implies

$$u(0, \beta_0) = c(0, \beta_0).^7 \tag{5.76}$$

The chain rule formula (5.75) is clearly meaningless here. Using the notation introduced in the paragraph following (5.22), we can represent the solution throughout the CRW as $\mathbf{U}(\alpha, \beta)$. In particular, along $r = 0$, the solution $\mathbf{U}(0, t)$ can be expressed as $\mathbf{U}(\alpha(t), \beta(t))$ and differentiated at $t = 0$.

Proposition 5.27 (The linear GRP in Eulerian coordinates – sonic case) *Assume the wave pattern of Figure 5.5(c) and let $r = 0$ be represented in the characteristic framework by $(\alpha(t), \beta(t))$, where β_0 satisfies (5.76).*
The solution $\left(\frac{\partial}{\partial t}\mathbf{U}\right)_0$ *to the linear GRP in this case is given by*

$$\left(\frac{\partial}{\partial t}\mathbf{U}\right)_0 = \frac{\partial}{\partial \alpha}\mathbf{U}(0, \beta_0) \cdot \alpha'(0) + \frac{\partial}{\partial \beta}\mathbf{U}(0, \beta_0) \cdot \beta'(0), \tag{5.77}$$

where

$$\alpha'(0) = -A(0)g_{\mathrm{L}}\beta_0^{1/2}, \tag{5.78}$$

$$\beta'(0) = \frac{1}{2}\beta_0^{1/2}A(0)\left[\frac{\partial(\rho c)}{\partial \alpha}(0, \beta_0) - \frac{\partial(\rho u)}{\partial \alpha}(0, \beta_0)\right]. \tag{5.79}$$

The derivatives $\frac{\partial}{\partial \beta}\mathbf{U}(0, \beta_0)$ *are obtained from the associated Riemann solution* $\mathbf{R}^{\mathrm{A}}(\mu; \mathbf{U}_{\mathrm{L}}, \mathbf{U}_{\mathrm{R}})$ *[see (5.3)] or its Lagrangian equivalent* $\mathbf{V}^{\mathrm{A}}(\xi, t)$ *[where*

---

[7] This equality explains the term "sonic case"; the material velocity coincides with the sonic speed.

β corresponds to the C_- characteristic $\xi(t) = -A(0)g_L\beta t]$. In the case of a γ-law gas use Equations (5.40) and (5.41).

The derivatives $\frac{\partial}{\partial\alpha}\mathbf{U}(0,\beta_0)$ are obtained in the process of the resolution of the CRW (Proposition 5.12 and Corollary 5.16). In the case of a γ-law gas see Remark 5.28 later in this subsection.

Proof Equation (5.77) follows from the chain rule, and it remains to prove (5.78), (5.79). Regarding $t(\alpha, \beta)$ as a function along $(\alpha(t), \beta(t))$ we have the obvious identity $t = t(\alpha(t), \beta(t))$. Differentiating at $t = 0$ and using (5.22) we get

$$1 = \frac{\partial t}{\partial\alpha}(0,\beta_0)\cdot\alpha'(0) + \frac{\partial t}{\partial\beta}(0,\beta_0)\cdot\beta'(0) = -A(0)^{-1}g_L^{-1}\beta_0^{-1/2}\alpha'(0),$$

which proves (5.78). The proof of (5.79) is considerably more intricate and the reader may skip it on first reading.

The line $r = 0$ is represented by $\xi = \xi(t) = \xi(\alpha(t), \beta(t))$, so by (5.22)

$$\xi'(t) = \frac{\partial\xi}{\partial\alpha}\alpha'(t) + \frac{\partial\xi}{\partial\beta}\beta'(t)$$

$$= \left[\beta(t)^{1/2} + 2\alpha(t)\epsilon(0,\beta)\right]\alpha'(t) + \frac{1}{2}\alpha(t)\beta(t)^{-1/2}\beta'(t) + O(t^2), \quad (5.80)$$

whereas from (5.71) we have

$$\xi'(t) = -A(0)\rho(\alpha(t), \beta(t))\,u(\alpha(t), \beta(t)),$$

so that

$$\xi'(t) = -A(0)\left\{(\rho u)(0,\beta_0) + \frac{\partial(\rho u)}{\partial\alpha}(0,\beta_0)\cdot\alpha(t)\right.$$

$$\left. + \frac{\partial(\rho u)}{\partial\beta}(0,\beta_0)\cdot(\beta(t) - \beta_0)\right\} + O(t^2). \quad (5.81)$$

However, the characteristic relation (5.26)(ii), evaluated for the associated Riemann solution (i.e., along the degenerate C_+ characteristic $\alpha = 0$), in conjunction with (5.76), yields[8]

$$c(0,\beta_0)^2\frac{\partial\rho}{\partial\beta}(0,\beta_0) + \rho(0,\beta_0)c(0,\beta_0)\frac{\partial u}{\partial\beta}(0,\beta_0) = c(0,\beta_0)\frac{\partial}{\partial\beta}(\rho u)(0,\beta_0) = 0.$$

$$(5.82)$$

---

[8] The CRW for the associated solution is isentropic, so that

$$\frac{\partial p}{\partial\beta}(0,\beta) = c(0,\beta)^2\frac{\partial\rho}{\partial\beta}(0,\beta).$$

Therefore (5.81) can be rewritten as

$$\xi'(t) = -A(0)\left\{(\rho u)(0, \beta_0) + \frac{\partial(\rho u)}{\partial \alpha}(0, \beta_0) \cdot \alpha(t)\right\} + O(t^2)$$

$$= -A(0)\left\{(\rho c)(0, \beta_0) + \frac{\partial(\rho u)}{\partial \alpha}(0, \beta_0) \cdot \alpha(t)\right\} + O(t^2). \quad (5.83)$$

Comparing zero-order terms in (5.80) and (5.83) we capture again (5.78), since $g(0, \beta_0) = g_L \beta_0$. For first-order terms we have

$$2\epsilon(0, \beta_0)\alpha'(0)^2 + \alpha'(0)\beta'(0)\beta_0^{-1/2} + \beta_0^{1/2}\alpha''(0)$$

$$= -A(0)\frac{\partial(\rho u)}{\partial \alpha}(0, \beta_0) \cdot \alpha'(0). \quad (5.84)$$

Similarly, differentiating the identity $t(\alpha(t), \beta(t)) = t$ and retaining first-order terms we have

$$2\eta(0, \beta_0)\alpha'(0)^2 + A(0)^{-1}g_L^{-1}\beta_0^{-3/2}\alpha'(0)\beta'(0) - A(0)^{-1}g_L^{-1}\beta_0^{-1/2}\alpha''(0) = 0. \quad (5.85)$$

Combining (5.84) and (5.85) we can eliminate $\alpha''(0)$ to obtain

$$2\alpha'(0)\left[A(0)g_L\beta_0\eta(0, \beta_0) + \epsilon(0, \beta_0)\right] + 2\beta_0^{-1/2}\beta'(0) = -A(0)\frac{\partial(\rho u)}{\partial \alpha}(0, \beta_0). \quad (5.86)$$

It remains to evaluate the expression in the square brackets in (5.86). To this end we use Equation (5.23)(i) along $(\alpha(t), \beta(t))$. Once again we expand up to first-order terms and we use $\tilde{A}(\alpha(t), \beta(t)) = A(0)$ to get, by (5.22),

$$\beta(t)^{1/2} + 2\alpha(t)\epsilon(0, \beta_0) = -A(0)g(0, \beta_0)\left[-A(0)^{-1}g_L^{-1}\beta(t)^{-1/2}\right.$$

$$+ 2\alpha(t)\eta(0, \beta_0)] - A(0)\frac{\partial g}{\partial \alpha}(0, \beta_0)$$

$$\times \left[A(0)^{-1}g_L^{-1}\beta_0^{-1/2}\alpha(t)\right] + O(t^2).$$

Using $g(0, \beta_0) = g_L \beta_0$ and equating first-order terms we get

$$2\left[A(0)g_L\beta_0\eta(0, \beta_0) + \epsilon(0, \beta_0)\right] = -\frac{\partial g}{\partial \alpha}(0, \beta_0)g_L^{-1}\beta_0^{-1/2}.$$

Inserting this expression in (5.86) and noting (5.78) we obtain (5.79). □

Remark 5.28 (γ-law gas) In the case of a γ-law gas the α derivatives in (5.79) are obtained by combining Equations (5.40), (5.41), and (5.47) and Corollaries 5.15 and 5.16. The value of β_0, obtained from (5.76) in conjunction with (5.41), is

$$\beta_0 = \left[\frac{\gamma - 1}{\gamma + 1} \left(\frac{u_L}{c_L} + \frac{2}{\gamma - 1} \right) \right]^{\frac{\gamma+1}{\gamma-1}}. \tag{5.87}$$

We have concluded the treatment of the linear GRP in all cases. As in the Lagrangian case (Summary 5.24), we include a short summary.

Summary 5.29 (Solving the linear GRP in Eulerian coordinates)

(I) *Obtain the Lagrangian solution (namely, ξ and t derivatives at the contact discontinuity) as in Summary 5.24.*

(II) *Determine the various (initial) speeds of the waves Γ_1, Γ_2, Γ_3 in Eulerian coordinates. In the case depicted in Figures 5.2 and 5.5, these are (see Table 5.5 for notation)*

$$u_L - c_L \text{ and } u^* - c_L^* \text{ for the head and tail } C_- \text{ characteristics of } \Gamma_1,$$

$$u^* \text{ for the speed of the contact discontinuity } \Gamma_2,$$

$$\sigma_E = \frac{\rho_R u_R - \rho_R^* u_R^*}{\rho_R - \rho_R^*} \text{ for the shock } \Gamma_3.$$

(III) *Locate the line $r = 0$ relative to these waves and use the associated Riemann solution $\mathbf{R}^A \left(\frac{r}{t}; \mathbf{U}_L, \mathbf{U}_R \right)$ [see (5.3)] to obtain the initial values*

$$\mathbf{U}_0 = \lim_{t \to 0+} \mathbf{U}(0, t).$$

(IV) *If the line $r = 0$ is not contained in a CRW (nonsonic case) evaluate the derivative $\left(\frac{\partial}{\partial t} \mathbf{U} \right)_0$ as in Proposition 5.26.*

(V) *If the line $r = 0$ is contained in a CRW (sonic case) evaluate the derivative $\left(\frac{\partial}{\partial t} \mathbf{U} \right)_0$ as in Proposition 5.27.*

Remark 5.30 (The "acoustic case") We emphasize [see Summary 5.24(IV)] that in the acoustic case the time derivatives $\left(\frac{\partial u}{\partial t} \right)^*$, $\left(\frac{\partial p}{\partial t} \right)^*$ are readily available (Proposition 5.9). It then follows [Summary 5.24(II) and (III)] that the full solution to the linear GRP in Lagrangian coordinates is reduced to simple algebraic expressions. In particular, these expressions do not depend on the equation of state.

Remark 5.31 (Directional derivatives) Note that the preceding discussion enables us to find the directional derivatives in any desired direction (instead of $r = 0$). In a nonsonic direction, one simply applies Equation (5.75), replacing the direction $\xi'(0) = -A(0)\rho_0 u_0$ by any other direction [expressed in (ξ, t) coordinates]. Compare this also with Remark 5.4(a).

5.2 The GRP Numerical Method for Quasi-1-D, Compressible, Inviscid Flow

The previous section dealt with the solution to the linear GRP. Given initial piecewise-linear data, we can find the value of the solution [to the system (4.45)] and its time derivative at the singularity. In this section we show how to implement this solution in the design of a suitable numerical scheme, and in fact a group of such schemes.

The schemes presented here range from the very basic one (Definition 5.37), which is just a straightforward, easy-to-implement, extension of the classical Godunov scheme, to the "full GRP" scheme (Definition 5.41), which requires the full power of the analysis presented in Section 5.1. However, we emphasize that, in all cases, the schemes are based on explicit formulas, derived on the basis of the Riemann solution and the equation of state. For a γ-law gas, these formulas are given in detail in Section 5.1. Once these formulas are incorporated into the numerical fluxes, the schemes prove to be robust and no intricate postprocessing procedures are needed (except for a simple "slope limiter", described in the final paragraph of this section). This claim will be amply demonstrated in Chapter 6 and by the numerical examples discussed in Part II of this monograph.

The basic methodology has already been introduced in Section 3.1. We take a uniform spatial grid[9] $r_j = j\Delta r$, $-\infty < j < \infty$, and uniformly spaced time levels $t_{n+1} = t_n + k$, $t_0 = 0$. We refer to the interval $(r_{j-1/2}, r_{j+1/2})$ as "cell j" and to its endpoints as the "cell boundaries."

At the time level t_n, the solution to (4.45) in cell j is approximated by an average $\mathbf{U}_j^n$. In analogy with Equation (3.1) we advance the averages $\{\mathbf{U}_j^n\}_j$ to the next time level by a general ("quasi-conservative") scheme

$$\mathbf{U}_j^{n+1} = \mathbf{U}_j^n - \frac{\Delta t}{(\Delta v)_j}\left[A(r_{j+1/2})\mathbf{F}_{j+1/2}^{n+1/2} - A(r_{j-1/2})\mathbf{F}_{j-1/2}^{n+1/2}\right]$$

$$- \frac{\Delta t}{\Delta r}\left[\mathbf{G}_{j+1/2}^{n+1/2} - \mathbf{G}_{j-1/2}^{n+1/2}\right], \tag{5.88}$$

---

[9] The fixed mesh size is taken here for convenience and can be modified to allow for nonuniform grids. Also, if the coordinate r is restricted (say, $r \geq 0$ in the spherical case), the same restrictions apply to the discrete grid.

where

$$(\Delta v)_j = \int_{r_{j-1/2}}^{r_{j+1/2}} A(r)\,dr$$

is the volume of the duct segment in cell j. The scheme (5.88) is a "finite-volume" scheme. It is obtained by integrating the quasi-conservation law (4.45) [after multiplication by $A(r)$] over the space–time rectangle $(r_{j-1/2}, r_{j+1/2}) \times (t_n, t_{n+1})$. The integral

$$\int_{r_{j-1/2}}^{r_{j+1/2}} A(r)\mathbf{U}(r, t_n)\,dr$$

is approximated by $\mathbf{U}_j^n(\Delta v)_j$, and similarly for $t = t_{n+1}$. The integral

$$\int_{r_{j-1/2}}^{r_{j+1/2}} A(r)\frac{\partial}{\partial r}\mathbf{G}(\mathbf{U}(r, t))\,dr$$

is approximated by $\frac{(\Delta v)_j}{\Delta r}\left[\mathbf{G}\left(\mathbf{U}(r_{j+1/2}, t)\right) - \mathbf{G}\left(\mathbf{U}(r_{j-1/2}, t)\right)\right]$. The "side" integrals

$$\int_{t_n}^{t_{n+1}} A(r_{j\pm1/2})\mathbf{F}\left(\mathbf{U}(r_{j\pm1/2}, t)\right)\,dt, \qquad \int_{t_n}^{t_{n+1}} \mathbf{G}\left(\mathbf{U}(r_{j\pm1/2}, t)\right)\,dt$$

are then approximated, respectively, by

$$A(r_{j\pm1/2})\mathbf{F}_{j\pm1/2}^{n+1/2}, \qquad \mathbf{G}_{j\pm1/2}^{n+1/2}, \tag{5.89}$$

which need to be determined. In fact, their evaluation in terms of the data $\left\{\mathbf{U}_j^n\right\}_j$[10] is what is commonly referred to as the "design of a scheme."

Definition 5.32 The terms $\mathbf{F}_{j\pm1/2}^{n+1/2}$, $\mathbf{G}_{j\pm1/2}^{n+1/2}$ are called the "numerical fluxes" for the quasi-conservative scheme (5.88).

In the rest of this section we follow closely the presentation in Section 3.1. In particular, we shall always assume that the ratio $\frac{k}{\Delta r}$ satisfies the CFL condition

---

[10] As in Chapter 3, we refer here only to "explicit" schemes.

in the sense of Corollary 3.2 and Remark 3.4. However, in the case at hand (and in the case of virtually any hyperbolic system of physical interest) there exists no simple bound on the "maximal wave speed" [analogous to (3.7)]. Thus, the restriction on the size of $\frac{k}{\Delta r}$ cannot be accomplished as in (3.6). Instead, we shall always assume that the CFL condition is satisfied in the sense that no wave issuing from a singularity $r_{j+1/2}$ at time $t = t_n$ reaches the adjacent cell boundaries $r_{j-1/2}, r_{j+3/2}$ during the time interval $(t_n, t_n + k)$. In practice, this is achieved by inspecting all such waves at time $t = t_n$ and taking their maximal speed S_n. Since (generally speaking) the wave speeds vary in time, an additional "safety" factor $\mu_{\text{CFL}} < 1$ is added to make up for a possible growth of the maximal speed. This factor is then labeled "CFL ratio", and the next time step k_n is set to be $k_n = \mu_{\text{CFL}} \cdot \frac{\Delta r}{S_n}$. For notational simplicity we omit henceforth the dependence of k on n, and we write $k_n = k$.[11]

The Godunov Scheme

Given the initial data $\mathbf{U}_0(r) = \mathbf{U}(r, 0)$, we define the initial set of cell averages by

$$\mathbf{U}_j^0 = \frac{1}{(\Delta v)_j} \int_{r_{j-1/2}}^{r_{j+1/2}} A(r) \mathbf{U}_0(r) \, dr, \qquad -\infty < j < \infty. \tag{5.90}$$

Next we assume that the cell averages $\{\mathbf{U}_j^n\}_j$ are known and determine the numerical fluxes in (5.88). To this end we assume that the cross-sectional area is "locally uniform" in cells $j-1, j, j+1$. The system (4.45) is then transformed into the "planar" one (4.47), near the cell boundaries $r_{j\pm1/2}$. We further assume that the flow distribution is piecewise (or "cellwise") constant, being equal to $\mathbf{U}_j^n$ throughout cell j (at time $t = t_n$). These assumptions imply that, owing to the CFL condition, the solution in the time interval $(t_n, t_n + k)$ consists of a "sequence of Riemann problems." Each cell boundary $r_{j+1/2}$ carries an initial discontinuity, separating two constant states $\mathbf{U}_L = \mathbf{U}_j^n$, $\mathbf{U}_R = \mathbf{U}_{j+1}^n$ [see (4.100)]. Translating the point $r = r_{j+1/2}$ to the origin, and using the notation introduced in Section 4.1 [see the paragraph following (4.100)], we conclude that the solution along the line $r = r_{j+1/2}$ is constant and equal to $\mathbf{R}(0; \mathbf{U}_j^n, \mathbf{U}_{j+1}^n)$. The CFL restriction on k prevents the waves emanating from $r_{j+1/2}$ from reaching either $r_{j-1/2}$ or $r_{j+3/2}$, as seen in Figure 5.6.

---

[11] Note that all numerical examples in Chapter 3, as well as those of Chapter 6, are indeed performed with a fixed time step k. This facilitates the use of such examples as test cases for various schemes.

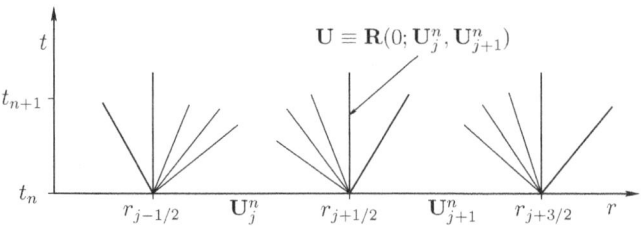

Figure 5.6. Structure of the solution for Godunov's scheme.

The situation now is completely analogous to that of the scalar conservation law. Indeed, integrating the system (4.47) over the space–time rectangle $(r_{j-1/2}, r_{j+1/2}) \times (t_n, t_n + k)$ we see that the average, over cell j, of the solution at time $t = t_{n+1}$ is *exactly* given by the formula for $\mathbf{U}_j^{n+1}$ in the following.

Definition 5.33 (The Godunov scheme) Given the initial distribution $\{\mathbf{U}_j^0\}_{-\infty < j < \infty}$, determine successively (for $n = 1, 2, \dots$) the cell averages by

$$\mathbf{U}_j^{n+1} = \mathbf{U}_j^n - \frac{k}{(\Delta v)_j} \left[A(r_{j+1/2}) \mathbf{F}_{j+1/2}^{G,n+1/2} - A(r_{j-1/2}) \mathbf{F}_{j-1/2}^{G,n+1/2} \right]$$
$$- \frac{k}{\Delta r} \left[\mathbf{G}_{j+1/2}^{G,n+1/2} - \mathbf{G}_{j-1/2}^{G,n+1/2} \right], \qquad -\infty < j < \infty, \quad (5.91)$$

where the numerical fluxes satisfy

$$\mathbf{F}_{j+1/2}^{G,n+1/2} = \mathbf{F}\left(\mathbf{R}(0; \mathbf{U}_j^n, \mathbf{U}_{j+1}^n)\right), \qquad -\infty < j < \infty,$$
$$\mathbf{G}_{j+1/2}^{G,n+1/2} = \mathbf{G}\left(\mathbf{R}(0; \mathbf{U}_j^n, \mathbf{U}_{j+1}^n)\right), \qquad -\infty < j < \infty, \qquad (5.92)$$

with $\mathbf{F}(\mathbf{U})$, $\mathbf{G}(\mathbf{U})$ as in (4.45).

Remark 5.34 It should be emphasized that in light of our assumptions concerning the local uniformity of the cross-sectional area and cellwise constant distribution of the flow variables at $t = t_n$, the formula (5.91) is the only possible choice. Indeed, the exact solution along $r = r_{j+1/2}$ is then given by $\mathbf{R}(0; \mathbf{U}_j^n, \mathbf{U}_{j+1}^n)$, for a suitably short time. Integrating (4.45) over cell j, between $t = t_n$ and $t = t_{n+1}$, leads to (5.91) for the cell average of the exact solution [compare (3.10)].

The Basic GRP Scheme

We have seen in Section 3.1 that (at least in the context of scalar conservation laws) the approximate solutions obtained by the Godunov scheme are quite adequate. This is true even in regards to the "capturing" of jump discontinuities. However, the excessive "dissipativity" of the scheme tends to "spread out" discontinuities (compare Figures 3.10 and 3.12) and to "clip out" extremal points (compare Figures 3.5 and 3.7). These numerical effects are even more pronounced in the case of systems, as we shall see in Chapter 6.

The remedy suggested in Section 3.1 was based on the fundamental observation by van Leer [112]: Replace the cellwise-constant distribution of variables at time $t = t_n$ by a "piecewise-linear" distribution, thus achieving second-order accuracy [see (3.15) and Claim 3.8]. At the same time, maintaining the "upwind" character of the scheme as in (5.88) enables the accurate capturing of jump discontinuities. Following this line of thought, we now assume, as in (3.14), that the flow variables are linearly distributed in each cell, so that at time $t = t_n$

$$\mathbf{U}^n(r) = \mathbf{U}_j^n + (r - r_j)\mathbf{L}_j^n, \qquad r_{j-1/2} < r < r_{j+1/2}. \qquad (5.93)$$

At the cell boundaries $r_{j\pm1/2}$ we therefore allow a jump of both the variables and their gradients, as in Figure 5.7. The problem is once again to determine the numerical fluxes (5.88). Furthermore, if we assume that all slopes $\mathbf{L}_j^n$ vanish and the cross-sectional area is (locally) uniform, we are reduced to the Godunov setup and we require that the fluxes coincide with those of the Godunov scheme (5.92).

The reasoning employed in the case of the Godunov scheme can be implemented here. The CFL condition implies that during the time interval $(t_n, t_n + k)$ the exact solution $\mathbf{U}(r, t)$ to the system (4.45), subject to initial data $\mathbf{U}^n(r)$, is not affected (along the line $r = r_{j+1/2}$) by waves issuing from the neighboring discontinuities at $r_{j-1/2}, r_{j+3/2}$. Shifting $r = r_{j+1/2}$ to $r = 0$ (see Remark 5.1) and the time t_n to $t = 0$, we see that the solution $\mathbf{U}(r_{j+1/2}, t), t_n \leq t \leq t_{n+1}$, is

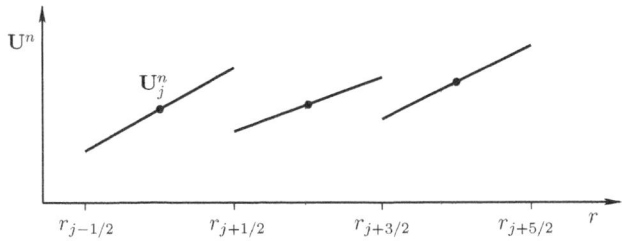

Figure 5.7. Distribution of flow variables at time $t = t_n$ (GRP setup).

that of the GRP where, as in (5.1),

$$\mathbf{U}_L = \mathbf{U}_j^n + \frac{\Delta r}{2}\mathbf{L}_j^n, \qquad \mathbf{U}_R = \mathbf{U}_{j+1}^n - \frac{\Delta r}{2}\mathbf{L}_{j+1}^n,$$

$$\mathbf{U}_L' = \mathbf{L}_j^n, \qquad\qquad \mathbf{U}_R' = \mathbf{L}_{j+1}^n. \tag{5.94}$$

Unlike the case of the Godunov scheme, where the waves emanating from the discontinuities propagate along straight lines (see Figure 5.6), these waves are now typically as shown in Figure 5.1(a). The solution $\mathbf{U}(r_{j+1/2}, t)$ cannot be obtained exactly and we must resort to appropriate approximations. The basic guideline here is to maintain the second-order accuracy, as in (3.15).

With an eye to a simple second-order extension of the Godunov scheme, we now try to design numerical fluxes based on the GRP solution. It should be analogous to the algorithm developed in the scalar case (Construction 3.10). It will be assumed that the CFL condition is satisfied and the ratio $\frac{k}{\Delta r}$ = *constant*.

As in (3.21), the first step is the evaluation of the "instantaneous" values of the solution $\mathbf{U}(r, t)$ at the jump discontinuities $(r_{j+1/2}, t_n)$. These values are obtained, in accordance with (5.3), as Riemann solutions related to the values of $\mathbf{U}^n(r)$ at the cell boundaries. Designating these values as

$$\mathbf{U}_{j+1/2,-}^n = \mathbf{U}_j^n + \frac{\Delta r}{2}\mathbf{L}_j^n, \qquad \mathbf{U}_{j+1/2,+}^n = \mathbf{U}_{j+1}^n - \frac{\Delta r}{2}\mathbf{L}_{j+1}^n \tag{5.95}$$

[these are $\mathbf{U}_L$, $\mathbf{U}_R$ in (5.94)] and using the notation of (5.3), we get

$$\mathbf{U}_{j+1/2}^n = \lim_{t \to t_n+} \mathbf{U}(r_{j+1/2}, t) = \mathbf{R}^A\left(0; \mathbf{U}_{j+1/2,-}^n, \mathbf{U}_{j+1/2,+}^n\right). \tag{5.96}$$

Next we examine the meaning of "second-order accuracy" in the present context. As was the case in Definition 2.21 and (3.15), this notion is applicable only in regions of smooth flow.[12] Suppose that $\tilde{\mathbf{U}}(r, t)$ is smooth in a neighborhood of the rectangle $[r_{j-1/2}, r_{j+1/2}] \times [t_n, t_{n+1}]$. Then the functions $\mathbf{F}(\tilde{\mathbf{U}})$, $\mathbf{G}(\tilde{\mathbf{U}})$ are smooth in the rectangle and may be compared to the numerical fluxes

---

[12] The restriction to smooth flows is because we measure the "truncation" error in "pointwise" terms, as in the proof of Claim 3.8. It can be relaxed if "integral" norms are used to measure this error. To a large extent, however, the conclusions concerning the accuracy of the scheme are norm independent.

$\mathbf{F}_{j+1/2}^{n+1/2}, \mathbf{G}_{j+1/2}^{n+1/2}$. The scheme is second-order accurate if

$$A(r_{j+1/2})\mathbf{F}_{j+1/2}^{n+1/2} - A(r_{j-1/2})\mathbf{F}_{j-1/2}^{n+1/2} = \frac{1}{k} \int_{t_n}^{t_{n+1}} \left[A(r_{j+1/2})\mathbf{F}(\widetilde{\mathbf{U}}(r_{j+1/2}, t)) \right.$$

$$\left. - A(r_{j-1/2})\mathbf{F}(\widetilde{\mathbf{U}}(r_{j-1/2}, t)) \right] dt + O(k^3),$$

$$\mathbf{G}_{j+1/2}^{n+1/2} - \mathbf{G}_{j-1/2}^{n+1/2} = \frac{1}{k} \int_{t_n}^{t_{n+1}} \left[\mathbf{G}(\widetilde{\mathbf{U}}(r_{j+1/2}, t)) - \mathbf{G}(\widetilde{\mathbf{U}}(r_{j-1/2}, t)) \right] dt + O(k^3).$$

$$(5.97)$$

Claim 3.8 and its proof can be repeated here verbatim.

Claim 5.35 (Second-order accuracy) *Let* $\widetilde{\mathbf{U}}(r, t)$ *be smooth, and let* $\mathbf{F}_{j+1/2}^{n+1/2}$, $\mathbf{G}_{j+1/2}^{n+1/2}$ *be the corresponding numerical fluxes. Then the scheme (5.88) is second-order accurate if*

$$\mathbf{F}_{j+1/2}^{n+1/2} = \mathbf{F}\left(\widetilde{\mathbf{U}}(r_{j+1/2}, t_n)\right) + \frac{k}{2} \frac{\partial}{\partial t} \mathbf{F}\left(\widetilde{\mathbf{U}}(r_{j+1/2}, t_n)\right),$$

$$\mathbf{G}_{j+1/2}^{n+1/2} = \mathbf{G}\left(\widetilde{\mathbf{U}}(r_{j+1/2}, t_n)\right) + \frac{k}{2} \frac{\partial}{\partial t} \mathbf{G}\left(\widetilde{\mathbf{U}}(r_{j+1/2}, t_n)\right).$$

$$(5.98)$$

Turning back to the setup of piecewise-linear data, we let $\mathbf{U}^n(r)$ [as in (5.93)] approximate the smooth flow $\widetilde{\mathbf{U}}(r, t_n)$. We need to express the right-hand sides of (5.98) in terms of $\mathbf{U}^n(r)$, which does not seem to be an easy task. However, taking in (5.95)

$$\mathbf{U}_j^n = \widetilde{\mathbf{U}}(r_j, t_n), \qquad \mathbf{L}_j^n = \frac{\partial}{\partial r} \widetilde{\mathbf{U}}(r_j, t_n), \qquad (5.99)$$

we have by Taylor's theorem

$$\mathbf{U}_{j+1/2,\pm}^n = \widetilde{\mathbf{U}}(r_{j+1/2}, t_n) + \mathbf{B}_\pm(r_{j+1/2}, t_n) \cdot k^2, \qquad (5.100)$$

where $\mathbf{B}_\pm(r, t)$ are smooth functions. The same conclusion applies therefore to the Riemann solutions $\mathbf{U}_{j+1/2}^n$ [see (5.96)] and we conclude that

$$\mathbf{U}_{j+1/2}^n - \mathbf{U}_{j-1/2}^n - \left[\widetilde{\mathbf{U}}(r_{j+1/2}, t_n) - \widetilde{\mathbf{U}}(r_{j-1/2}, t_n)\right] = O(k^3). \qquad (5.101)$$

The handling of $\frac{\partial}{\partial t} \widetilde{\mathbf{U}}(r_{j+1/2}, t_n)$ is the central goal of the numerical aspect of this monograph. It leads us naturally to the linear GRP as posed in Definition 5.3. In fact, given the linear distribution (5.94) on the two sides of the discontinuity at

$r = r_{j+1/2}$, the results of Section 5.1 (see Summary 5.29) allow us to determine the time derivative of the evolving solution. However, the algebraic calculations can be considerably reduced by turning to the acoustic case. Indeed, by (5.100) the limiting values $\mathbf{U}^n_{j+1/2,\pm}$ differ from $\widetilde{\mathbf{U}}(r_{j+1/2}, t_n)$ only by an $O(k^2)$ term, whereas the slopes $\mathbf{L}^n_j$, $\mathbf{L}^n_{j+1}$ differ from $\frac{\partial}{\partial r}\widetilde{\mathbf{U}}(r_{j+1/2}, t_n)$ by an $O(k)$ term. It follows that the evolution of the time derivatives based on the acoustic approximation [Proposition 5.9 and Summary 5.24(IV)] entails only an $O(k)$ smooth error relative to the exact time derivative. If the acoustic approximation of the time derivative is designated by $\left(\frac{\partial}{\partial t}\mathbf{U}^{ac}\right)^n_{j+1/2}$, the foregoing remarks yield

$$\left(\frac{\partial}{\partial t}\mathbf{U}^{ac}\right)^n_{j+1/2} - \left(\frac{\partial}{\partial t}\mathbf{U}^{ac}\right)^n_{j-1/2} = O(k^2). \tag{5.102}$$

Inspecting the right-hand sides of (5.98) we see that the time derivatives are multiplied by k; hence, as in (5.101), the differences between values at the two cell boundaries $r_{j\pm1/2}$ are again $O(k^3)$. These observations may therefore be summarized in the following key statement for the GRP method.

Theorem 5.36 (Numerical fluxes for the basic GRP scheme) *Consider the piecewise-linear distribution* $\mathbf{U}^n(r)$ *as in (5.93). At the point* $r_{j+1/2}$ *define the values* $\mathbf{U}^n_{j+1/2}$ *as the Riemann solution (5.96) and* $\left(\frac{\partial}{\partial t}\mathbf{U}^{ac}\right)^n_{j+1/2}$ *as the acoustic time derivative, based on the linear profiles (5.94) (see Construction 5.39 in the following). Define the numerical fluxes by*

$$\mathbf{F}^{n+1/2}_{j+1/2} = \mathbf{F}\left(\mathbf{U}^n_{j+1/2} + \frac{k}{2}\left(\frac{\partial}{\partial t}\mathbf{U}^{ac}\right)^n_{j+1/2}\right),$$

$$\mathbf{G}^{n+1/2}_{j+1/2} = \mathbf{G}\left(\mathbf{U}^n_{j+1/2} + \frac{k}{2}\left(\frac{\partial}{\partial t}\mathbf{U}^{ac}\right)^n_{j+1/2}\right). \tag{5.103}$$

Using these fluxes in (5.88), we find that the resulting scheme is of second-order accuracy.

Definition 5.37 The scheme presented in Theorem 5.36 is the basic GRP scheme, and we label it the E_1 scheme.

Remark 5.38 (The E_1 scheme as a generalization to Godunov's scheme) Godunov's scheme serves as the foundation of the E_1 scheme. As has been observed in Sections 3.1 and 3.2 (and will be further demonstrated in Chapter 6) it carries the main burden of the "upwinding." As a result, it serves as the basis for the "shock-capturing" capability of the scheme. The refinement involved in the E_1 scheme contributes (as has already been amply illustrated in Sections 3.2 and 3.3) to the high resolution (i.e., "sharpness") of the captured discontinuities.

Inasmuch as "computation time" is considered, the construction of the time derivatives $\left(\frac{\partial}{\partial t}\mathbf{U}^{ac}\right)^n_{j+1/2}$ (which we recall later in Construction 5.39 for the reader's convenience) adds, in practical cases, only 2–5% to the time required for the computation of the Riemann solution $\mathbf{U}^n_{j+1/2}$. Note that no information concerning the equation of state is needed for the time derivatives. A further simplification can obviously be achieved by replacing the exact Riemann solution (5.96) by an "approximate" one (see Footnote 1 in Section 5.1).

Note in particular that if the slopes $\mathbf{L}^n_j$, $\mathbf{L}^n_{j+1}$ vanish, then $\mathbf{U}^n_{j+1/2,+} = \mathbf{U}^n_{j+1}$, $\mathbf{U}^n_{j+1/2,-} = \mathbf{U}^n_j$. If, in addition, $A'(r_{j+1/2}) = 0$, then $\left(\frac{\partial}{\partial t}\mathbf{U}^{ac}\right)^n_{j+1/2} = 0$ and, by (5.92),

$$\mathbf{F}^{n+1/2}_{j+1/2} = \mathbf{F}^{G,n+1/2}_{j+1/2}, \qquad \mathbf{G}^{n+1/2}_{j+1/2} = \mathbf{G}^{G,n+1/2}_{j+1/2}. \tag{5.104}$$

In fact, because of (5.100) the choice (5.96) for $\mathbf{U}^n_{j+1/2}$ as the Riemann solution can be replaced, say, by $\frac{1}{2}\left(\mathbf{U}^n_{j+1} + \mathbf{U}^n_j\right) = \tilde{\mathbf{U}}(r_{j+1/2}, t_n) + O(k^2)$, without losing the second-order accuracy expressed in (5.101). However, the reducibility (5.104) to Godunov's scheme is then lost.

We recall the simple steps required in the evaluation of $\left(\frac{\partial}{\partial t}\mathbf{U}^{ac}\right)^n_{j+1/2}$, using the simplified notation of (5.94). It should be emphasized that, unlike the continuity assumption $\mathbf{V}_L = \mathbf{V}_R$ in Proposition 5.9, we have here in general $\mathbf{U}_L \neq \mathbf{U}_R$. This is the reason for keeping the two-sided notation in (5.13). Evidently, the result for $\left(\frac{\partial}{\partial t}\mathbf{U}^{ac}\right)^n_{j+1/2}$ based on Proposition 5.9 will only approximate the exact derivative $\frac{\partial}{\partial t}\mathbf{U}(r_{j+1/2}, t_n)$, within an $O(k)$ error bound, as already discussed. At the same time, in view of Remark 5.38, we retain the Riemann solution $\mathbf{U}^n_{j+1/2}$ as in (5.96). The full wave pattern at the singularity is determined by this Riemann solution. This includes the type and initial speeds of all three waves Γ_1, Γ_2, Γ_3 (see Figure 5.1). As in (5.72) we introduce $\mathbf{U}_0 = \mathbf{U}^n_{j+1/2}$ for the limiting values along $r = r_{j+1/2}$ and $\mathbf{U}^*_L$, $\mathbf{U}^*_R$ (see Table 5.5) for the limiting values along the contact discontinuity Γ_2. We recall that $\lambda = \frac{A'(r_{j+1/2})}{A(r_{j+1/2})}$ and the initial Lagrangian derivatives used in (5.13) are obtained from the Eulerian ones $\mathbf{U}'_L$, $\mathbf{U}'_R$ via (5.7). To keep the notation in the following construction in line with the foregoing notation, we take $r_{j+1/2}$ as the location of the singularity (replacing $r = 0$ in the context of Section 5.1).

Construction 5.39 (The acoustic time derivative $\left(\frac{\partial}{\partial t}\mathbf{U^{ac}}\right)^n_{j+1/2}$)

$$\mathbf{U}_0 = \mathbf{U}^n_{j+1/2} = \mathbf{R}^A\left(0; \mathbf{U}_L, \mathbf{U}_R\right) \qquad (\textit{see (5.96)}).$$

$\mathbf{U}^*_L, \ \mathbf{U}^*_R = $ *limiting values on the two sides*
of the contact discontinuity Γ_2.

The full wave pattern is determined by the Riemann solution $\mathbf{R}^A\left(\mu; \mathbf{U}_L, \mathbf{U}_R\right)$.

(A) *If the line* $r = r_{j+1/2}$ *is to the right of* Γ_3 *[as in Figure 5.5(a)] or to the left of* Γ_3, *the derivative* $\left(\frac{\partial}{\partial t}\mathbf{U^{ac}}\right)^n_{j+1/2}$ *is simply determined from the initial data by (4.45).*

(B) *Suppose that the line* $r = r_{j+1/2}$ *is between* Γ_1 *and* Γ_3 *(on either side of the contact discontinuity* Γ_2*). Both* Γ_1 *and* Γ_3 *are now viewed as characteristics* C_- *and* C_+ *(see Figure 5.3). The initial slopes of* C_- *(viewed from its two sides) are* $u_L - c_L$, $u^* - c^*_L$, *whereas those of* C_+ *are* $u_R + c_R$, $u^* + c^*_R$.

(C) *Denote by* $\left(\frac{\partial Q}{\partial r}\right)^n_L$, $\left(\frac{\partial Q}{\partial r}\right)^n_R$ *the initial slopes in cells* j, $j+1$, *respectively [these are the components of* $\mathbf{L}^n_j$, $\mathbf{L}^n_{j+1}$, *respectively, in (5.94)]. Define the set of coefficients in (5.13) by*

$$a_L = 1, \qquad b_L = (g^*_L)^{-1},$$

$$d_L = -(g^*_L)^{-1}\left[c_L g_L \left(\frac{\partial u}{\partial r}\right)^n_L + c_L \left(\frac{\partial p}{\partial r}\right)^n_L - \lambda u_L c_L g_L\right],$$

$$a_R = -1, \qquad b_R = (g^*_R)^{-1},$$

$$d_R = -(g^*_R)^{-1}\left[c_R g_R \left(\frac{\partial u}{\partial r}\right)^n_R - c_R \left(\frac{\partial p}{\partial r}\right)^n_R + \lambda u_R c_R g_R\right],$$

where

$$\lambda = \frac{A'(r_{j+1/2})}{A(r_{j+1/2})}.$$

Obtain the time derivatives $\left(\frac{\partial u}{\partial t}\right)^*$, $\left(\frac{\partial p}{\partial t}\right)^*$ *along the contact discontinuity (Theorem 5.7) by solving the pair of linear equations*

$$a_L\left(\frac{\partial u}{\partial t}\right)^* + b_L\left(\frac{\partial p}{\partial t}\right)^* = d_L,$$

$$a_R\left(\frac{\partial u}{\partial t}\right)^* + b_R\left(\frac{\partial p}{\partial t}\right)^* = d_R,$$

and then

$$\left(\frac{\partial \rho}{\partial t}\right)_L^* = (c_L^*)^{-2}\left(\frac{\partial p}{\partial t}\right)^*, \quad \left(\frac{\partial \rho}{\partial t}\right)_R^* = (c_R^*)^{-2}\left(\frac{\partial p}{\partial t}\right)^*.$$

(D) Evaluate the ξ derivatives at the contact discontinuity as in Summary 5.24:

$$A(r_{j+1/2})\left(\frac{\partial p}{\partial \xi}\right)^* = -\left(\frac{\partial u}{\partial t}\right)^*,$$

$$A(r_{j+1/2})\left(\frac{\partial u}{\partial \xi}\right)_R^* = -(g_R^*)^{-2}\left(\frac{\partial p}{\partial t}\right)^* - \lambda(\rho_R^*)^{-1}u^*,$$

$$A(r_{j+1/2})\left(\frac{\partial u}{\partial \xi}\right)_L^* = -(g_L^*)^{-2}\left(\frac{\partial p}{\partial t}\right)^* - \lambda(\rho_L^*)^{-1}u^*.$$

$$A(r_{j+1/2})\left(\frac{\partial \rho}{\partial \xi}\right)_L^* = (g_L^*)^{-1}\left[\rho_L\left(\frac{\partial u}{\partial r}\right)_L^n\right.$$

$$\left. + c_L\left(\frac{\partial \rho}{\partial r}\right)_L^n + \lambda\rho_L u_L + (c_L^*)^{-2}\left(\frac{\partial p}{\partial t}\right)^*\right],$$

$$A(r_{j+1/2})\left(\frac{\partial \rho}{\partial \xi}\right)_R^* = (g_R^*)^{-1}\left[-\rho_R\left(\frac{\partial u}{\partial r}\right)_R^n + c_R\left(\frac{\partial \rho}{\partial r}\right)_R^n\right.$$

$$\left. - \lambda\rho_R u_R - (c_R^*)^{-2}\left(\frac{\partial p}{\partial t}\right)^*\right]. \tag{5.105}$$

(E) As in Proposition 5.26 define

$$[\partial_t^0 \mathbf{U}, \partial_\xi^0 \mathbf{U}] = \begin{cases} \left[\left(\frac{\partial}{\partial t}\mathbf{U}\right)_L^*, \left(\frac{\partial}{\partial \xi}\mathbf{U}\right)_L^*\right] & \text{if } r = r_{j+1/2} \text{ is between } \Gamma_1 \text{ and } \Gamma_2, \\ \left[\left(\frac{\partial}{\partial t}\mathbf{U}\right)_R^*, \left(\frac{\partial}{\partial \xi}\mathbf{U}\right)_R^*\right] & \text{if } r = r_{j+1/2} \text{ is between } \Gamma_2 \text{ and } \Gamma_3, \end{cases}$$

and evaluate finally [see (5.75)]

$$\left(\frac{\partial}{\partial t}\mathbf{U}^{\mathrm{ac}}\right)_{j+1/2}^n = \partial_t^0 \mathbf{U} - A(r_{j+1/2})\rho_0 u_0 \partial_\xi^0 \mathbf{U}. \tag{5.106}$$

(The acoustic case is always "nonsonic".)

Remark 5.40 (L_1 scheme) This construction describes the E_1 scheme. For a numerical approximation in the Lagrangian framework, the process is even simpler; the evaluation of the time derivatives in step (C) of the construction suffices to construct the numerical fluxes. Indeed, the grid lines $\xi = constant$ now coincide with the contact discontinuities issuing from all boundaries.

In both cases, the E_1 and L_1 schemes constitute the *simplest* extensions of the Godunov scheme that combine second-order accuracy with high-resolution properties in capturing discontinuities. The majority of the numerical examples in Chapter 6 are carried out using these schemes, where only a few grid points with large jumps are treated by the E_∞ algorithm.

The E_∞ and L_∞ Schemes, Intermediate Schemes, and MUSCL

In Theorem 5.36 we defined the numerical fluxes based on the acoustic approximation. However, starting from the piecewise-linear distribution (5.93) we can evaluate, by means of the full solution to the linear GRP, the exact time derivative $\left(\frac{\partial}{\partial t}\mathbf{U}\right)^n_{j+1/2}$. This procedure corresponds to that described in Summary 5.29 for the determination of $\left(\frac{\partial}{\partial t}\mathbf{U}\right)_0$.[13] Equation (5.103) for the numerical fluxes is then replaced by

$$\mathbf{F}^{n+1/2}_{j+1/2} = \mathbf{F}\left(\mathbf{U}^n_{j+1/2} + \frac{k}{2}\left(\frac{\partial}{\partial t}\mathbf{U}\right)^n_{j+1/2}\right),$$

$$\mathbf{G}^{n+1/2}_{j+1/2} = \mathbf{G}\left(\mathbf{U}^n_{j+1/2} + \frac{k}{2}\left(\frac{\partial}{\partial t}\mathbf{U}\right)^n_{j+1/2}\right).$$

(5.107)

Definition 5.41 (E_∞ scheme) The scheme (5.88), with numerical fluxes given by (5.107), is called the E_∞ Scheme.

As in Remark 5.40, there is a corresponding L_∞ scheme; the fluxes are then obtained as in the first step of Summary 5.29 (detailed in Summary 5.24).

The reason for the indices in labeling the E_1 and E_∞ schemes lies in the order (in k) in which the numerical fluxes are evaluated. In E_∞, they are obtained from the exact solution to the linear GRP, and no approximation is involved [once the linear distributions $\mathbf{U}^n(r)$ are given]. In contrast, in the case of E_1, the numerical

---

[13] In this case Construction 5.39 is modified as follows:

(a) Steps (C) and (D) for the evaluation of the Lagrangian solutions are now performed as in Summary 5.24(I)–(III).

(b) In the sonic case Equation (5.106) is replaced by Equations (5.77)–(5.79).

fluxes are based on the time derivative $\left(\frac{\partial}{\partial t}\mathbf{U}^{\mathrm{ac}}\right)_{j+1/2}^{n}$, which approximates the exact solution to the linear GRP only within $O(k)$. By this convention, if the computed time derivative approximates the exact one $\left(\frac{\partial}{\partial t}\mathbf{U}\right)_{j+1/2}^{n}$ within $O(k^2)$ error, we refer to the resulting scheme as an E_2 scheme. Such an approximation results if the acoustic approximation is replaced by the assumption that the waves Γ_1, Γ_3 are always *shocks* satisfying the Rankine–Hugoniot condition, which means that a CRW is replaced by a "rarefaction shock." This assumption and the resulting approximation is the one used by van Leer [112] in his pioneering work.[14] Bram van Leer named his Lagrangian scheme "MUSCL" (Monotonic Upstream-centered Scheme for Conservation Laws). In Appendix D we show that the MUSCL scheme is in fact an L_2 scheme. In particular, we derive there the MUSCL expressions for the approximation of the time derivatives $\left(\frac{\partial}{\partial t}\mathbf{V}^*\right)_{j+1/2}^{n}$ (see Table 5.5), namely, the time derivatives along the contact discontinuity.

Updating the Slopes

Once the numerical fluxes $\mathbf{F}_{j+1/2}^{n+1/2}$, $\mathbf{G}_{j+1/2}^{n+1/2}$ have been determined, the new averages $\left\{\mathbf{U}_j^{n+1}\right\}_j$ at $t = t_{n+1}$ are obtained by (5.88). To construct the new piecewise-linear profiles $\mathbf{U}^{n+1}(r)$ [see (5.93)] it is necessary to determine the new slopes $\mathbf{L}_j^{n+1}$. This is done in exactly the same way as in the scalar case (Step 3 of Construction 3.10). Thus, let $\left(\frac{\partial}{\partial t}\mathbf{U}^{\mathrm{app}}\right)_{j+1/2}^{n}$ be the approximation to the time derivative of the solution to the linear GRP (which is $\left(\frac{\partial}{\partial t}\mathbf{U}^{\mathrm{ac}}\right)_{j+1/2}^{n}$ in the E_1 scheme and is $\left(\frac{\partial}{\partial t}\mathbf{U}\right)_{j+1/2}^{n}$ in the E_∞ scheme). We set

$$\mathbf{U}_{j+1/2}^{n+1} = \mathbf{U}_{j+1/2}^{n} + k \cdot \left(\frac{\partial}{\partial t}\mathbf{U}^{\mathrm{app}}\right)_{j+1/2}^{n}, \qquad (5.108)$$

$$\mathbf{L}_j^{n+1} = \frac{1}{\Delta r}\left(\mathbf{U}_{j+1/2}^{n+1} - \mathbf{U}_{j-1/2}^{n+1}\right). \qquad (5.109)$$

In particular, we observe that the solution to the linear GRP plays a double role. First, it is used in the evaluation of the numerical fluxes (at "half time step" $t = t_n + \frac{k}{2}$), and second, it serves to determine the new values $\mathbf{U}_{j+1/2}^{n+1}$ at cell boundaries, from which the slopes $\mathbf{L}_j^{n+1}$ are obtained.

---

[14] Actually, van Leer's scheme was Lagrangian, and therefore it corresponds to an L_2 scheme. In his work, the Eulerian profile was obtained by "remapping" the Lagrangian grid back into the Eulerian one.

Remark 5.42 Second-order accuracy is achieved even if the updated slopes are derived from the updated cell averages, for example,

$$\mathbf{L}_j^{n+1} = \frac{1}{2\Delta r}\left(\mathbf{U}_{j+1}^{n+1} - \mathbf{U}_{j-1}^{n+1}\right).$$

However, our point here is that the GRP solution is used not only to get the numerical fluxes (and thus $\mathbf{U}_j^{n+1}$ by the conservation laws) but also, *independently,* the cell-boundary values $\mathbf{U}_{j+1/2}^{n+1}$. This suggests the interpretation that, in the GRP approach, not only the new cell averages but also the new slopes are based on a direct discretization of the underlying differential equations.

Concluding the GRP Algorithm

We can now sum up the GRP algorithm, which is a second-order, upwind, finite-difference scheme for the system (4.45) governing quasi-1-D, compressible flow. Referring to this system, we have the vector of unknown flow variables

$$\mathbf{U} = \begin{bmatrix} \rho \\ \rho u \\ \rho E \end{bmatrix}.$$

At the cell boundaries $\{r_{j+1/2}\}$ we obtain the values $\{\mathbf{U}_{j+1/2}^n\}$ by solving the (planar) Riemann problem (5.96). Next we evaluate the "instantaneous" time derivatives $\left(\frac{\partial}{\partial t}\mathbf{U}^{\mathrm{app}}\right)_{j+1/2}^n$. A detailed description of their evaluation in the acoustic case (E_1) is given in Construction 5.39. The necessary modification for the E_∞ scheme (the full solution to the linear GRP) is then given in Footnote 13. Following the evaluation of $\{\mathbf{U}_{j+1/2}^n\}$ and the corresponding time derivatives, the numerical fluxes are given by

$$\mathbf{F}_{j+1/2}^{n+1/2} = \mathbf{F}\left(\mathbf{U}_{j+1/2}^n + \frac{k}{2}\left(\frac{\partial}{\partial t}\mathbf{U}^{\mathrm{app}}\right)_{j+1/2}^n\right),$$

$$\mathbf{G}_{j+1/2}^{n+1/2} = \mathbf{G}\left(\mathbf{U}_{j+1/2}^n + \frac{k}{2}\left(\frac{\partial}{\partial t}\mathbf{U}^{\mathrm{app}}\right)_{j+1/2}^n\right),$$
(5.110)

and the new values $\{\mathbf{U}_j^{n+1}\}$ are given by (5.88). The new cell boundary values $\{\mathbf{U}_{j+1/2}^{n+1}\}$ are determined according to (5.108) and the new slopes $\{\mathbf{L}_j^{n+1}\}$ by (5.109).

Finally, the slopes in all cells are subjected to a "slope-limiter" algorithm, without ever changing the average values $\{\mathbf{U}_j^{n+1}\}$. The fact that slope-limiting is indispensable has already been demonstrated in Chapter 3 (see Example 3.14). The final slopes of the "primitive" variables ρ, p, u[15] are determined by the GRP "slope limiter" given in Construction 3.15. The slopes of all other flow variables required in the GRP algorithm (on both sides of cell boundaries $r_{j+1/2}$) are determined from their functional dependence on (ρ, p, u).

---

[15] This was the practice followed in all the examples of Chapter 6. Numerical experiments have shown, however, that applying the "slope-limiter" restriction to other sets of flow variables (e.g., S, p, u) did not produce any noticeable change in the results.

6

Analytical and Numerical Treatment of Fluid Dynamical Problems

Here the fluid dynamical theory and GRP schemes of Chapters 4 and 5 are applied to one-dimensional test cases. The problems are aimed primarily at demonstrating the capabilities of the scheme, but they are also revealing of nontrivial fluid dynamical phenomena that arise even at the relatively simple one-dimensional settings considered here. In Section 6.1 we treat a shock tube problem, using several scheme options to solve it. An interesting class of fluid dynamical problems is that of wave interactions, to which Section 6.2 is devoted. We selected four different cases in this class, shock–shock, shock–contact, shock–rarefaction, and rarefaction–contact interactions. In each case the GRP solution is compared to either an exact one or to a solution of a Riemann problem that approximates the exact one in some "asymptotic" sense. In the remainder of the chapter we employ the quasi-one-dimensional ("duct flow") scheme, solving three different problems, comparing each numerical solution to the corresponding exact one. Section 6.3 treats a spherically converging flow of cold gas, and Section 6.4 is devoted to the flow induced by an expanding sphere. Finally, in Section 6.5 we present a detailed treatment of the steady flow in a converging–diverging nozzle, obtained numerically as a large-time solution by the GRP scheme.

The notation for the fluid dynamical variables here is identical to that of Chapters 4 and 5.

6.1 The Shock Tube Problem

A shock tube problem is a particular type of Riemann problem, having two initial states that comprise quiescent fluids of different pressures and densities, separated by a diaphragm. This problem serves as an idealized model of the flow commencing upon the burst of a diaphragm that separates a high-pressure ("driver") gas from a low-pressure ("driven") gas in a shock tube (see *Shock Wave Handbook* [16]). Following the survey article by Sod [104], the shock tube problem with the data proposed by Sod has become a standard test case for the evaluation of numerical schemes. We selected this problem as a basic demonstration of the GRP fluid dynamical scheme presented in Chapter 5. The initial value problem is solved here in the two frameworks discussed in Chapter 5 – Eulerian and Lagrangian – as well as in a hybrid Eulerian–Lagrangian system.

The data of Sod's shock tube problem (at $t = 0$) consist of a left state $[\rho, p, u]_L = [1, 1, 0]$ for $x < 50$ and a right state $[\rho, p, u]_R = [0.125, 0.100, 0]$ for $x > 50$, with a perfect gas of $\gamma = 1.40$ on both sides. The resulting IVP is a Riemann problem with an initial discontinuity located at $x = 50$. Its (self-similar) solution consists of three waves (compare Construction 4.46):

(a) a right-propagating shock wave moving along the line $x = 1.752156\,t$,
(b) a contact discontinuity moving along the line $x = 0.9274526\,t$, and
(c) a left-propagating CRW with head and tail characteristics given, respectively, by $x = -1.183216\,t$, $x = -0.6514463\,t$.

At the contact discontinuity the (continuous) velocity and pressure are $[u^*, p^*] = [0.9274526, 0.3031302]$, and the densities on either side are $\rho_L^* = 0.4263194$, $\rho_R^* = 0.2655737$ (see Figure 4.16).

In the numerical simulation the spatial domain is $0 \leq x \leq 100$, which is divided into equal-length cells of $\Delta x = 1$. The boundary conditions are $u = 0$ at the domain endpoints, which implies that the only nonvanishing flux component there is the pressure term in the momentum flux. The computation is performed in the time interval $[0, 20]$, with a constant time step Δt chosen so that $\mu_{CFL} < 1$.[1]

Two numerical solutions are calculated in each case, one with the Godunov (first-order-accurate) scheme and the other with the GRP (second-order-accurate) scheme. For both schemes the results are presented as spatial distributions of velocity and density, with the numerical solution plotted as discrete points, and the corresponding exact solution shown as solid lines.

Consider first the Eulerian case, with a time step $\Delta t = 0.20$, which produces a value $\mu_{CFL} \approx 0.45$. Both Godunov and GRP results are shown in Figure 6.1. The most noticeable feature here is the dramatic improvement of the second-order solution relative to the first-order one. It is particularly interesting to note that the improved accuracy of the GRP solution is not merely in smooth regions of the

---

[1] As in Definition 2.27 this is a restriction on the ratio $\frac{\Delta t}{\Delta x}$, namely, $\mu_{CFL} = \frac{S_{max}\Delta t}{\Delta x} < 1$, where S_{max} is the maximum wave speed in absolute value. Compare (3.6) and the discussion following Definition 5.32.

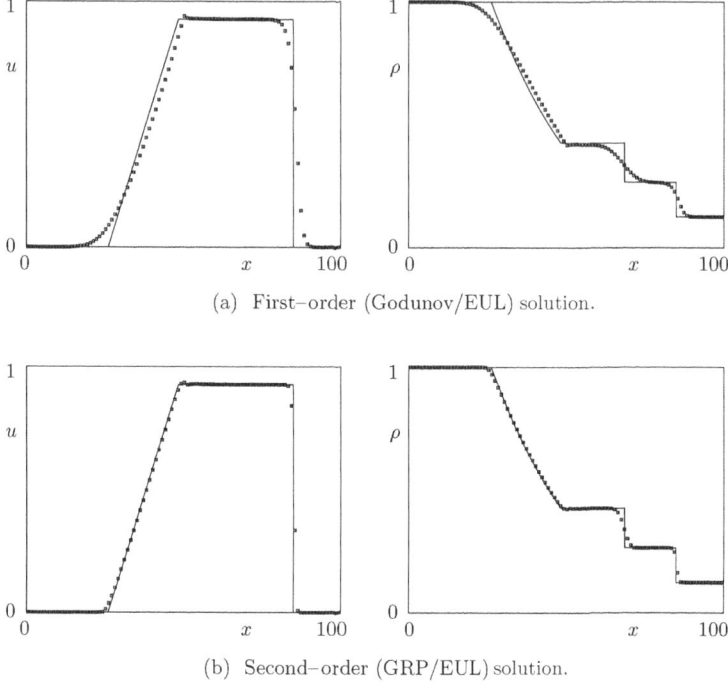

(a) First–order (Godunov/EUL) solution.

(b) Second–order (GRP/EUL) solution.

Figure 6.1. Eulerian solution of shock tube flow at $t = 20$ (100 cells).

flow (the rarefaction wave) but also at flow singularities – the shock and contact discontinuities and the gradient discontinuities at the head and tail points of the rarefaction wave. This property of the numerical scheme is commonly referred to as a "high-resolution" capability, which is taken to mean that singular flow features are "sharply resolved" by the finite-difference approximation. From an analytic point of view, this capability is evidence to an "optimal" combination of the second-order-accurate scheme with the mandatory "limiter" [see the discussion following Equations (5.110)] that effectively reduces the scheme to a first-order level of accuracy at points of discontinuity.

The Lagrange scheme follows directly from the basic Lagrangian analysis of the generalized Riemann problem presented in Summary 5.24. Thus, each cell interface (grid point) follows the particle path and hence coincides with the contact discontinuity, and we have $u_{i+1/2}^{n} = u^{*}$, $u_{i+1/2}^{n+1/2} = u^{*} + \frac{\Delta t}{2}\left(\frac{du}{dt}\right)^{*}$, where here "$i + 1/2$" refers to the Lagrangian grid point $\xi_{i+1/2}$ [see Equation (4.79)]. We remark that the only nonzero flux terms at a Lagrange cell interface are the pressure-related flux components of the momentum and energy equations [see Equations (4.83)].

The Lagrangian computation was performed with identical initial data and boundary conditions; the only difference is the time step, which was reduced to $\Delta t = 0.160$, producing $\mu_{\text{CFL}} \approx 0.43$. Both Godunov and GRP results are shown in Figure 6.2. As in the Eulerian case, we notice a remarkable improvement of the second-order solution relative to the first-order one. It is also interesting to compare the two solutions – the Eulerian and the Lagrangian – which differ only in the choice of coordinates (and, consequently, grid and finite-difference scheme). Clearly, the Eulerian solution is superior in accuracy to the Lagrangian one in all features but for the sharper resolution of the contact discontinuity in the Lagrangian case. This may be interpreted as indicating the significance of grid spacing. In the Eulerian case the grid spacing is uniform, whereas the Lagrangian grid points are attached to (moving) "material points." Consequently, fluid compression produces narrower cells, whereas fluid expansion produces wider cells. For example, inspection of Figure 6.1 and 6.2 shows that the rarefaction wave is spread over 23 cells in the Eulerian case, compared to 16 cells in the Lagrangian case.

Because this sample problem shows dependence on the coordinate type and the corresponding finite-difference scheme, could a better performance be

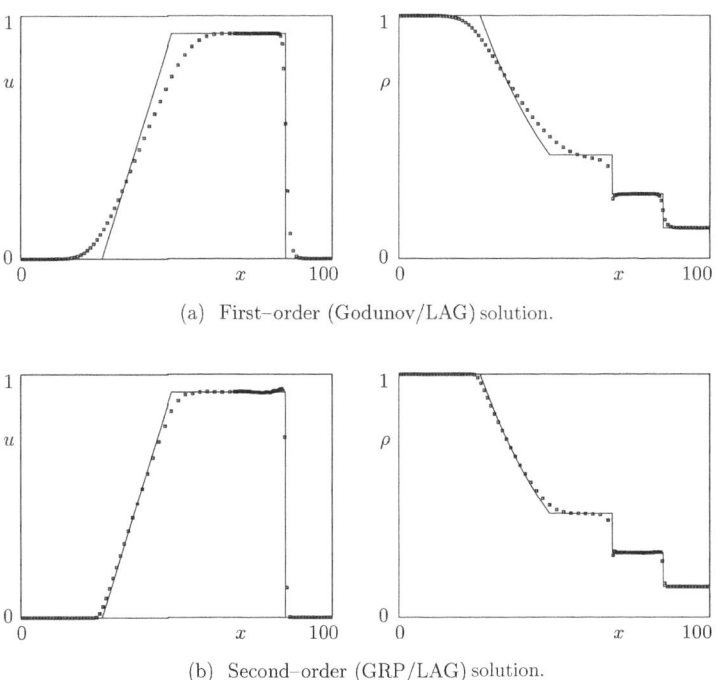

(a) First–order (Godunov/LAG) solution.

(b) Second–order (GRP/LAG) solution.

Figure 6.2. Lagrangian solution of shock tube flow at $t = 20$ (100 cells).

obtained with an "intermediate" type of grid? Specifically, we refer to the so-called ALE (arbitrary Euler Lagrange) grid option, where grid points move at an arbitrary (time-dependent) velocity (Hirt and Amsden [62]). The ALE grid in our sample problem is chosen to maintain a sharp contact discontinuity, by specifying the grid point initially at $x = 50$ (the initial position of the discontinuity) as a Lagrangian point. Thus, this midpoint maintains the separation between the left-state and the right-state fluids at all times. The grid points on the left side and on the right side are assigned velocities designed to maintain a uniformly spaced grid on either side. These moving grid points are neither "Eulerian" nor "Lagrangian"; they are in fact "hybrid" (i.e., "ALE") points. The fluxes at these points are readily obtained from the full GRP analysis (see Remark 5.31).

The ALE time step was the same as that of the Lagrangian computation, producing nearly the same value of the CFL ratio ($\Delta t = 0.160$ and $\mu_{CFL} \approx 0.43$). Indeed, the results in Figure 6.3 do show an "intermediate" accuracy level between the Eulerian and the Lagrangian cases. In particular, we note that the tail of the rarefaction wave is better resolved in the ALE solution than in the Lagrangian solution, but not as well as in the Eulerian solution.

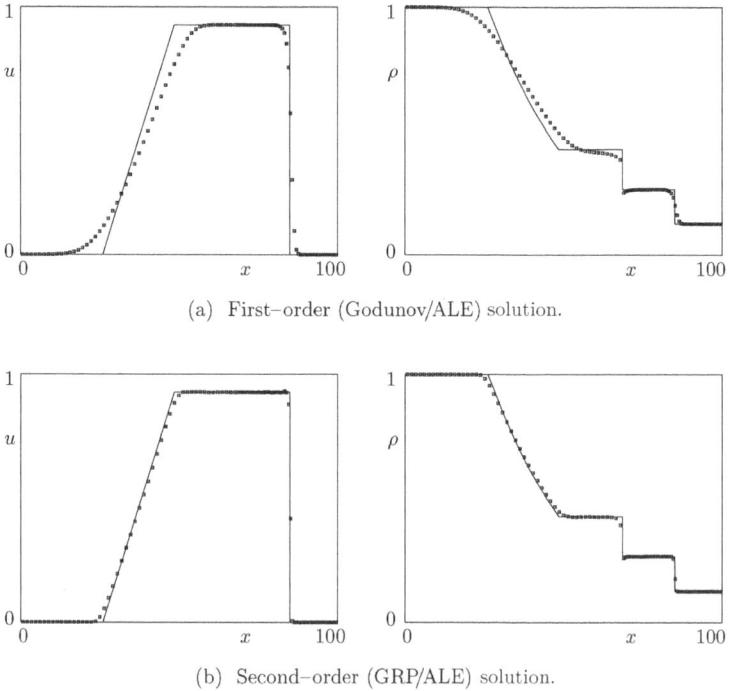

(a) First–order (Godunov/ALE) solution.

(b) Second–order (GRP/ALE) solution.

Figure 6.3. ALE solution of shock tube flow at $t = 20$ (100 cells).

In summary, we have demonstrated that the Eulerian GRP scheme, with a relatively modest grid (of 100 cells), produces quite an accurate high-resolution solution of a typical shock tube problem. The improvement relative to the Godunov scheme is dramatic, which demonstrates the significant progress achieved by increasing the order of accuracy from first to second order. We also take notice of the dependence of the numerical solution on the coordinate type and the related grid. The quality of the Lagrangian solution is lower than that of the corresponding Eulerian solution, with the ALE scheme producing a solution of intermediate quality.

Remark 6.1 As in the ALE approach referred to here, one can use the full GRP analysis to "track" various singularities (shock, contact discontinuity, edge characteristics of a rarefaction wave, etc.) in the flow. We refer to Section 8.2 for further details.

6.2 Wave Interactions

The interaction between a pair of single waves (taken here as one-dimensional shock, CRW, or contact discontinuity) is a problem of fundamental theoretical and practical significance in fluid dynamics. For a theoretical introduction to the fluid dynamics of wave interaction we refer the reader to the classical book by Courant and Friedrichs [30, Chapter III, Section D]. Although here our interest is naturally focused at the theoretical and numerical aspects of the topic, we point out that analysis of wave interactions is motivated by the need to understand related phenomena in diverse applications, such as shock tubes, shock tunnels, and internal combustion engines. In our consideration of various wave interaction cases, we point out that each case (in a sense to be explained in the following) may be related to a respective "asymptotic Riemann problem."

The initial data for the shock tube problem considered in Section 6.1 consist of two adjacent constant states $[\mathbf{U}_L, \mathbf{U}_R]$. They lead to a self-similar solution, which in general comprises three waves emanating from the discontinuity. The next natural step would be the consideration of the IVP for which the initial data consist of three constant states $[\mathbf{U}_L, \mathbf{U}_M, \mathbf{U}_R]$ [see Figure 6.4(a)]. In general, each discontinuity would lead to three waves, and the resulting wave structure would evolve into a complex pattern because of wave interactions. In this section we assume therefore that the initial data are selected so as to produce a *single wave* emanating from each one of the two discontinuities.[2] This is always

---

[2] In other words, we consider a reduced set of three-state IVPs, namely, IVPs producing two single waves. In current publications, "wave interaction" commonly refers to interactions of two single waves (see, e.g., *Shock Wave Handbook* [16, Chapter 7]).

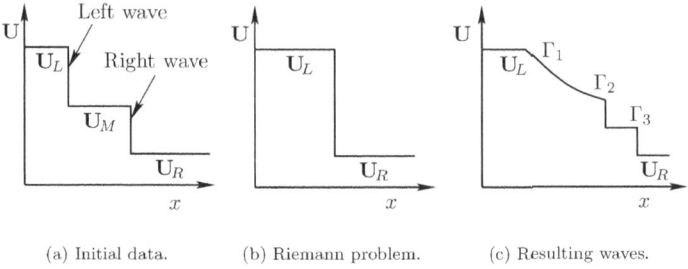

(a) Initial data.   (b) Riemann problem.   (c) Resulting waves.

Figure 6.4. Wave interaction – A schematic description.

possible since a particular Riemann problem may be designed to produce just a single wave (see discussion and Footnote 3, following Definition 5.2). To fix ideas, consider the procedure for selecting U_L, given the "middle" state U_M. Let the left wave [see Figure 6.4(a)] be a k-wave ($k = 1, 2, 3$). Using the notation introduced for "interaction curves" (see Section 4.2, Summary 4.45) we see that U_L lies on the curve I_k^r, namely, the curve representing all *left* states connected to the *right* state U_M by a k-wave. To be more specific, we shall henceforth denote this curve by $I_k^r(U_M)$. Similarly, if the right wave is a k-wave, then U_R lies on the curve $I_k^l(U_M)$, which represents all *right* states connected to the *left* state U_M by a k-wave. Equivalently, in this case U_M lies on the curve $I_k^r(U_R)$.

Considering the wave interaction in terms of a time evolution, we refer to the (x, t) diagram in Figure 6.5 (where for simplicity the incident waves are taken as sharp jumps – shocks or contact discontinuities). The interaction takes place ("abruptly") at the point where the two trajectories intersect, which we shall always take as $(x, t) = (0, 0)$. The waves Γ_1, Γ_2, Γ_3 emanating from that point have the same pattern as a solution to a Riemann problem [compare Figures 4.16

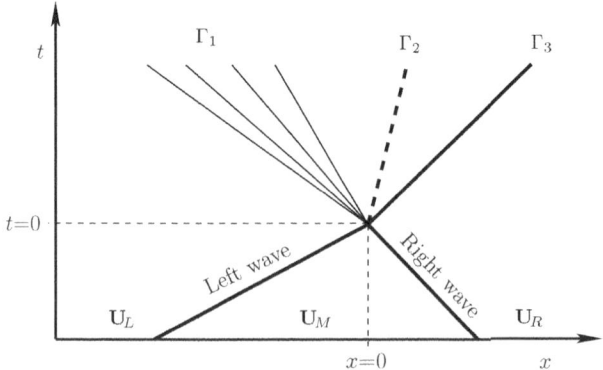

Figure 6.5. Schematic (x, t) diagram of the interaction of two-waves.

and 5.1(b)]. Indeed, at $(x, t) = (0, 0)$ the entire flow profile reduces to the single discontinuity $[\mathbf{U}_L, \mathbf{U}_R]$, as also shown in Figure 6.4(b), where it is appropriately labeled a "Riemann problem." The postinteraction flow depicted in Figure 6.4(c) thus corresponds to a profile of the waves Γ_1, Γ_2, Γ_3 in Figure 6.5, taken at some fixed time $t > 0$. Summarizing the case where each of the interacting waves is a sharp jump, we see that the IVP always reduces to a Riemann problem[3] at $(x, t) = (0, 0)$; it may be solved as such, and, in particular, the solution is self-similar relative to the origin [see Equation (4.100) and the discussion around it].

We now turn to the case where one of the waves (or even both) is a CRW, and we designate by "wave front" the extreme (head or tail) characteristics bounding the CRW on either side. The interaction then commences at $(x, t) = (0, 0)$ where the earliest intersection between the two wave fronts takes place. Obviously, this is also the moment at which the length of the intermediate state $\mathbf{U}_M$ in the flow profile just vanishes. In this case the flow profile at $t = 0$ includes (in addition to the extreme states $\mathbf{U}_L$, $\mathbf{U}_R$) segments of continuous variation (within the CRW) and thus does not correspond to a Riemann problem, and its solution at $t > 0$ is non-self-similar.

It is instructive to consider the Riemann problem $[\mathbf{U}_L, \mathbf{U}_R]$ obtained by eliminating the middle state $\mathbf{U}_M$. Loosely speaking, its self-similar solution is related to the full foregoing wave interaction in some "asymptotic" sense. It is shown that at large times $t > 0$, the full interaction solution approaches the corresponding Riemann solution. However, owing to some "fine features" of the solution to the interaction IVP that persist for all times, the convergence to the Riemann solution is only in some "overall" sense. The analysis of such situations, sometimes referred to as "perturbed" Riemann problems, is known as stability analysis. There are no rigorous results in this direction for the case of nonisentropic, compressible flow [i.e., the system (4.47)], but we refer to Liu [86] for the analysis of a simpler case.

As for the numerical solution, we use the same (Eulerian) GRP scheme employed in Section 6.1. Moreover, as in that case, the computation domain is taken to be wide enough so that the waves emanating from the considered interaction do not reach the domain endpoints. We then impose the respective state ($\mathbf{U}_L$ or $\mathbf{U}_R$) as the state defining the flux at those endpoints. Additionally, in all the wave interactions studied here we assume that the fluid is a perfect gas with $\gamma = 1.4$.

In the rest of this section we deal with four cases of wave interaction. The first two involve only discontinuous waves (shocks and contacts), and the other

---

[3] In a two-wave IVP the (x, t) trajectories do not necessarily intersect, in which case the IVP is not a wave-interaction problem.

two involve a CRW. In Subsection 6.2.1 we consider the physically common case of shock transition through a contact discontinuity, and in Subsection 6.2.2 the interaction is between two 3-shocks. An interaction between a 3-shock and a 1-CRW is then presented in Subsection 6.2.3, followed by the interaction of a 3-CRW and a contact discontinuity in Subsection 6.2.4.

6.2.1 Shock–Contact Interaction

Here we consider the physically common case of shock wave interaction with a contact discontinuity. We take the left wave to be a 3-shock, while the right wave is a contact discontinuity. Let the three states $[U_L, U_M, U_R]$ be as in Figure 6.4(a). Then U_L lies on the shock branch of $I_3^r(U_M)$, and $[U_M, U_R]$ is a contact discontinuity (i.e., $\rho_M \neq \rho_R$ while the pressure and velocity are continuous across this surface). When crossing the contact discontinuity from left to right, the density may either increase or decrease, and we select the latter case. As we shall see in a moment, the interaction will then produce a "transmitted" 3-shock and a "reflected" 1-CRW. We specify the shock by the postshock pressure $p_L/p_M = 5$, and the contact discontinuity by $\rho_R/\rho_M = 0.3$. Using the Rankine–Hugoniot jump conditions for a perfect gas (Summary 4.49) gives the initial states

$$U_L = [\rho, p, u]_L = [2.8182, 5, 1.6064],$$

$$U_M = [\rho, p, u]_M = [1, 1, 0], \qquad (6.1)$$

$$U_R = [\rho, p, u]_R = [0.3, 1, 0].$$

The exact solution to this wave interaction (at $t = 0+$) is readily obtained by solving the Riemann problem having the initial data $[U_L, U_R]$. This solution (recall Construction 4.46) is obtained by the intersection of the interaction curves $I_1^l(U_L)$, $I_3^r(U_R)$ in the (u, p) plane, as depicted graphically (to true scale) in Figure 6.6. This figure also shows that U_L lies on the shock branch of $I_3^r(U_M)$ and that the (u, p) values of U_M and U_R coincide. However, since the densities $\rho_M \neq \rho_R$, the curves $I_3^r(U_M)$, $I_3^r(U_R)$ are different. The state U_L is situated in the (u, p) plane *above* $I_3^r(U_R)$; consequently, the intersection point (u^*, p^*) is obtained on the *rarefaction branch* of $I_1^l(U_L)$. The Riemann solution thus consists of a (left-facing) 1-CRW and a (right-facing) 3-shock. The velocity and pressure at the contact discontinuity are $(u^*, p^*) = (2.059973, 3.301911)$, and the corresponding densities on either side are $\rho_L^* = 2.095325$, $\rho_R^* = 0.6711996$. In obtaining this solution we have followed the procedure described in Appendix C.

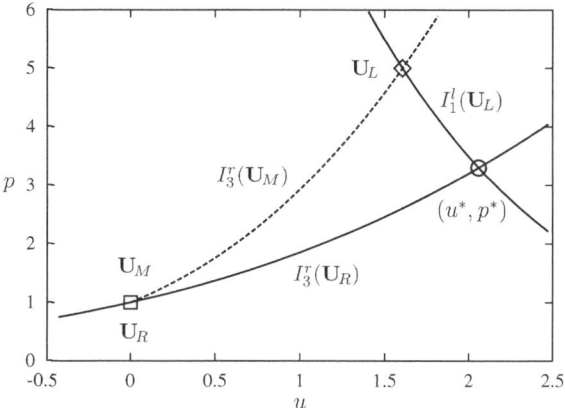

Figure 6.6. Riemann problem for the shock–contact interaction ($\gamma = 1.4$). Left state: 3-shock with $p_L = 5$. Middle state $[\rho, p, u]_M = [1, 1, 0]$. Right state $[\rho, p, u]_R = [0.3, 1, 0]$.

Another representation of this wave interaction is the (true-scale) wave diagram shown in (Figure 6.7). The "incident" shock moves at speed $\sigma_L = 2.490$. When the shock passes through the contact surface at $t = 0$, the flow splits into a 1-CRW (facing left), a contact discontinuity, and a 3-shock (compare Figure 4.16). The speeds of the extreme characteristic lines of the 1-CRW (shown in Figure 6.7) are 0.0304085292 (head) and 0.325695142 (tail), and the 3-shock moves at speed $\sigma_3 = 3.72482334$.

We now turn to the numerical solution of this two-shock interaction, using the same (Eulerian) GRP scheme as previously used in Section 6.1. We choose

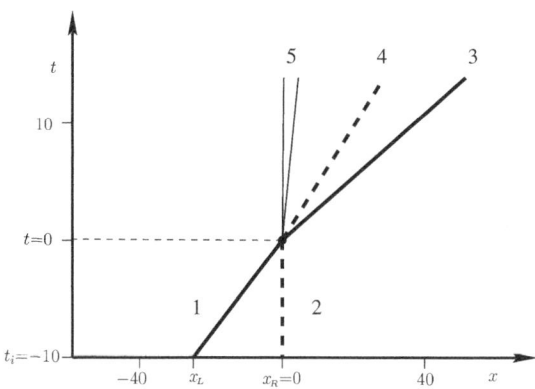

Figure 6.7. Wave diagram of the shock–contact interaction. 1 – "Incident" 3-shock, 2 – contact discontinuity, 3 – "transmitted" 3-shock, 4 – contact discontinuity, 5 – "reflected" 1-CRW.

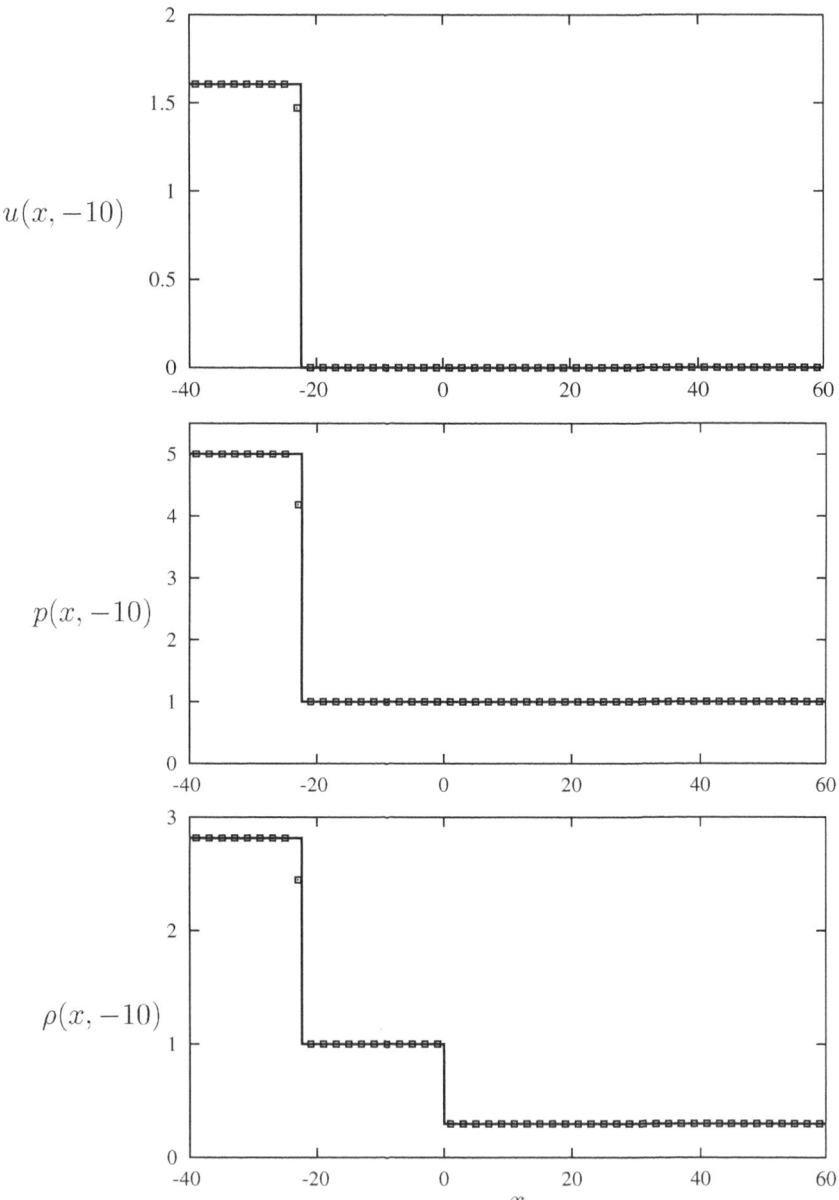

Figure 6.8. Initial data for shock–contact interaction. Solid line – exact data; points – GRP (50 cells shown).

the initial time as $t_i = -10$, so that the initial data are as follows:

$$U(x, -10) = \begin{cases} U_L = [\rho, p, u]_L = [2.8182, 5, 1.6064], & x < x_L, \\ U_M = [\rho, p, u]_M = [1, 1, 0], & x_L < x < x_R, \\ U_R = [\rho, p, u]_R = [0.3, 1, 0], & x_R < x, \end{cases}$$

(6.2)

where the left and right discontinuities are at $x_L = -10\sigma_L = -24.90$ and $x_R = 0$, respectively.

The grid occupies the interval $[-40, 100]$ and is divided into 70 cells of length $\Delta x = 2$ each. The integration was performed from $t = t_i = -10$ to the final time $t = 20$ with a constant time step $\Delta t = 0.20$, corresponding to the CFL ratio $\mu_{CFL} \approx 0.49$. In accordance with the schematic description in Figure 6.4, the exact and numerical results are displayed at the initial time ($t_i = -10$) in Figure 6.8,[4] at the interaction moment ($t = 0$) in Figure 6.9, and at the postinteraction time ($t = 20$) in Figure 6.10. In each case the results are displayed over a subinterval of length 100 (50 cells), containing the relevant wave pattern (i.e., bounded by the initial states U_L, U_R).

At the interaction moment $t = 0$ the exact distribution is simply the two-state jump $[U_L, U_R]$, with the discontinuity at $(x, t) = (0, 0)$. The computed distribution agrees quite well with the exact one, considering the relatively coarse grid (Figure 6.9).

The postinteraction results (Figure 6.10), are shown at a moment where the entire wave pattern is resolved by merely 38 cells. It is remarkable that the GRP scheme resolves this wave structure so accurately, considering the relative coarseness of the grid.

6.2.2 Shock–Shock Interaction

In this case both waves are 3-shocks; that is, U_L lies on the shock branch of $I_3^r(U_M)$, while U_R lies on the shock branch of $I_3^l(U_M)$. Observe that this also implies that U_M lies on the shock branch of $I_3^r(U_R)$. For both shocks the preshock state is on the right, so we select the right state as the quiescent state

$$U_R = [\rho, p, u]_R = [1, 1, 0].$$

The first shock ("right wave") is then obtained by specifying a postshock pressure of $p_M = 5$; across the second shock ("left wave") the pressure jumps tenfold

---

[4] Since x_L does not coincide with a grid point, the values in the cells containing it are appropriately interpolated. This is done for all wave-interaction cases here.

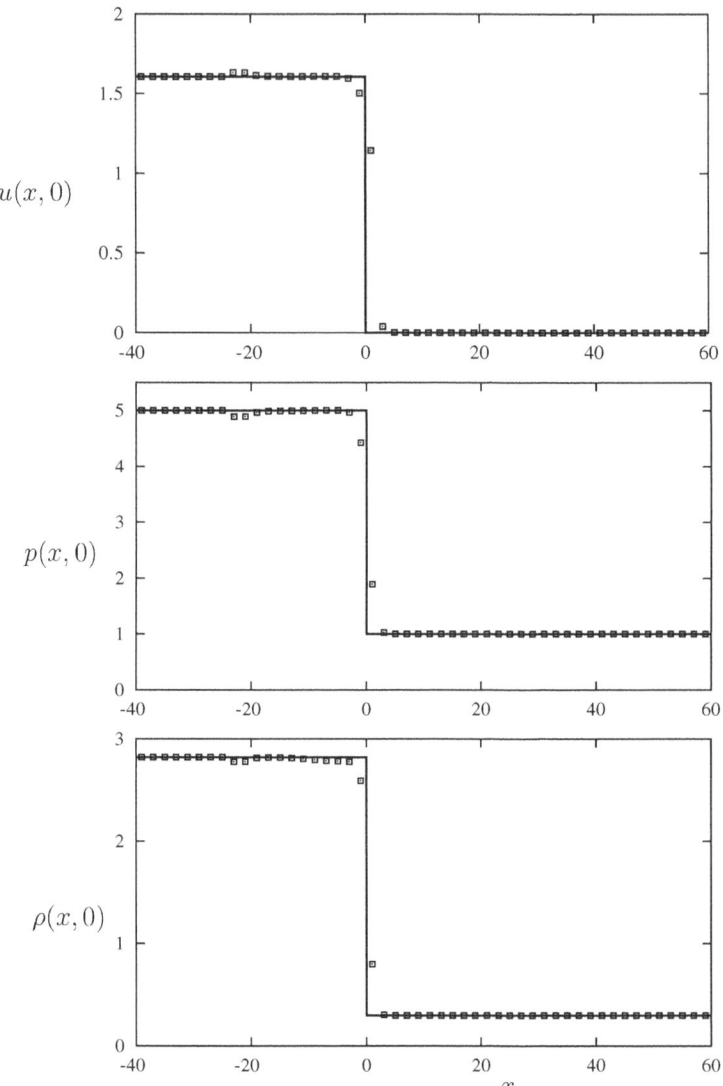

Figure 6.9. Shock–contact interaction. Shock and contact merge at $(x, t) = (0, 0)$. Solid line – exact solution; points – GRP (50 cells shown).

to $p_L = 50$. Using the Rankine–Hugoniot jump conditions for a perfect gas (Summary 4.49), we find that the postshock states are given by

$$
\begin{aligned}
\mathbf{U}_M &= [\rho, p, u]_M = [2.8182, 5, 1.6064], \\
\mathbf{U}_L &= [\rho, p, u]_L = [10.744, 50, 5.0386].
\end{aligned}
\tag{6.3}
$$

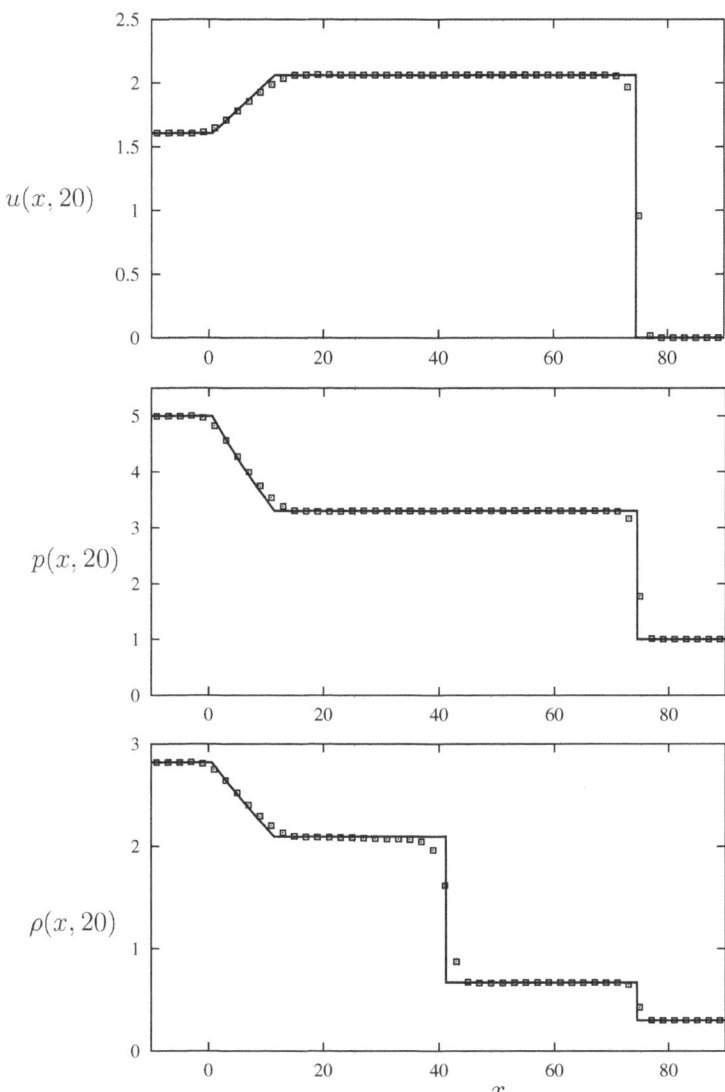

Figure 6.10. Shock–contact postinteraction results at $t = 20$. Solid line – exact solution; points – GRP (50 cells shown).

Recall that with respect to the state $\mathbf{U}_M$ the left shock is supersonic whereas the right shock is subsonic [see the discussion following Equation (4.78)]. The latter is therefore overtaken by the former. At that moment the flow profile is reduced to the single jump discontinuity $[\mathbf{U}_L, \mathbf{U}_R]$, that is, to a Riemann problem

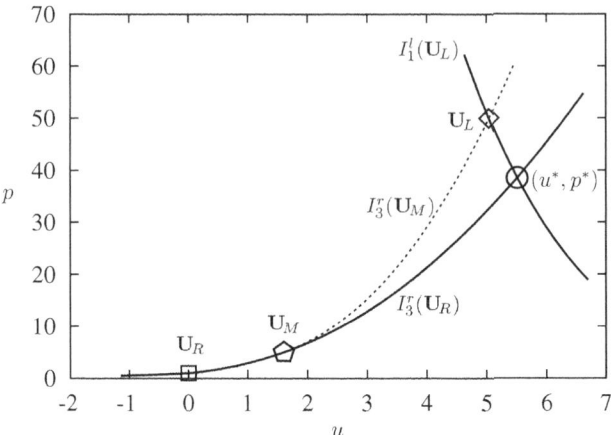

Figure 6.11. Riemann problem for the shock–shock interaction. Right state $[\rho, p, u]_R = [1, 1, 0]$, $\gamma = 1.4$; first shock: $p_M = 5$; second shock: $p_L = 50$.

[as in Figure 6.4(b)]. The exact solution to this wave interaction is thus obtained by solving the corresponding Riemann problem.

The solution to a Riemann problem (recall Construction 4.46) is obtained by the intersection of the interaction curves $I_1^l(\mathbf{U}_L)$ and $I_3^r(\mathbf{U}_R)$ in the (u, p) plane, as shown (to true scale) in Figure 6.11. Additionally, this figure shows the two-shock process leading to the initial data (6.3): $\mathbf{U}_M$ is on the shock branch of $I_3^r(\mathbf{U}_R)$, and $\mathbf{U}_L$ is on the shock branch of $I_3^r(\mathbf{U}_M)$. The state $\mathbf{U}_L$ is situated in the (u, p) plane *above* $I_3^r(\mathbf{U}_R)$; consequently, the intersection point (u^*, p^*) is obtained on the *rarefaction branch* of $I_1^l(\mathbf{U}_L)$. The solution to this Riemann problem thus consists of a (left-facing) 1-CRW and a (right-facing) 3-shock. In obtaining this solution we have followed the procedure described in Appendix C. The velocity and pressure at the contact discontinuity are $(u^*, p^*) = (5.5059, 38.509)$, and the corresponding densities on either side are $\rho_L^* = 8.9161$, $\rho_R^* = 5.2136$.

Another representation of this wave interaction is the (true-scale) wave diagram shown in Figure 6.12. The right and left shocks move at speeds $\sigma_R = 2.490$ and $\sigma_L = 6.259$, respectively. The latter overtakes the former at $t = 0$, and as already discussed, the flow then splits into a 1-CRW (facing left), a contact discontinuity, and a 3-shock. The speeds of the extreme characteristic lines of the 1-CRW (shown in Figure 6.12) are 2.486 (head) and 3.047 (tail), and the 3-shock moves at speed $\sigma_3 = 6.813$.

We now turn to the numerical solution of this two-shock interaction, using the GRP scheme. We choose the initial time as $t_i = -9$, so that the initial data

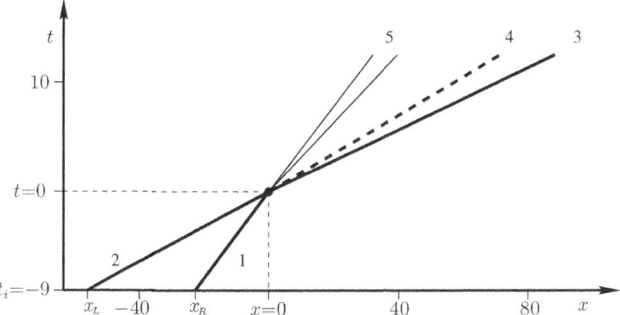

Figure 6.12. Wave diagram of the shock–shock interaction. 1 – Right "incident" 3-shock; 2 – left "incident" 3-shock; 3 – "transmitted" 3-shock; 4 – contact discontinuity, 5 – "reflected" 1-CRW.

are as follows:

$$U(x, -9) = \begin{cases} U_L = [\rho, p, u]_L = [10.744, 50, 5.0386], & x < x_L, \\ U_M = [\rho, p, u]_M = [2.8182, 5, 1.6064], & x_L < x < x_R, \\ U_R = [\rho, p, u]_R = [1, 1, 0], & x_R < x, \end{cases}$$

(6.4)

where the left and right jump discontinuities are positioned at $x_L = -9\sigma_L = -56.3299$ and $x_R = -9\sigma_R = -22.4098$.

The grid occupies the interval $[-100, 140]$ and is divided into 120 cells of length $\Delta x = 2$ each. The integration was performed from $t = t_i = -9$ to the final time $t = 18$ with a constant time step $\Delta t = 0.125$, corresponding to the CFL ratio $\mu_{CFL} \approx 0.48$. In accordance with the schematic description in Figure 6.4, the exact and numerical results are displayed at the initial time ($t_i = -9$) in Figure 6.13, at the interaction moment ($t = 0$) in Figure 6.14, and at the post-interaction time ($t = 18$) in Figure 6.15. In each case the results are displayed over a subinterval of length 100 (50 cells), containing the relevant wave pattern (i.e., bounded by the initial states U_L, U_R). Note that this choice is suggested by the fact that in the present case all wave speeds are positive (see Figure 6.12).

At the interaction moment $t = 0$ the exact distribution is simply the two-state jump $[U_L, U_R]$, with the discontinuity at $(x, t) = (0, 0)$. The computed distribution agrees quite well with the exact one, considering the relatively coarse grid (Figure 6.14).

The postinteraction results (Figure 6.15) are shown at a moment where the entire wave pattern is resolved by merely 40 cells. It is remarkable that the GRP scheme resolves this wave structure so accurately, considering the relative coarseness of the grid.

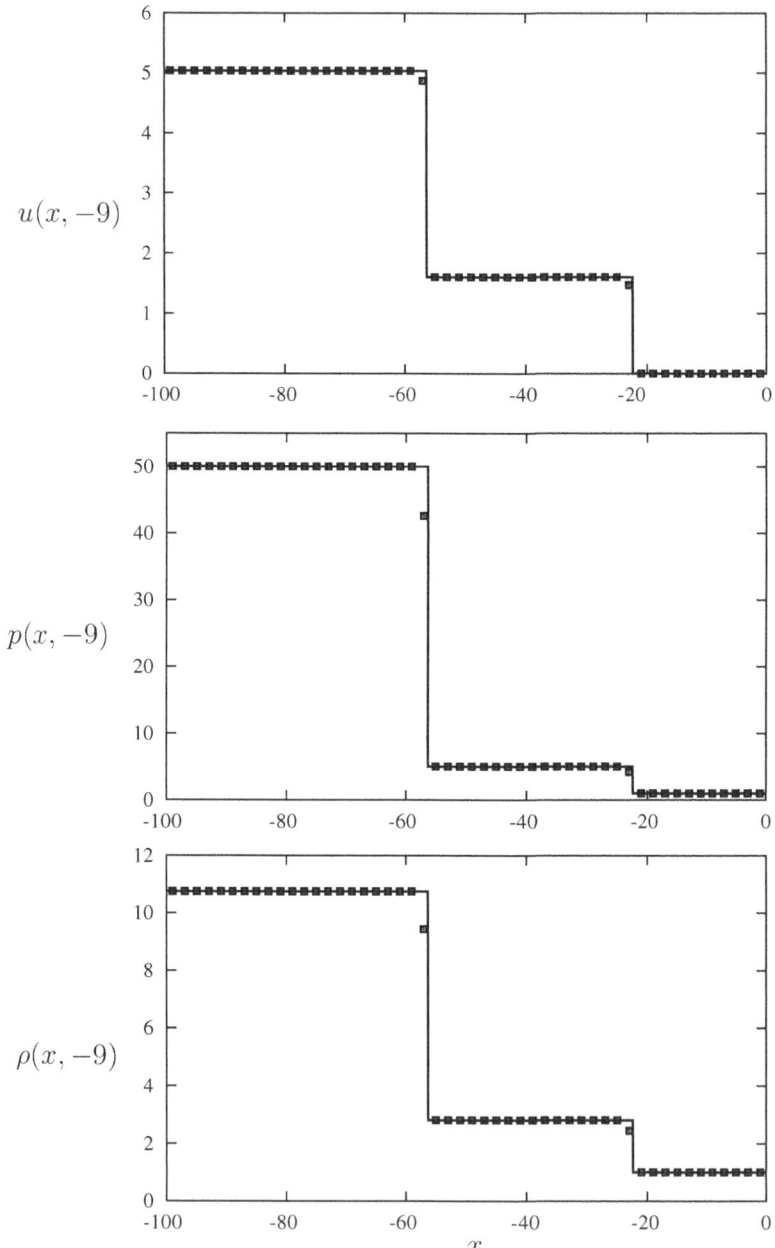

Figure 6.13. Initial data for shock–shock interaction. Solid line – exact data; points – GRP (50 cells shown).

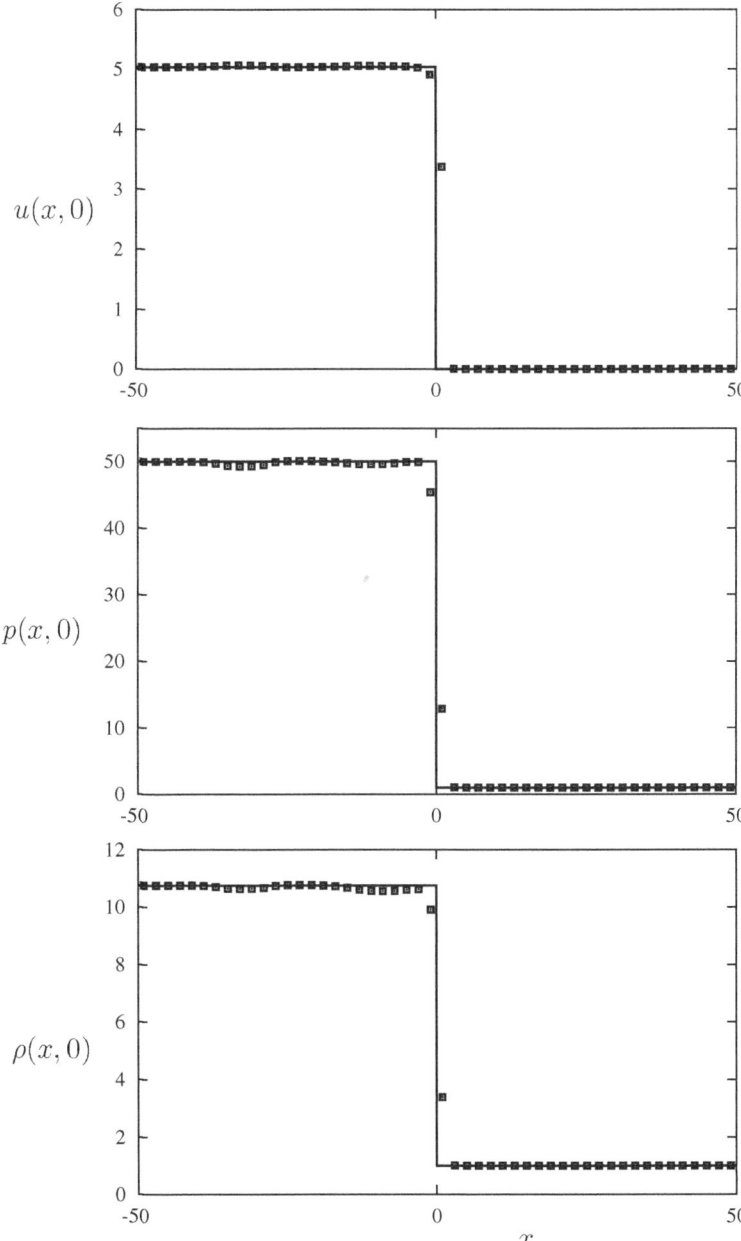

Figure 6.14. Shock–shock interaction. Shocks merge at $(x, t) = (0, 0)$. Solid line – exact solution; points – GRP (50 cells shown).

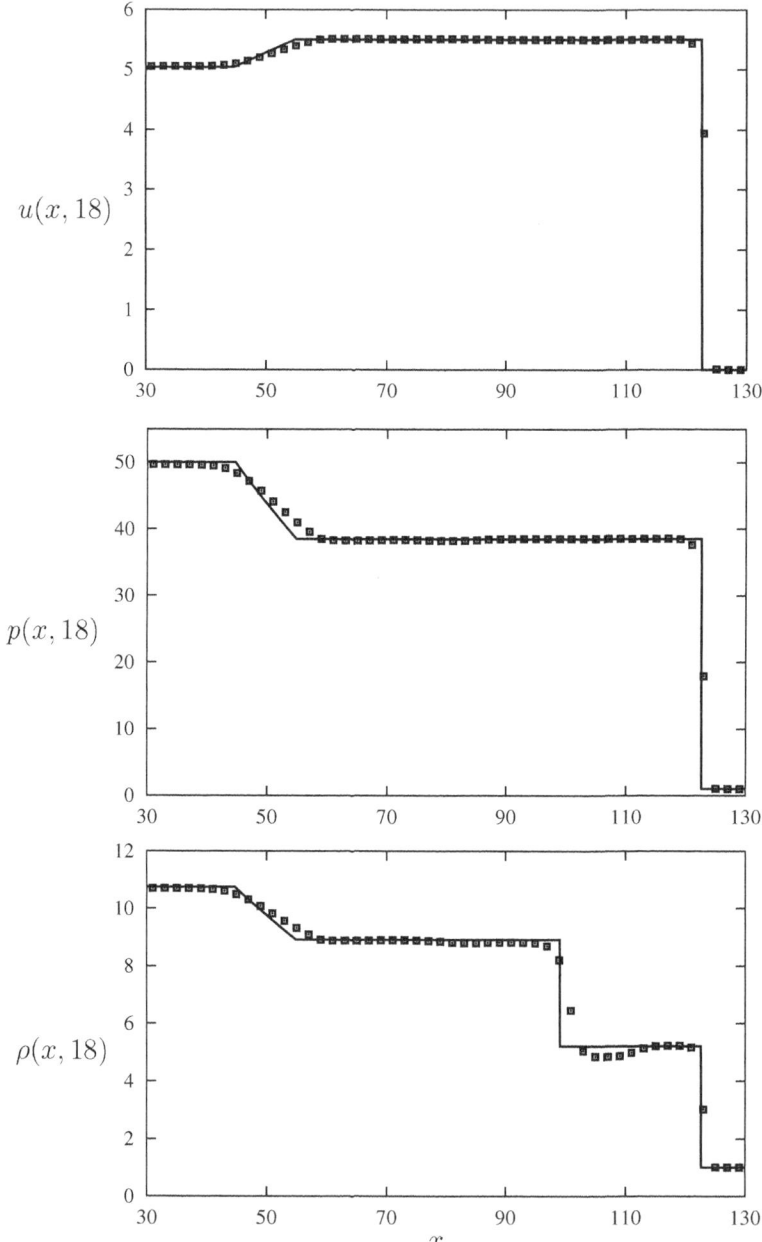

Figure 6.15. Shock–shock postinteraction results at $t = 18$. Solid line – exact solution; points – GRP (50 cells shown).

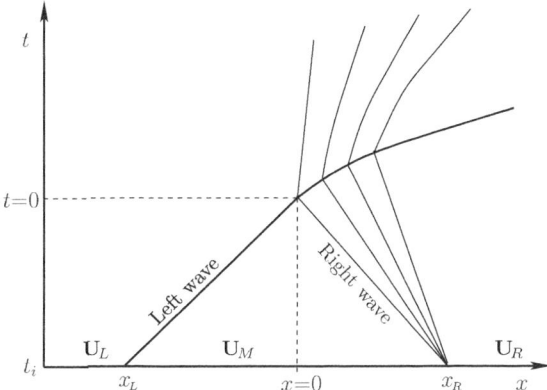

Figure 6.16. Diagram of the interaction of a 3-shock and a 1-CRW.

6.2.3 Shock–CRW Interaction

The former two examples dealt with the interaction between a pair of sharp discontinuities (shock–contact, shock–shock), resulting in a self-similar wave pattern. In the present case we consider the interaction between a 3-shock (facing right) and a 1-CRW (facing left), as depicted schematically in Figure 6.16. This interaction produces a non-self-similar solution (see the discussion at the beginning of this section).

As before, the two waves are specified by the postwave pressure. Assuming $p_M = \rho_M = 1$, we take the pressure behind the shock as $p_L = 5$ and the pressure behind the CRW as $p_R = 1/5$. Using the shock and rarefaction relations for a perfect gas (Summary 4.49) gives the initial data at time $t = t_i = -10$:

$$\mathbf{U}(x, -10) = \begin{cases} \mathbf{U}_L = [\rho, p, u]_L = [2.8182, 5.0, 1.6064], & x < x_L, \\ \mathbf{U}_M = [\rho, p, u]_M = [1, 1, 0], & x_L < x < x_R, \\ \mathbf{U}_R = [\rho, p, u]_R = [0.31676, 0.2, 1.2152], & x_R < x. \end{cases}$$

(6.5)

The shock speed is $\sigma_L = 2.4900$, and the speed of the head of the CRW is $-\sqrt{\gamma p_M / \rho_M} = -1.1832$. The initial discontinuities are thus positioned at $x_L = -10\sigma_L = -24.900$ and $x_R = 11.832$.

At $(x, t) = (0, 0)$ the left shock intersects the head characteristic of the CRW. The flow profile at $t = 0$ consists of a constant left state $\mathbf{U}_L$ $(x < 0)$ and a smoothly varying right state, with limiting value at $x = 0+$ of $\mathbf{U}_M$. Referring to the discussion of the GRP in Section 5.1, we infer that the wave pattern for $t > 0$ consists of a shock discontinuity (which is the only wave in the Riemann

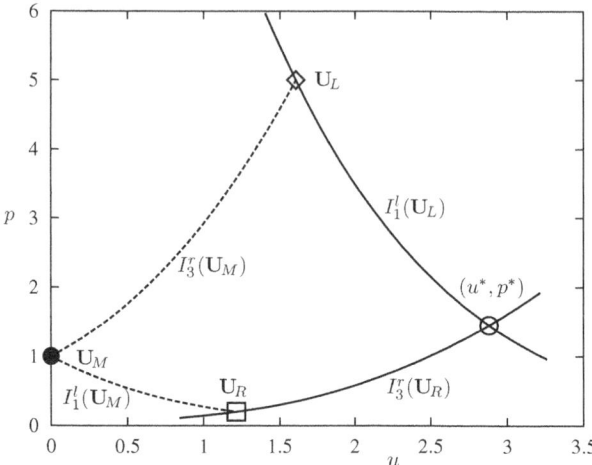

Figure 6.17. Riemann problem for shock–rarefaction interaction. Zero state $[\rho, p, u]_0 = [1, 1, 0]$, $\gamma = 1.4$. Right shock: $p_L = 5$. Left rarefaction: $p_R = 1/5$.

solution for initial data $[\mathbf{U}_L, \mathbf{U}_M]$). This wave separates two regions of smooth flow, as shown in Figure 6.16.

We observe, however, that the solution to the three-state IVP (6.5) should approach at large times the solution to the Riemann problem corresponding to the initial data $[\mathbf{U}_L, \mathbf{U}_R]$. The procedure for obtaining the Riemann solution has already been discussed in the preceding subsections and is presented graphically (to true scale) in Figure 6.17, which also shows the shock branch of $I_3^r(\mathbf{U}_M)$ on which $\mathbf{U}_L$ lies and the rarefaction branch of $I_1^l(\mathbf{U}_M)$ on which $\mathbf{U}_R$ lies. The intersection point of the interaction curves $I_1^l(\mathbf{U}_L)$ and $I_3^r(\mathbf{U}_R)$ is $(u^*, p^*) = (2.8819, 1.4527)$. The waves Γ_1 and Γ_3 resolving the initial discontinuity $[\mathbf{U}_L, \mathbf{U}_R]$ are a 1-CRW and a 3-shock (same type as the respective interacting waves). The corresponding densities on either side of the contact discontinuity are $\rho_L^* = 1.1656$, $\rho_R^* = 1.0647$.

The GRP numerical solution of the IVP (6.5) was performed in the wide computation domain $[-50, 400]$, which was divided into a grid of 450 equal cells. The time integration proceeded from $t_i = -10$ to the final time $t = 100$ in constant steps of $\Delta t = 0.125$, resulting in a CFL ratio of $\mu_{CFL} = 0.54$.

The initial data are shown in Figure 6.18. The early time evolution, corresponding to $t = -2$, (i.e., just before the start of the interaction), is displayed in Figure 6.19. Here we clearly see the shock discontinuity approaching the head of the CRW. From this time on, we consider the postinteraction solution. Here we examine the solution at the time sequence $t = 10, 20, 50, 100$. Qualitatively, the evolution of the numerical solution is as shown in Figure 6.16.

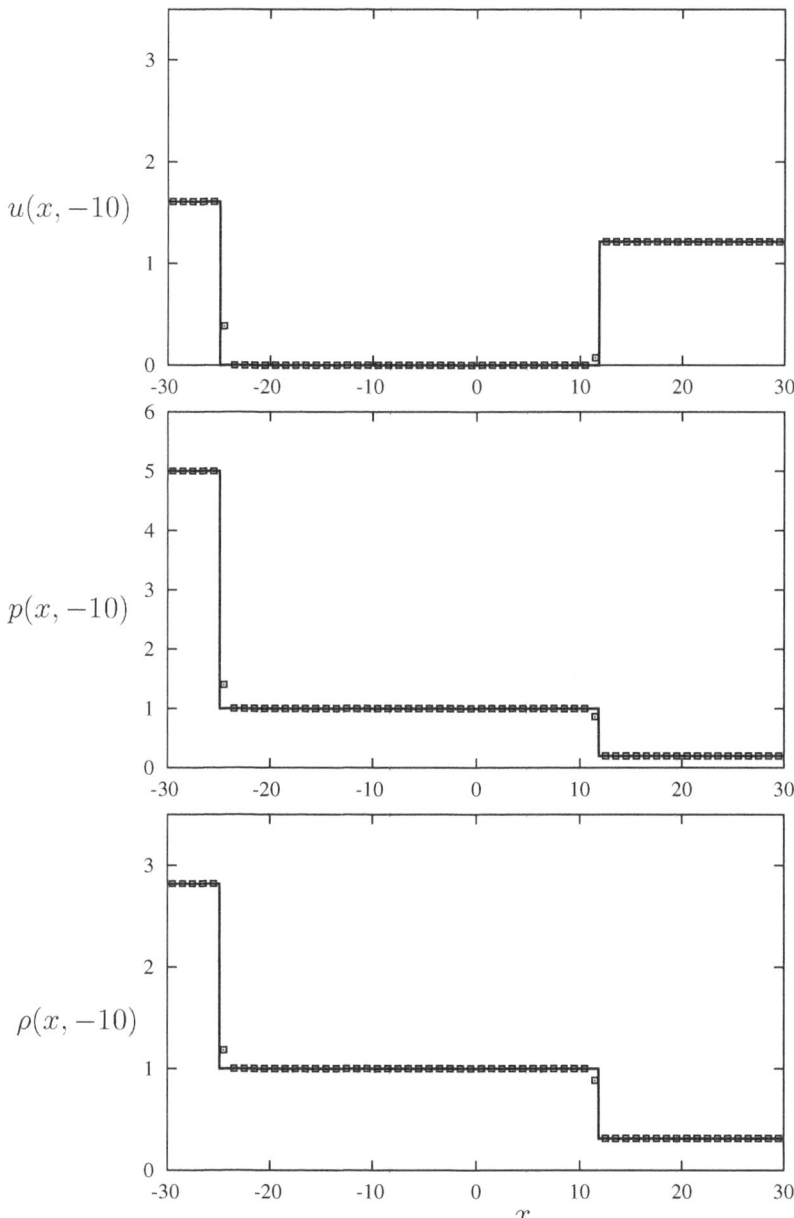

Figure 6.18. Initial data for shock–CRW interaction. Middle state $[\rho, p, u]_M = [1, 1, 0]$. Perfect gas ($\gamma = 1.4$). 3-shock: $p_L = 5$. 1-CRW: $p_R = 1/5$. Solid line – exact data; points – GRP (60 cells).

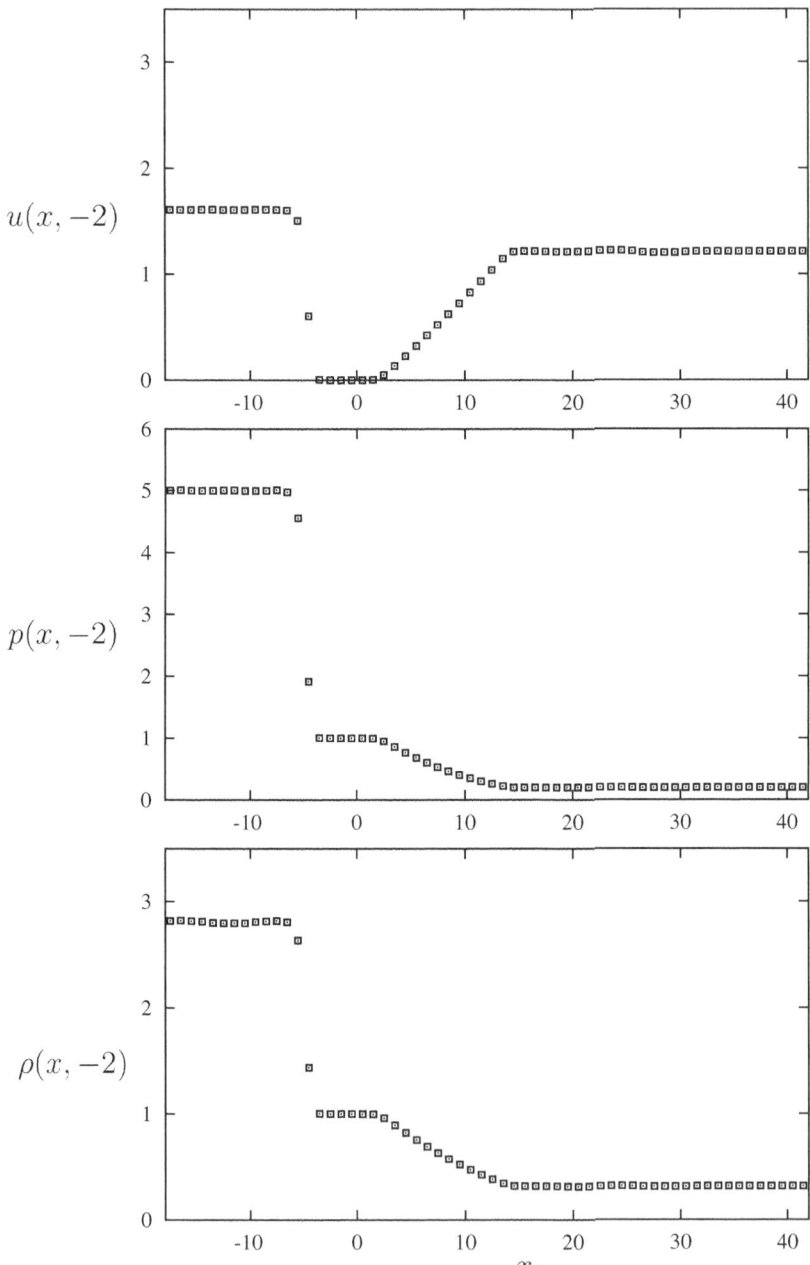

Figure 6.19. Shock–CRW results at preinteraction time $t = -2$. Middle state $[\rho, p, u]_M = [1, 1, 0]$. Perfect gas ($\gamma = 1.4$). 3-shock: $p_L = 5$. 1-CRW: $p_R = 1/5$. Points – GRP (60 cells).

To inspect its asymptotic behavior, we plot in Figure 6.20 the exact Riemann solution corresponding to the initial data $[\mathbf{U}_L, \mathbf{U}_R]$ at $t = 0$. At this relatively early time ($t = 10$) we see that the disparity between the two solutions is rather large. In Figure 6.21 we display density profiles for the time sequence $t = 10, 20, 50, 100$. The results exhibit an asymptotic convergence to the Riemann solution. However, in the IVP solution there is no contact discontinuity. Rather, the steep variation in density observed near the Riemann contact is due to a fluid layer of smoothly varying entropy generated by the shock as it passed through the CRW.

6.2.4 CRW–Contact Interaction

In Subsection 6.2.1 we considered the case of shock–contact interaction. Here we take up the analogous case of CRW–contact interaction.

Let the left wave, which initially separates the states $[\mathbf{U}_L, \mathbf{U}_M]$, be a 3-CRW [see Figure 6.4(a)], while the right wave is a contact discontinuity separating the states $[\mathbf{U}_M, \mathbf{U}_R]$, with $\rho_M > \rho_R$. It is known (Courant and Friedrichs [30, Section 79]) that the interaction in this case produces a left-facing compression wave and a right-facing rarefaction wave. Specifically, we assume the following initial data for the time $t = t_i = -10$:

$$\mathbf{U}(x, -10) = \begin{cases} \mathbf{U}_L = [\rho, p, u]_L = [0.42317, 0.3, -0.93484], & x < x_L, \\ \mathbf{U}_M = [\rho, p, u]_M = [1, 1, 0], & x_L < x < x_R, \\ \mathbf{U}_R = [\rho, p, u]_R = [0.3, 1, 0], & x_R < x. \end{cases}$$
$$(6.6)$$

The initial discontinuities are positioned so that the interaction will commence at $(x, t) = (0, 0)$; that is, $x_L = -10 \times 1.1832 = -11.832$ and $x_R = 0$.

As in the case of the CRW–shock interaction, it will prove instructive to consider the "asymptotic Riemann problem" having initial data $[\mathbf{U}_L, \mathbf{U}_R]$. The interaction curves for this case are shown (to true scale) in Figure 6.22. The intersection of $I_1^1(\mathbf{U}_L)$ and $I_3^r(\mathbf{U}_R)$ produces the contact values $(u^*, p^*) = (-1.2083, 0.43585)$ and the corresponding left and right densities $\rho_L^* = 0.55173$, $\rho_R^* = 0.16577$. We note that the states $\mathbf{U}_M$ and $\mathbf{U}_R$ are represented by the same (u, p) point. Yet, since $\rho_M \neq \rho_R$ the respective interaction curves $I_3^r(\mathbf{U}_M)$, $I_3^r(\mathbf{U}_R)$ are different. In fact, because of this difference, the intersection point between $I_3^r(\mathbf{U}_R)$ and $I_1^1(\mathbf{U}_L)$ lies on the rarefaction branch of the former and on the shock branch of the latter. The solution to this Riemann problem thus consists of a 1-shock (facing left) and a 3-CRW (facing right).

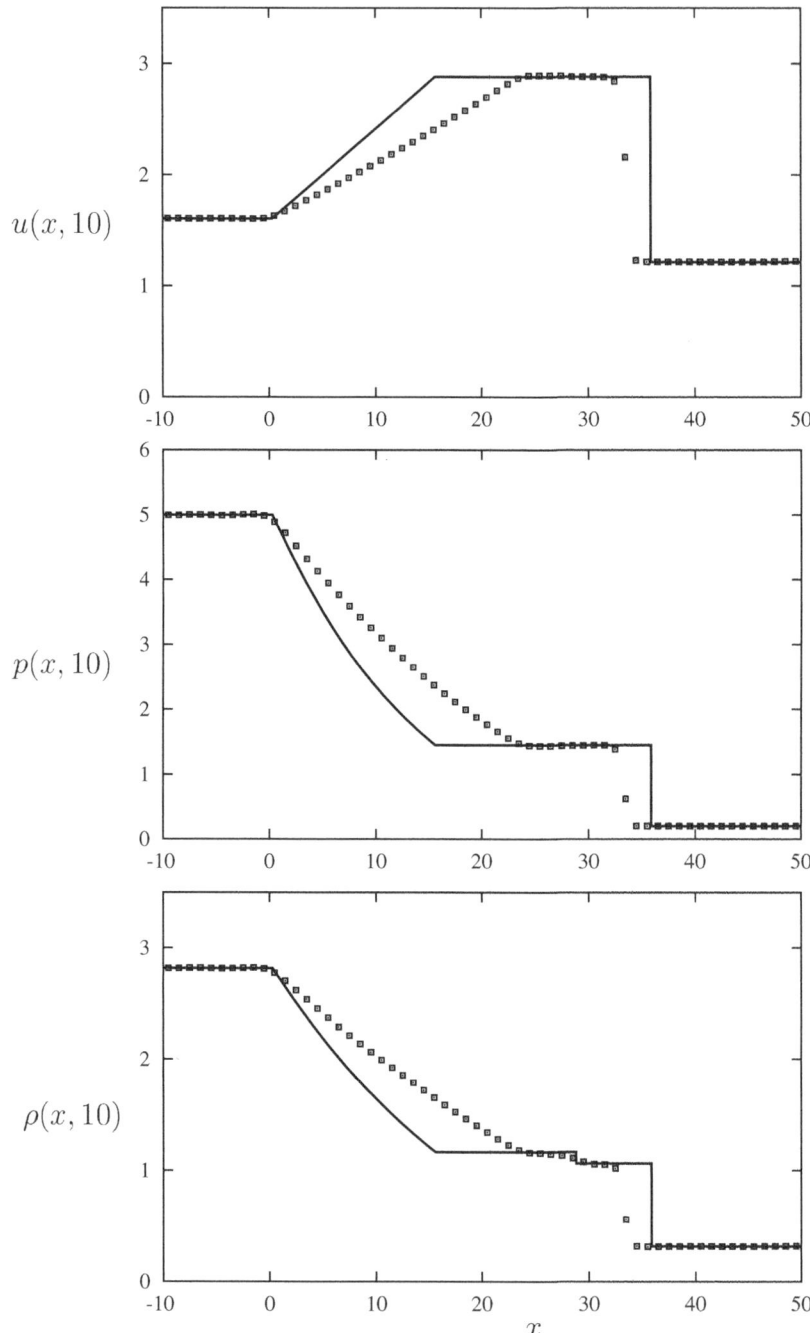

Figure 6.20. Shock–CRW postinteraction results at $t = 10$. Middle state $[\rho, p, u]_M = [1, 1, 0]$. Perfect gas ($\gamma = 1.4$). 3-shock: $p_L = 5$. 1-CRW: $p_R = 1/5$. Solid line – Riemann solution; points – GRP (60 cells).

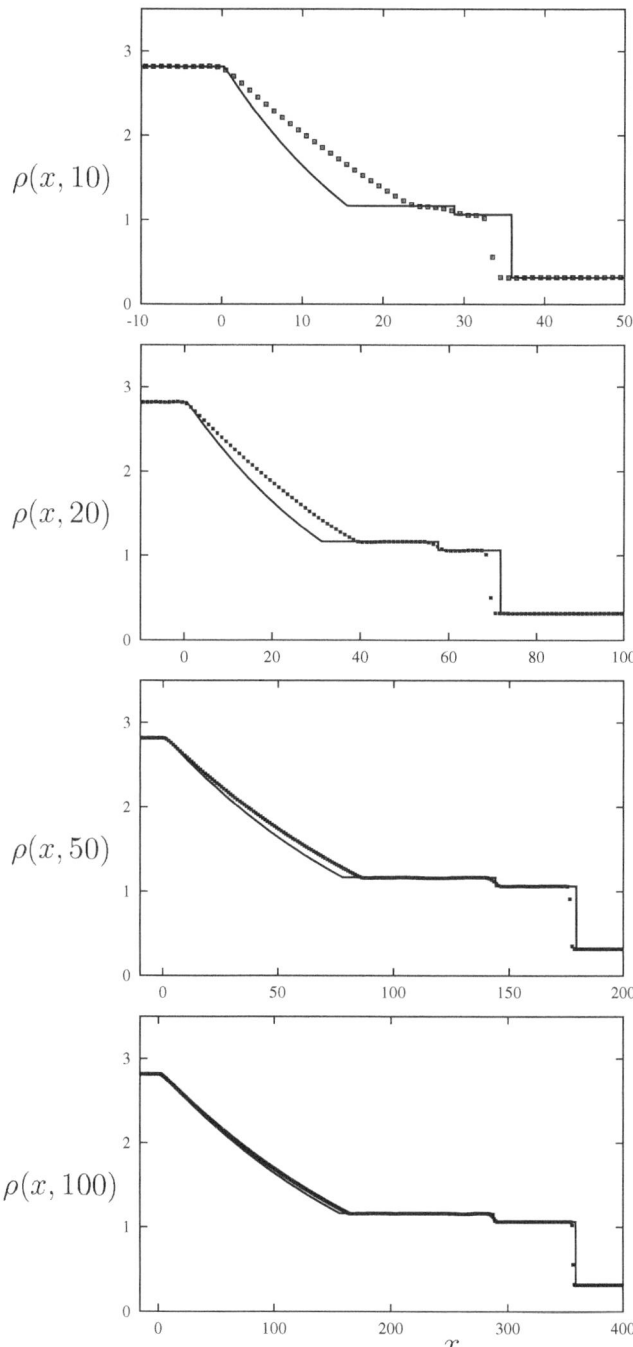

Figure 6.21. Shock–CRW interaction. Time sequence of density. Middle state: $[\rho, p, u]_M = [1, 1, 0]$ Perfect gas ($\gamma = 1.4$). 3-shock: $p_L = 5$. 1-CRW: $p_R = 1/5$. Solid line – Riemann solution; Points – GRP.

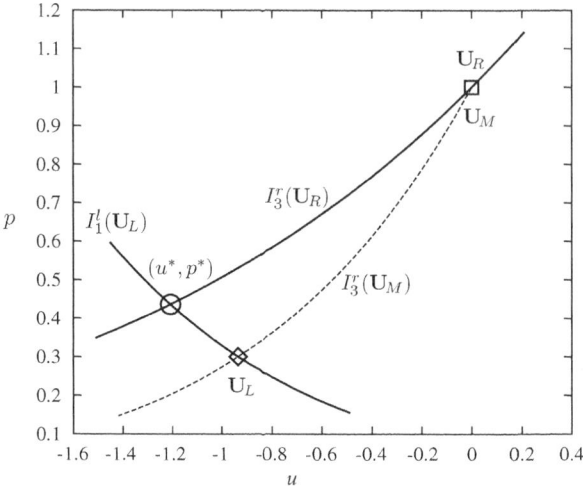

Figure 6.22. Interaction of CRW with a contact discontinuity. Middle state: $[\rho, p, u]_0 = [1, 1, 0]$, $\gamma = 1.4$. Right state: $[\rho, p, u]_R = [0.3, 1, 0]$. 3-CRW: $[\rho, p, u]_L = [0.42317, 0.3, -0.93484]$.

Approximate Analysis of the Interaction

Consider the interaction as depicted qualitatively in the (x, t) wave diagram of Figure 6.23. At the beginning of the interaction $(x, t) = (0, 0)$, the flow profile consists of a smoothly varying CRW $(x < 0)$ and a quiescent state $\mathbf{U}_R$ $(x > 0)$.

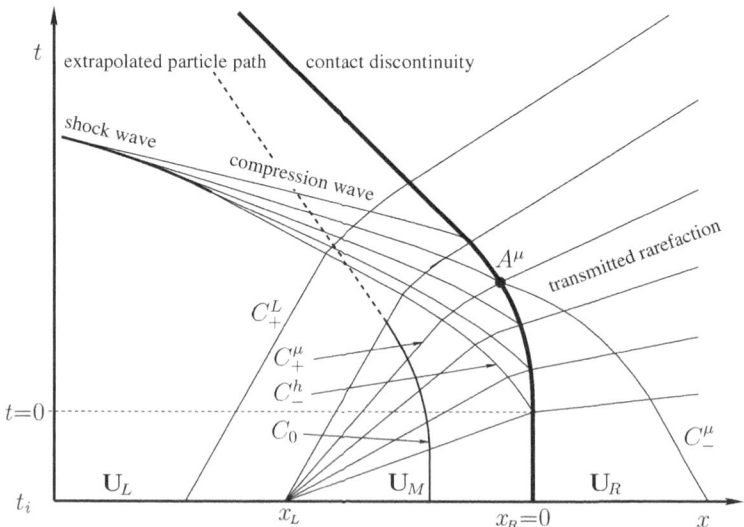

Figure 6.23. CRW–contact interaction schematic.

Recalling the discussion of the GRP in Section 5.1, we infer that the contact discontinuity persists for $t > 0$. In fact, it will separate two regions of smoothly varying flow, at least for some short time $t > 0$. We wish to understand the gradual evolution of the flow toward the aforementioned "asymptotic Riemann solution." To this end we invoke some approximate analytic reasoning, since an exact solution is not available.[5] When $\rho_R < \rho_M$ the interaction produces a left-facing compression wave and a right-facing rarefaction wave. The compression wave evolves into a sharp shock wave, forming at some time $t_c > 0$. In the following, the analytic arguments supporting this claim are briefly outlined.

Referring to Figure 6.23, we let C_+^μ denote a characteristic line within the CRW, having a slope $\mu = u_\mu + c_\mu$, where μ, u_μ, c_μ are constant along C_+^μ up to its intersection with the head characteristic C_-^h of the reflected wave. It then extends (as a curved line) and intersects the contact surface at a point A_μ. We assume that as the extended part of C_+^μ traverses the interaction zone, it does not intersect a shock front, so that the entropy $S_\mu = S_M$ is constant on the entire line C_+^μ. Denote by Q_1, Q_2, respectively, the left and right values of any flow variable at the point A_μ. The velocity and pressure are continuous across the contact; that is, $u_1 = u_2$, $p_1 = p_2$. But $c_1 \neq c_2$, $\rho_1 \neq \rho_2$, and by the isentropic assumption, $S_1 = S_M$, $S_2 = S_R$ (however, $S_1 \neq S_2$). Invoking the γ-law isentropic relations, $c_1/c_M = (p_1/p_M)^{\frac{\gamma-1}{2\gamma}}$, $c_2/c_R = (p_2/p_R)^{\frac{\gamma-1}{2\gamma}}$, and using $p_M = p_R$, we infer that

$$c_1/c_M = c_2/c_R, \qquad c_M/c_R = \sqrt{\rho_R/\rho_M}. \qquad (6.7)$$

Now, recall that the Riemann invariants $R_\pm$ are constant along $C_\pm$ in isentropic regions [see Equation (4.108)]. We apply these relations three times: First, $R_- = \frac{2}{\gamma-1}c_\mu - u_\mu = \frac{2}{\gamma-1}c_M$ is uniformly constant in the CRW up to the intersection with C_-^h [compare (4.99)]; second, $R_+^\mu = \frac{2}{\gamma-1}c_\mu + u_\mu = \frac{2}{\gamma-1}c_1 + u_1$ is constant along C_+^μ up to the point A_μ; third, $R_-^A = \frac{2}{\gamma-1}c_2 - u_2 = \frac{2}{\gamma-1}c_R$ is constant along the characteristic curve C_-^μ on the right side of the contact surface. Solving these equations for u_1 (which is equal to u_2), and using (6.7), we get

$$u_1 = \frac{2}{1 + \sqrt{\rho_R/\rho_M}} u_\mu. \qquad (6.8)$$

Since $\rho_R < \rho_M$ (6.8) implies that $|u_1| > |u_\mu|$, which must bring about a compression wave, as explained by the following reasoning. Let C_0 be a particle path starting at (x, t_i), $x_L < x < 0$ (Figure 6.23), and let this particle complete

---

[5] The method of characteristics can produce an accurate solution, but it has to be "tailor fitted" to the specific features of this case – a rather complex undertaking.

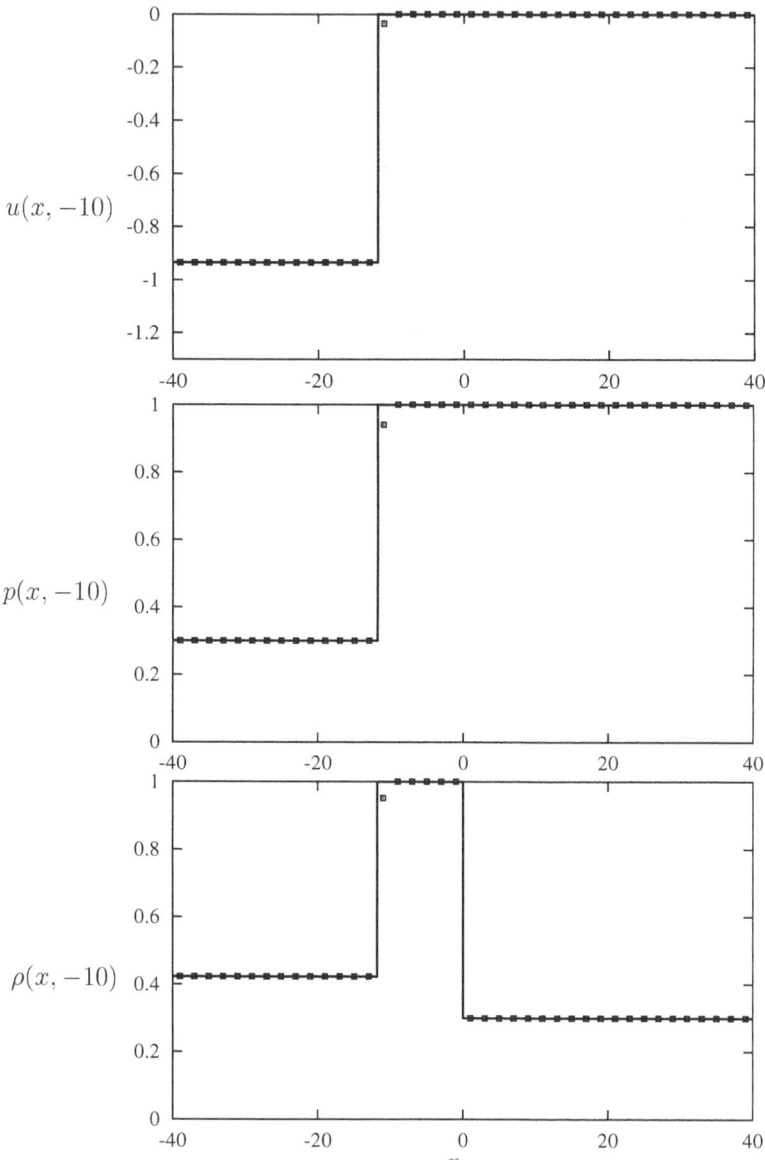

Figure 6.24. CRW–contact interaction. Initial data. Middle state: $[\rho, p, u]_0 = [1, 1, 0]$, $\gamma = 1.4$. $\mathbf{U}_L$: $[\rho, p, u]_L = [0.42317, 0.3, -0.93484]$. $\mathbf{U}_R$: $[\rho, p, u]_R = [0.3, 1, 0]$. Solid line – initial data; points – GRP (40 cells shown).

its passage through the CRW and continue unaffected by the interaction. If we take $\mu = u_L + c_L$ (the tail characteristic of the CRW), Equation (6.8) implies that the extrapolated straight part of C_0 (shown in Figure 6.23 as a dashed line) intersects the contact discontinuity since $|u_1| > |u_\mu|$. Evidently, such intersection is impossible and will indeed be avoided by the formation of a compression wave between C_0 and the contact trajectory. Similarly, the contact discontinuity accelerates away from the fluid on its right, thereby producing a right-facing rarefaction wave. These wave-generating effects are analogous to those of an accelerated piston moving into, or away from, the adjacent fluid (Courant and Friedrichs [30, Sections 43 and 49]).

Consider the nature of the foregoing approximation by taking (instead of C_+^μ) a characteristic curve C_+^L starting at (x, t_i), with $x < x_L$ (see Figure 6.23). With $x_L - x$ taken to be sufficiently large, C_+^L intersects the shock front prior to its intersection point with the contact trajectory. Clearly, the former analysis cannot apply at this point since the entropy is not constant along C_+^L. Thus, the contact velocity u_1 obtained by (6.8) for the tail characteristic differs from the large-time velocity at the contact surface. Yet, it may be a reasonable *approximation* to that velocity. For example, in the present case the large-time velocity is presumably that obtained from the "asymptotic Riemann solution" $u^* = -1.208341$. By (6.8) we obtain $u_1 = -1.208026$, with a relative error of about 0.0003. This indicates that the analytic approximation is reasonably accurate in the present case.

Numerical (GRP) Solution

We now turn to the numerical solution of the CRW–contact interaction, using the same GRP scheme employed in the previous cases. The grid occupies the interval $[-250, 250]$ and is divided into 250 cells of length $\Delta x = 2$ each. The initial data are as given in (6.6). The constant time step is $\Delta t = 0.4$, and the integration is performed from $t_i = -10$ to the final time $t = 100$ (275 integration steps), with the corresponding value of the CFL ratio $\mu_{CFL} \approx 0.63$.

The initial profiles are shown in Figure 6.24, along with the exact data. In Figure 6.25 we show the profiles obtained at the beginning of the interaction ($t = 0$), and also the initial data for the corresponding "asymptotic Riemann problem" $[U_L, U_R]$ (to which we refer to in the following as the "Riemann solution"). At time $t = 10$ (Figure 6.26) there is already an extremal point in the velocity and density profiles, indicating an incipient formation of the compression wave. This trend becomes more pronounced at $t = 20$ (Figure 6.27).

The full formation of the compression/shock wave is evident in Figure 6.28, where a time sequence of density profiles is shown. Clearly, the density peaks at a point that approaches the contact discontinuity of the Riemann solution.

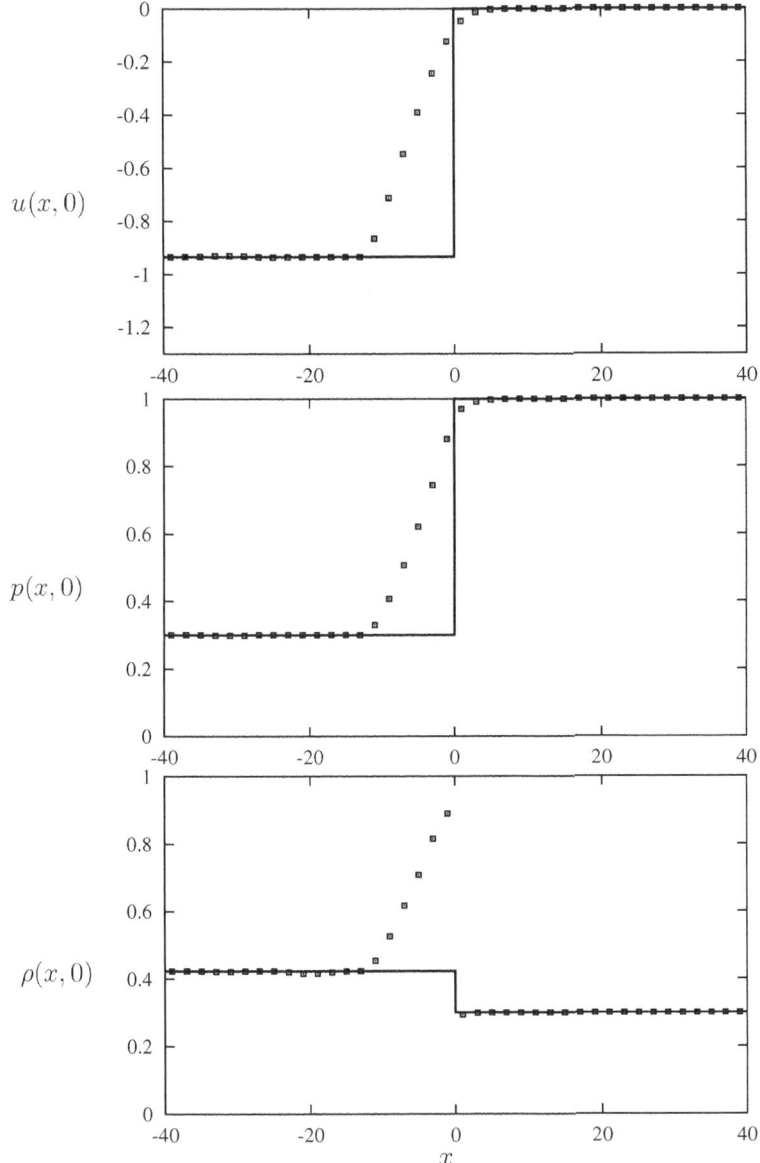

Figure 6.25. CRW–contact interaction. Middle state: $[\rho, p, u]_0 = [1, 1, 0]$, $\gamma = 1.4$. $\mathbf{U}_L$: $[\rho, p, u]_L = [0.42317, 0.3, -0.93484]$. $\mathbf{U}_R$: $[\rho, p, u]_R = [0.3, 1, 0]$. Solid line – initial data for "asymptotic RP"; points – GRP (40 cells shown).

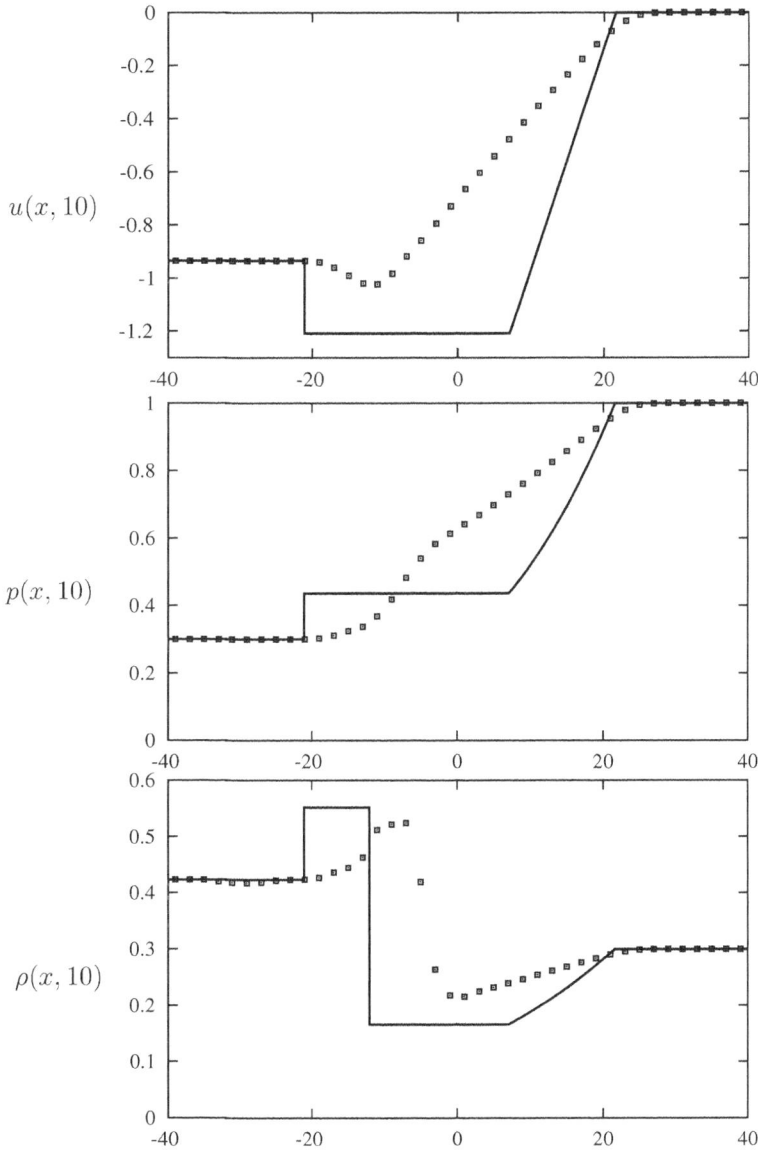

Figure 6.26. CRW–contact interaction. $t = 10$. Middle state: $[\rho, p, u]_0 = [1, 1, 0]$, $\gamma = 1.4$. $\mathbf{U}_L$: $[\rho, p, u]_L = [0.42317, 0.3, -0.93484]$. $\mathbf{U}_R$: $[\rho, p, u]_R = [0.3, 1, 0]$. Solid line – Riemann solution; points – GRP (40 cells shown).

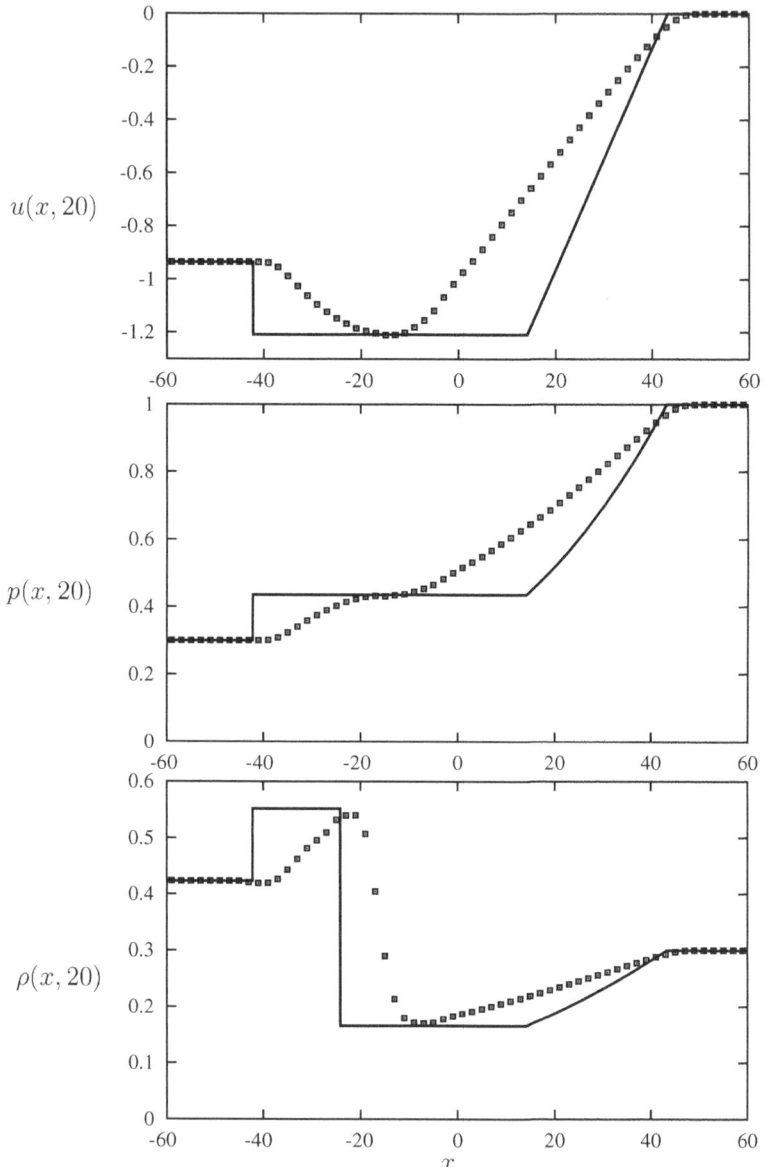

Figure 6.27. CRW–contact interaction. $t = 20$. Middle state: $[\rho, p, u]_0 = [1, 1, 0]$, $\gamma = 1.4$. $\mathbf{U}_{\mathrm{L}}$: $[\rho, p, u]_{\mathrm{L}} = [0.42317, 0.3, -0.93484]$. $\mathbf{U}_{\mathrm{R}}$: $[\rho, p, u]_{\mathrm{R}} = [0.3, 1, 0]$. Solid line – Riemann solution; points – GRP (40 cells shown).

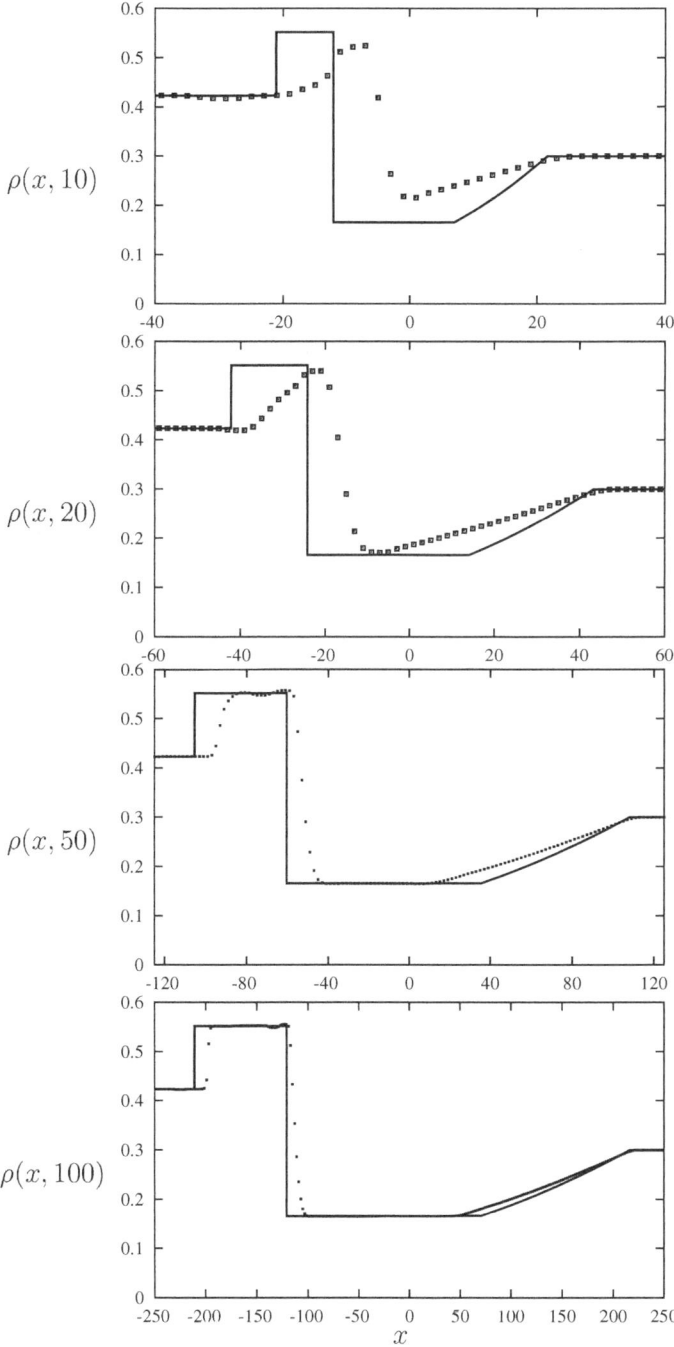

Figure 6.28. CRW–contact interaction. Time sequence of density. Middle state: $[\rho, p, u]_0 = [1, 1, 0]$, $\gamma = 1.4$. $\mathbf{U}_L$: $[\rho, p, u]_L = [0.42317, 0.3, -0.93484]$. $\mathbf{U}_R$: $[\rho, p, u]_R = [0.3, 1, 0]$. Solid line – Riemann solution; points – GRP.

To the left of the peak (corresponding to the fluid on the left side of the contact surface) we see a left-facing compression wave, which steepens and seems to have fully developed into a shock wave at $t = 100$. On the right side, the density profile spreads out, approaching at the latest time shown ($t = 100$) the CRW of the Riemann solution. Here the GRP method clearly demonstrates the capability of capturing a gradually forming shock wave.

Notwithstanding the apparent convergence to the Riemann solution (Figure 6.28), there are some fine differences between the IVP and Riemann solution that persist for all times. Consider in particular the fluid between the contact discontinuity and the shock front. In the Riemann solution the contact discontinuity corresponds to the resolution of the initial discontinuity $[\mathbf{U}_L, \mathbf{U}_R]$, whereas in the IVP it evolves "isentropically" from the contact discontinuity $[\mathbf{U}_M, \mathbf{U}_R]$. Furthermore, a layer of smoothly varying entropy is obtained in the IVP solution owing to the gradual formation of the shock. This gradual density distribution (seven points) appears in the GRP solution at late times ($t \geq 50$) as a captured contact discontinuity (Figure 6.28), which agrees well with the densities on either side of the contact in the Riemann solution. We attribute this agreement to the rather small entropy jump across the shock (relative magnitude ≈ 0.001). In cases of different initial data the "fine features" disagreement may become conspicuous.

6.3 Spherically Converging Flow of Cold Gas

The former examples in this chapter involved only planar flow. Here we consider a duct flow with spherical symmetry, governed by Equations (4.45), with cross-sectional area $A(r) = r^2$. This unique problem has been proposed by Noh [94] as a test case having an exact (self-similar) solution. The initial conditions consist of a spherically converging flow of a "cold" (zero-pressure) perfect gas with $\gamma = 5/3$, having finite density and uniform velocity:

$$\mathbf{U}(r, 0) = [\rho, p, u]_o = [1, 0, -1], \qquad 0 \leq r. \tag{6.9}$$

The solution consists of an expanding spherical shock (starting from the origin at $t = 0$). The fluid behind the shock is quiescent with uniform pressure p_- and density ρ_-. The pressure jumps across the shock front from its zero preshock value to p_-; thus it is an "infinite shock" as discussed in Corollary 4.52. The density ratio is therefore $\rho_-/\rho_+ = (\gamma + 1)/(\gamma - 1) = 4$. The pressure ahead of the shock is zero; hence the (spherical) velocity retains its initial value -1. The density, however, increases by (isentropic) spherical compression. If we denote the (constant) shock speed by σ_3, the mass flux through the shock front is

$\sigma_3 \rho_- = (\sigma_3 + 1)\rho_+$. It follows that $\sigma_3 = 1/3$. To find the density profile ahead of the shock (i.e., $r \geq t/3$) we observe that the fluid located at point r at time t was initially at the point $r + t$. This implies that the compression ratio is $\left(\frac{r+t}{r}\right)^2$; hence

$$\rho(r, t) = (1 + t/r)^2, \quad r \geq \frac{t}{3}. \tag{6.10}$$

In particular, at the shock front $r = t/3$, we find $\rho_+ = 16$; hence $\rho_- = 64$. The uniform pressure behind the shock is obtained from Equation (4.89)(i), $(p_+ - p_-)/(u_+ - u_-) = \sigma_3 \rho_-$; hence $p_- = 64/3$. The exact profiles are shown as solid lines in Figure 6.29.

The computation is performed with the GRP scheme for duct flows (Chapter 5), in the interval $[0 < r < 100]$, which is divided into a grid of 100 equal cells. The boundary conditions are zero velocity at the origin ($r = 0$) and the exact solution given by (6.10) at the outer boundary $r = 100$. A constant time step $\Delta t = 0.25$ is assumed, and the integration is performed up to $t = 225$. The corresponding CFL ratio is $\mu_{\text{CFL}} = 0.31$.

The results are shown in Figure 6.29, where the distributions of the flow velocity, pressure, and density are shown along with the exact solution. The agreement with the exact solution is very good, with discrepancies occurring primarily for the density distribution near the origin. This error is due to the "startup" of the captured shock near the origin, where the numerical dissipation generates an entropy higher than the exact value. At the final time the shock wave is very sharply resolved, with the postshock flow variables close to the exact values. It is also interesting to note, in this unique problem, that the GRP method can correctly capture a shock wave of infinite intensity.

For further numerical studies of Noh's problem by the GRP method, we refer to Ben-Artzi and Birman [6] and to Birman et al. [17]. An extensive adaptive-mesh computational analysis of this problem was recently performed by Gehmeyr, Cheng, and Mihalas [48], with results quite similar to those obtained here.

6.4 The Flow Induced by an Expanding Sphere

When a spherical surface, initially at the origin, expands at a constant speed s into a quiescent compressible fluid, a spherically expanding shock wave is formed. The shock moves at a constant speed $\sigma_3 > s$. This problem has been studied by Taylor [108] in terms of the self-similar fluid dynamical equations for a spherically symmetric flow. As in the former example, the numerical solution here is also obtained by the GRP scheme for a (spherical) duct flow. In this case, however, a new feature is introduced – the boundary condition at the expanding

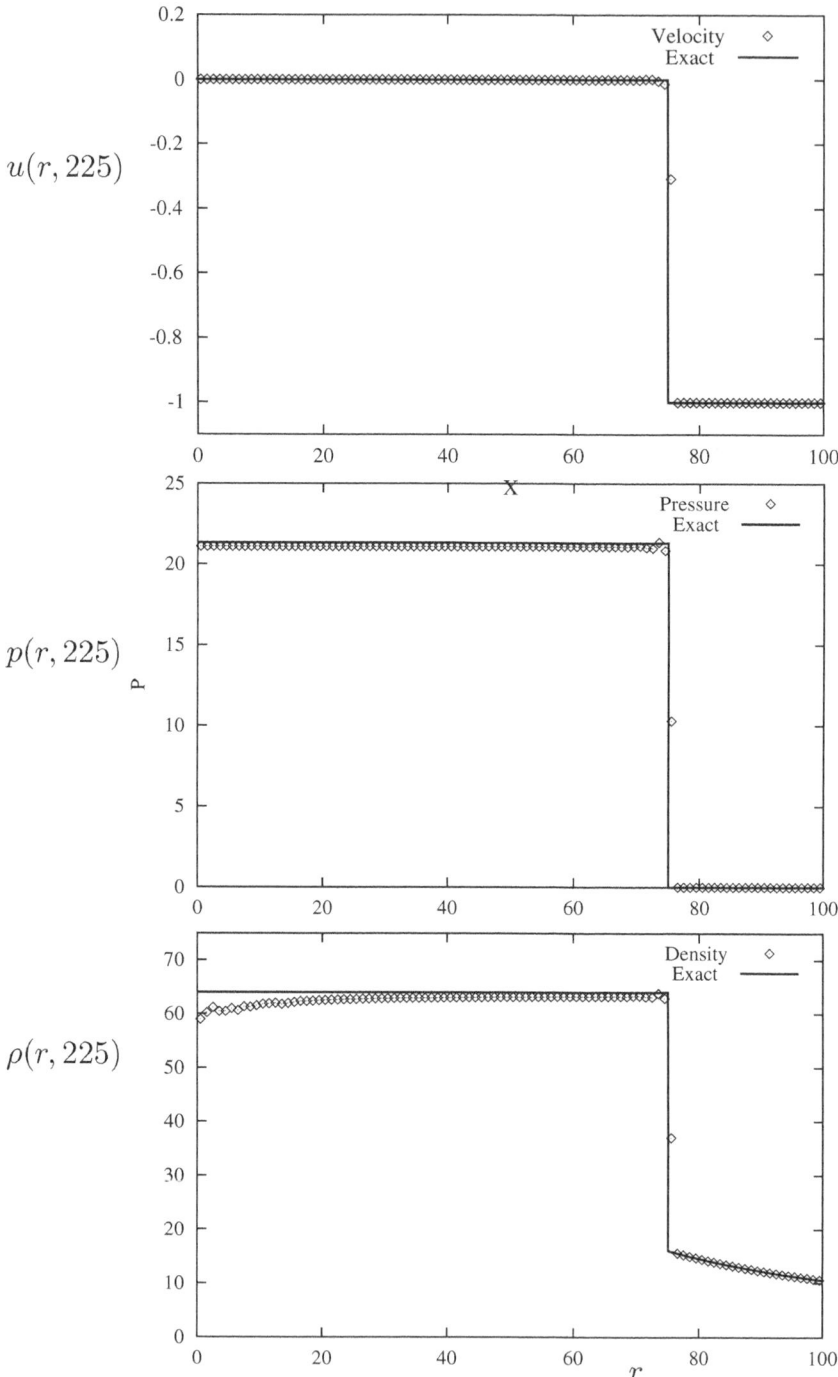

Figure 6.29. Distribution of flow variables for Noh's problem. Solid line – exact solution; points – GRP (100 cells shown).

spherical surface, which is treated by employing a moving grid. The fluid is a perfect gas with $\gamma = 1.4$, having the uniform initial state $\mathbf{U}_0 = [\rho, p, u]_0 = [1.4, 1, 0]$ and $s = 1$. The corresponding value of the shock speed is $\sigma_3 = 1.328$.

Taylor [108] has shown that the (self-similar isentropic) flow in the region between the surface and the shock is governed by a pair of ordinary differential equations for the fluid velocity and the speed of sound as functions of the similarity coordinate r/t. These equations are derived from the first two equations in (4.45) with $A(r) = r^2$ (and assuming that the flow variables are functions of r/t for $st \leq r \leq \sigma_3 t$). The exact solution is obtained by integrating these equations, requiring that at the spherical surface the fluid velocity equals s and that across the shock front (the constant speed of which, σ_3, is not a priori known) the Rankine–Hugoniot jump conditions hold.

The solution by the GRP scheme makes use of the ALE capability (see Sections 6.3 and 8.2). The initial conditions are $\mathbf{U}(r, 0) = \mathbf{U}_0$, with a grid extending initially from $r = 0$ to $r = 140$ and divided into 40 equal cells. The leftmost grid point (corresponding to the spherical surface) moves at the constant velocity of $s = 1$, while the rightmost point is stationary; all intermediate grid points move at a linearly interpolated velocity, keeping the (shrinking) grid intervals uniform at all time levels.

Special attention is given to the leftmost cell; its left boundary coincides with the expanding surface and the fluxes there cannot be obtained by solving a standard GRP. This is a typical situation in high-resolution computations, where moving boundaries (or complex geometric settings in multidimensional cases) make it necessary to deal in much greater detail with a small number of "special" cells, compared to the "standard" treatment accorded to the bulk of the mesh.

Let $\overline{\mathbf{U}} = [\overline{\rho}, \overline{p}, \overline{u}]$ and $\overline{\Delta \mathbf{U}} = [\overline{\Delta \rho}, \overline{\Delta p}, \overline{\Delta u}]$ be, respectively, the average state and linear variation in the leftmost cell. Thus, the state attributed to the left endpoint is $\overline{\mathbf{U}} - \frac{1}{2}\overline{\Delta \mathbf{U}}$. However, the expanding surface forces a velocity s at that point, which in general is different from $\overline{u} - \frac{1}{2}\overline{\Delta u}$. We regard this situation as an IVP with initial discontinuity at that point. Employing the notation of Section 6.2 (see Figure 6.6) we let $I_3^r(\overline{\mathbf{U}} - \frac{1}{2}\overline{\Delta \mathbf{U}})$ be the interaction curve (namely, the ensemble of all left states connected to the right state $\overline{\mathbf{U}} - \frac{1}{2}\overline{\Delta \mathbf{U}}$ by a 3-wave). We then take the state $\mathbf{U}^* \in I_3^r(\overline{\mathbf{U}} - \frac{1}{2}\overline{\Delta \mathbf{U}})$ so that $u^* = s$. The density ρ^* and pressure p^* are thereby uniquely determined. We take the state $\mathbf{U}^*$ as constant at that point throughout the time interval $[t_n, t_{n+1}]$, and we determine the fluxes accordingly. At the rightmost grid point we take the quiescent state $\mathbf{U}_0$ as the solution to the local GRP, assuming that for the time interval considered, the shock will not reach that point. The time integration is performed with a constant time step $\Delta t = 0.5$, which corresponds to a CFL ratio of $\mu_{CFL} = 0.686$, up to the final time $t = 100$.

The results shown in Figure 6.30 display a very good agreement with the exact solution, considering that the fluid in the region between the surface and the shock is resolved by merely 33 cells. It is noted that the shock wave is sharply captured, which is of special interest since shocks in spherical flow computation are generally less sharp than in planar flow.

In this self-similar flow the shock wave is of constant intensity, so that even though the flow between the surface and the shock front is not uniform, the entropy is. We therefore show (in Figure 6.30) the profile of $S = p/\rho^{\gamma}$, and indeed it is nearly uniform behind the shock front. The increase in entropy near the spherical surface is a "starting error" of the captured shock computation, which typically arises when a shock wave passes through a jump discontinuity or departs from a "piston-like" surface (as in the present case).

6.5 Converging–Diverging Nozzle Flow

In the former two examples we considered a special case of duct flow, namely, one-dimensional, spherically symmetric flow. The present example is also a duct flow, but it is different from the former two cases in several aspects. The quasi-one-dimensional flow here takes place in a duct where $A(r)$ is a nonmonotonic function; it serves as an *approximate* model to the fully two-dimensional flow in a nozzle of finite width (a comparison between the one-dimensional and the two-dimensional solutions of flow in a duct will be considered in Chapter 10 of this monograph). Moreover, since in most cases we are concerned with a *steady* compressible flow in such nozzles, our example is formulated as an IVP with boundary conditions designed to produce a steady flow at large times. Our presentation starts with a brief outline of one-dimensional steady flow in nozzles. We then specify a particular nozzle geometry and two cases of boundary conditions. In each case the exact steady solution is compared to the numerical (GRP) solution, and a strikingly good agreement is found.

A converging–diverging duct (the so-called Laval nozzle) is widely used for achieving steady supersonic flow in a variety of systems such as rocket motors and wind tunnels. The simplest analytic model for compressible flow in a Laval nozzle is the quasi-one-dimensional duct flow approximation (see Section 4.2). For a duct of finite lateral extent, this approximation is widely recognized as a basic "engineering model" for nozzle flow analysis, particularly when the flow is steady. The steady duct flow is governed by a system of ordinary differential

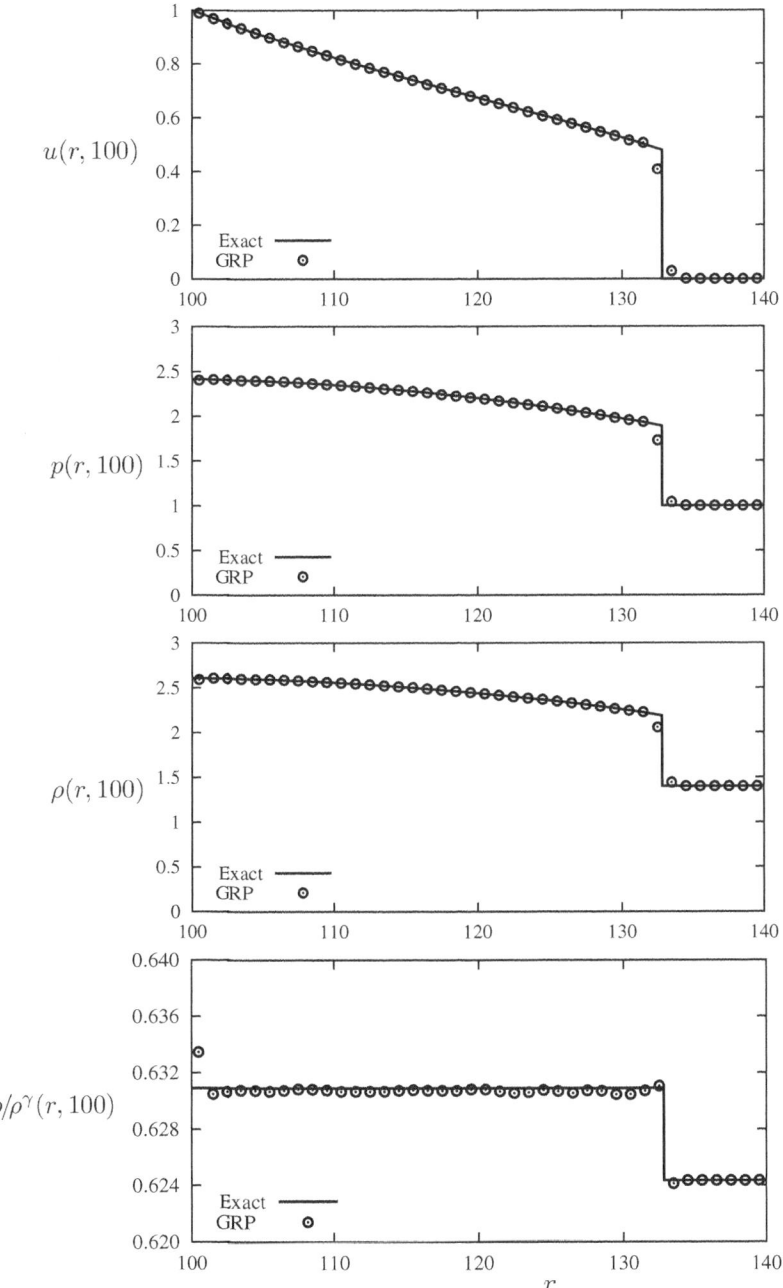

Figure 6.30. Flow variables in the fluid surrounding an expanding sphere. Perfect gas ($\gamma = 1.4$), $[u, p, \rho]_0 = [0, 1, 1.4]$. Sphere expansion speed $s = 1$. Solid line – exact solution; points – GRP (40 points shown).

equations [derived from (4.45)] that is readily integrated. For a comprehensive account of the steady one-dimensional theory we refer the reader to Shapiro [102, Chapter 5] or Liepmann and Roshko [84, Chapter 5]. In the following we present two cases of steady, quasi-one-dimensional nozzle flow of a perfect gas with a polytropic index $\gamma = 1.4$, which are compared to the large-time solution of the corresponding IVP, with suitably defined boundary conditions at the nozzle inlet and exit planes.

Nozzle Geometry and Steady Flow

Here we consider a steady flow in a converging–diverging nozzle, which occupies the interval $0 \le r \le 1$ and has a smooth cross-sectional area function[6] $A(r)$ given by the following expression:

$$
A(r) = \begin{cases} A_{\text{in}} \exp\left[-\log(A_{\text{in}})\sin^2(2\pi r)\right], & 0 \le r \le 0.25, \\[2em] A_{\text{ex}} \exp\left[-\log(A_{\text{ex}})\sin^2\left(\frac{2\pi(1-r)}{3}\right)\right], & 0.25 \le r \le 1, \end{cases}
$$

(6.11)

$$
A_{\text{in}} = A(0) = 4.8643, \qquad A_{\text{ex}} = A(1) = 4.2346
$$

(see the symmetric nozzle contour in Figures 6.31–6.34). It is noted that the function $A(r)$ for the converging part of the nozzle is different from $A(r)$ of the diverging part. The two contour parts are joined smoothly at the throat ($r = 0.25$), where $A(r) = 1$, $A'(r) = 0$, but $A''(r)$ is discontinuous.

From the theory of steady duct flow of a perfect gas (Shapiro [102, Chapter 5], Liepmann and Roshko [84, Chapter 5]) it follows that the Mach number $M(r) = u(r)/c(r)$ is fully determined by $A(r)$ through the algebraic relation[7]

$$
[A(r)]^2 = \frac{1}{[M(r)]^2}\left[\frac{2}{\gamma+1}\left(1 + \frac{\gamma-1}{2}[M(r)]^2\right)\right]^{(\gamma+1)/(\gamma-1)}.
$$

(6.12)

Furthermore, one can specify two arbitrary constants ("stagnation" pressure and density) p_0 and ρ_0 so that the steady flow profiles in the nozzle are given

---

[6] The values of the inlet and exit cross-sectional area A_{in} and A_{ex} correspond to a steady isentropic (shock-free) flow, where the inlet Mach number is $M_{\text{in}} = 0.12$ and (assuming a supersonic flow in the diverging part of the nozzle) the exit Mach number is $M_{\text{ex}} = 3$.

[7] This relation is restricted to smooth (isentropic) choked flows, meaning that the flow at the throat (point of minimal cross-sectional area) is sonic ($M = 1$).

by

$$p(r) = p_0 \left(1 + \frac{\gamma - 1}{2}[M(r)]^2 \right)^{-\frac{\gamma}{\gamma-1}},$$

$$\rho(r) = \rho_o \left(1 + \frac{\gamma - 1}{2}[M(r)]^2 \right)^{-\frac{1}{\gamma-1}}, \tag{6.13}$$

$$u(r) = M(r)\sqrt{\gamma p(r)/\rho(r)},$$

as long as the flow is smooth (shock free). This is shown by the solid lines in Figures 6.31 and 6.32. In fact, in this case we note that fixing p_0, ρ_0, there is only one possible value for the exit pressure $p(1)$, which is given by (6.13).

Now, it turns out that one can obtain, for fixed values of p_0, ρ_0, another class of steady state solutions; they consist of two intervals of smooth flow, separated by a steady (left-facing) shock wave (see the solid lines in Figs. 6.33–6.34). This set of solutions is obtained by specifying the exit pressure $p(1)$ within a certain range of values. Expressions (6.12) and (6.13) remain valid in the smooth region upstream of the shock (i.e., between the inlet $r = 0$ and the stationary shock). The pressure jump across the shock then serves to adjust the downstream pressure profile to its specified exit value $p(1)$.[8]

Two cases are considered here. In both we take $p_0 = \rho_0 = 1$ and $A(r)$ as in (6.12).

(A) A smooth flow where $p(1) = 0.0272237$ is obtained from (6.13) by taking $r = 1$ in (6.12), leading to $M(1) = 3$.

(B) Setting $p(1) = 0.4$ leads to a discontinuous steady-state solution, as shown by the solid lines in Figures 6.33 and 6.34. The location of the shock is $r = 0.76986$, and the pressure at the shock jumps from $p_- = 0.044707$ to $p_+ = 0.36548$. Observe that, in compliance with the previous discussion, the profiles of the steady flow upstream of the shock are identical to those of the smooth flow in this interval.

It is also interesting to notice the discontinuity in $M'(r)$ and $p'(r)$ at the throat ($r = 0.25$), caused by the discontinuity in $A''(r)$ there.

---

[8] Physically speaking, this corresponds to a "matching" requirement $p(1) = p_b$ between $p(1)$ and the "background pressure" p_h, which is the pressure of the ambient gas outside the nozzle exit (Shapiro [102, Chapter 5], Liepmann and Roshko [84, Chapter 5]). This matching is possible when p_b is not lower than the value corresponding to a shock wave at the exit (in our example $p_b = 0.28131$). At lower background pressure the flow throughout the nozzle is shock free and supersonic in the diverging part of the nozzle. In that case $M(1) = 3$ and $p(1) \neq p_b$.

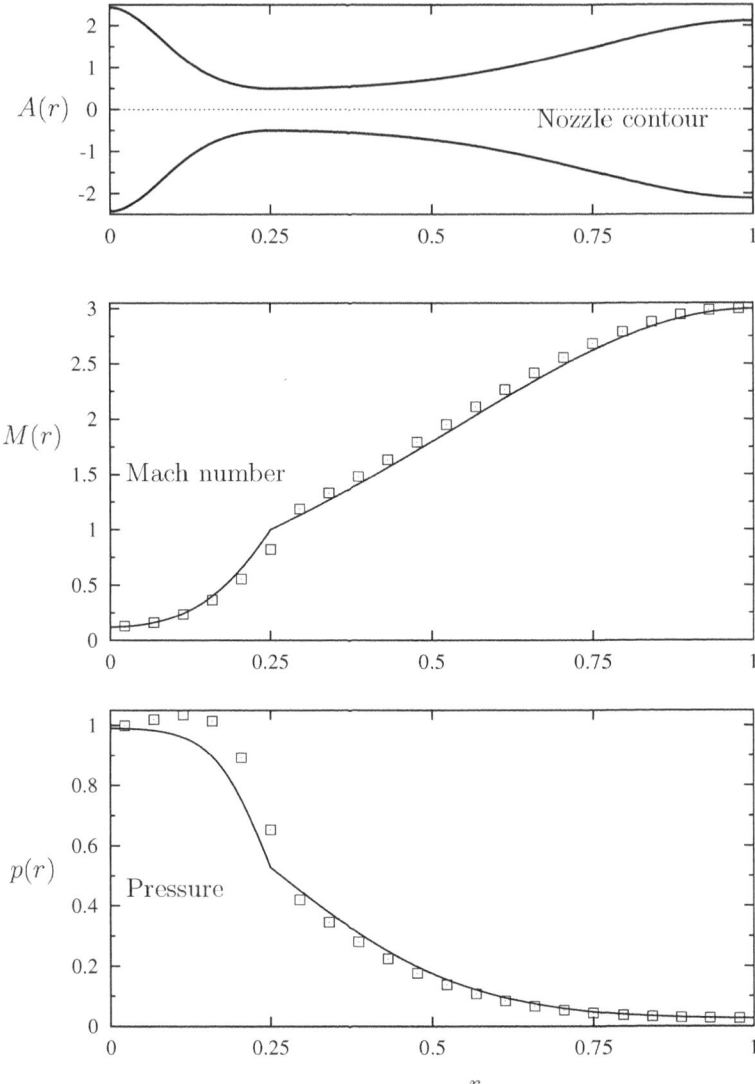

Figure 6.31. Large-time flow in Laval nozzle. First-order (Godunov) solution. $p_0 = 1$; $p_b = 0.02722$. Nominal (isentropic) $M_{ex} = 3$. Solid line – steady solution; points – Godunov (22 cells).

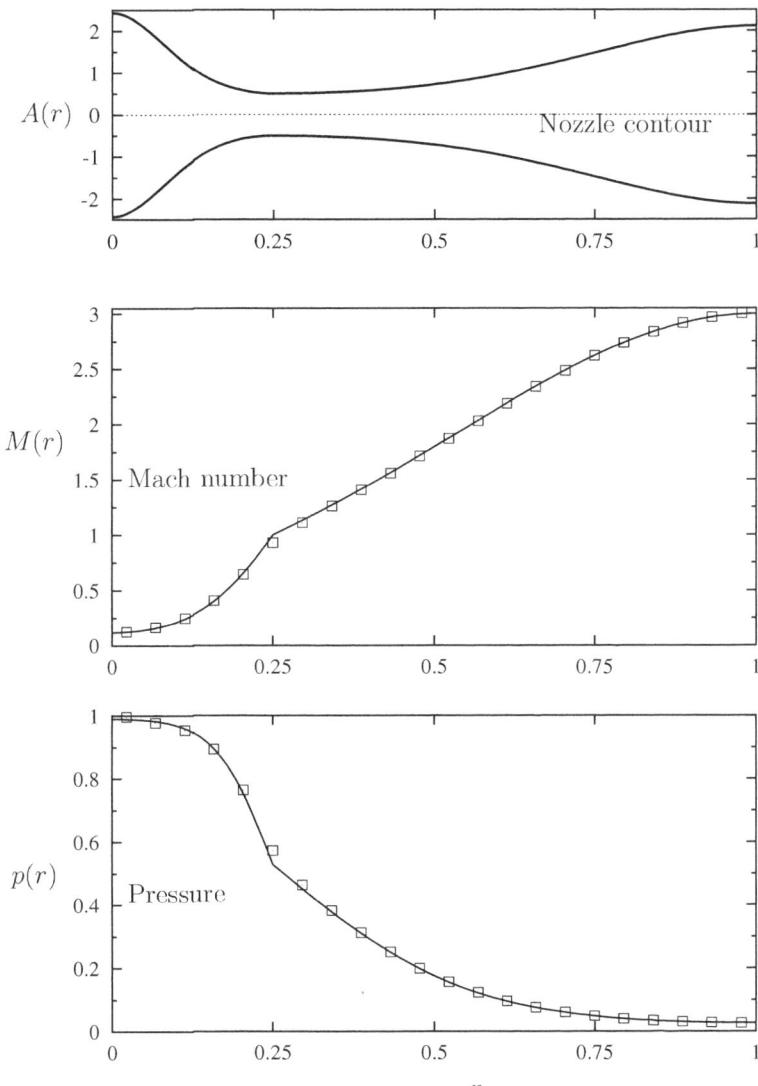

Figure 6.32. Large-time flow in Laval nozzle. Second-order (GRP) solution. $p_0 = 1$; $p_b = 0.02722$. Nominal (isentropic) $M_{ex} = 3$. Solid line – steady solution; points – GRP (22 cells).

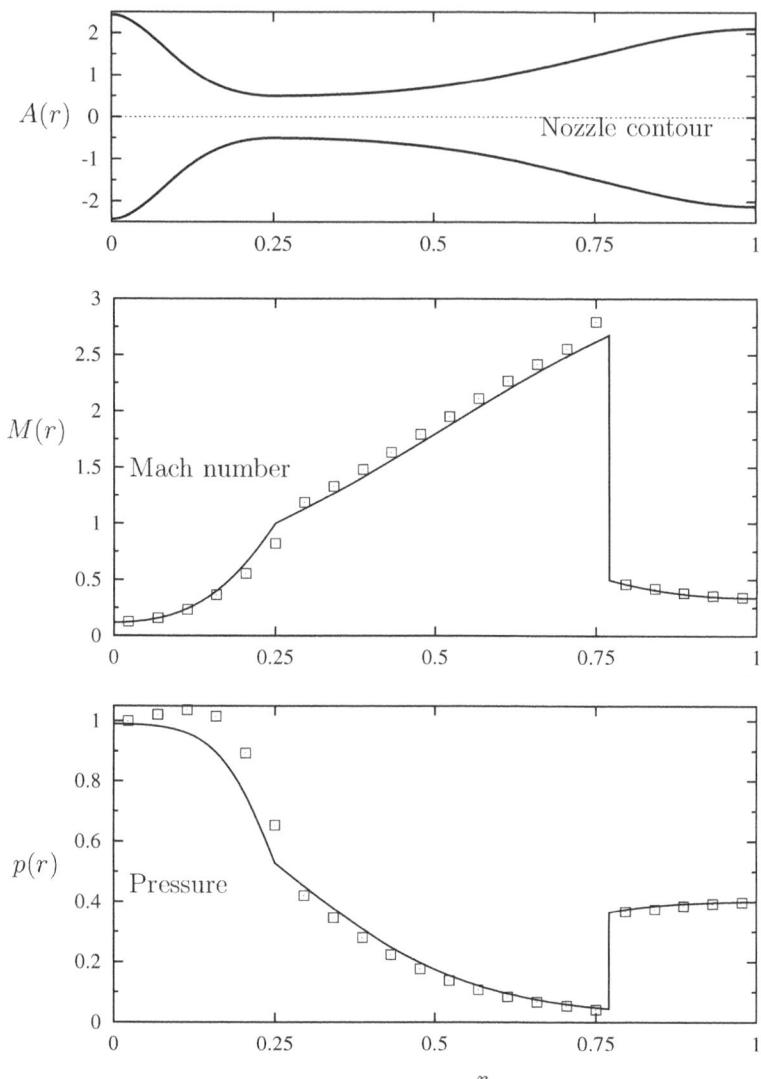

Figure 6.33. Large-time flow in Laval nozzle. First-order (Godunov) solution. $p_0 = 1$; $p_b = 0.4$. Nominal (isentropic) $M_{ex} = 3$. Solid line – steady solution; points – Godunov (22 cells).

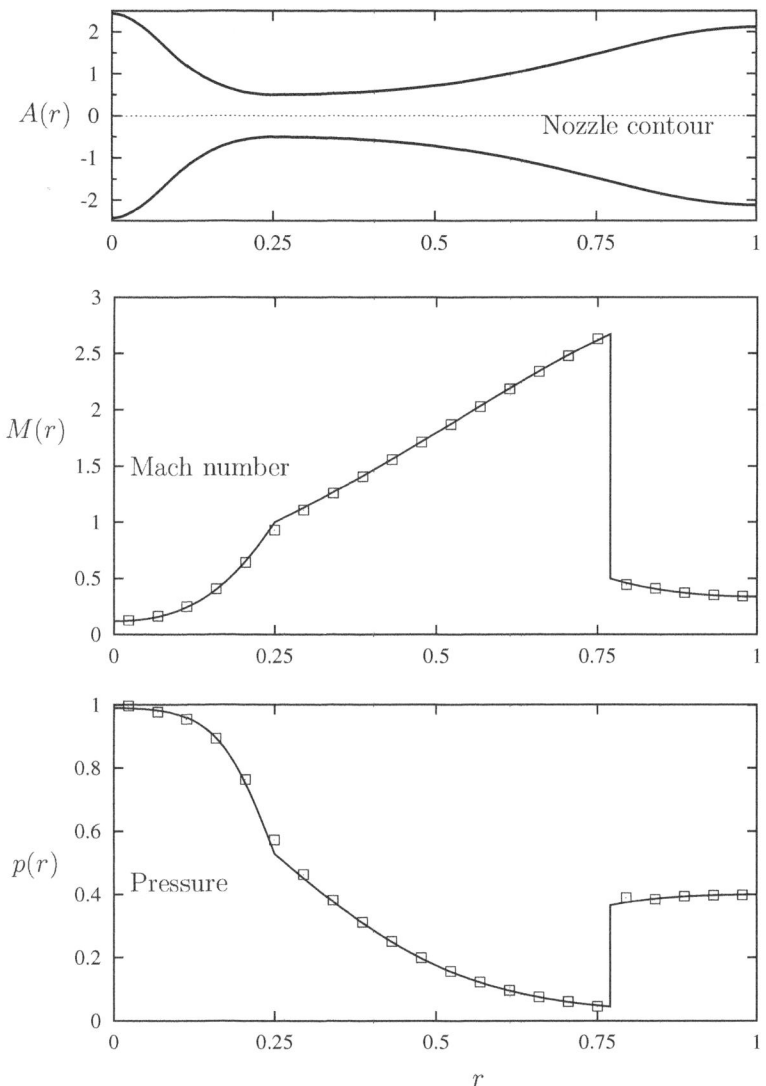

Figure 6.34. Large-time flow in Laval nozzle. Second-order (GRP) solution. $p_0 = 1$; $p_b = 0.4$. Nominal (isentropic) $M_{ex} = 3$. Solid line – steady solution; points – GRP (22 cells).

The Finite-Difference Solution

We seek a finite-difference solution to the IVP for the flow in the duct (6.11), with boundary conditions designed to produce a flow that approaches a steady state at large times. The initial data are discontinuous, as in a Riemann problem with the discontinuity at the throat point,

$$\mathbf{U}(r, 0) = \begin{cases} \mathbf{U}_{\mathrm{L}} = [\rho_0, \, p_0, \, 0], & r < 0.25, \\ \mathbf{U}_{\mathrm{R}} = [\rho_0(p_{\mathrm{b}}/p_0)^{1/\gamma}, \, p_{\mathrm{b}}, \, 0] & r > 0.25, \end{cases} \tag{6.14}$$

where p_{b} is the value designated for $p(1)$ at the steady-state solution [as in the preceding cases (A) and (B)]. Note that in both initial states the velocity is zero; in $\mathbf{U}_{\mathrm{L}}$ the values of pressure and density are those specified at the inlet ($r = 0$), and in $\mathbf{U}_{\mathrm{R}}$ the pressure has the value specified at the exit ($r = 1$). Both $\mathbf{U}_{\mathrm{L}}$ and $\mathbf{U}_{\mathrm{R}}$ lie on the same isentropic curve.

The boundary conditions are instrumental in obtaining a finite-difference solution that at large times approximates the exact steady flow in the nozzle. The main idea is to obtain the numerical fluxes at the boundary points as solutions to a suitable GRP, analogous to the treatment of regular ("internal") grid points. This is done by specifying a virtual initial state for the GRP at each boundary point, that is, at $r = 0-$ just "outside" the inlet, and at $r = 1+$ just "outside" the exit.

Consider first the inlet boundary point $r = 0$. Let M_{in} be the steady-flow Mach number at the inlet point [by (6.12) $M_{\mathrm{in}} = M(0) = 0.12$]. The steady-flow variables at the inlet are then given by

$$p_{\mathrm{in}} = p_0 \left(1 + \frac{\gamma - 1}{2} M_{\mathrm{in}}^2\right)^{-\frac{\gamma}{\gamma - 1}},$$

$$\rho_{\mathrm{in}} = \rho_0 \left(1 + \frac{\gamma - 1}{2} M_{\mathrm{in}}^2\right)^{-\frac{1}{\gamma - 1}}, \tag{6.15}$$

$$u_{\mathrm{in}} = M_{\mathrm{in}} \sqrt{\gamma p_{\mathrm{in}}/\rho_{\mathrm{in}}}.$$

Denote, respectively, by $\mathbf{U}_1$ and $\Delta \mathbf{U}_1$ the average state and linear variation in the cell adjacent to the inlet point (which we appropriately label as $1/2$). Let $\mathbf{U}_{1/2,-} = [\rho_{\mathrm{in}}, \, p_{\mathrm{in}}, \, u_{\mathrm{in}}]$, $\mathbf{U}_{1/2,+} = \mathbf{U}_1 - \frac{1}{2}\Delta \mathbf{U}_1$. The GRP at this point is then set up with the initial data

$$\mathbf{U}(r, 0) = \begin{cases} \mathbf{U}_{1/2,-}, & r < 0, \\ \mathbf{U}_{1/2,+} + r\,(\Delta \mathbf{U}_1/\Delta r), & r > 0, \end{cases} \tag{6.16}$$

where the corresponding initial slopes are taken as zero for the left state $(1/2, -)$ and $(\Delta \mathbf{U}_1/\Delta r)$ for the right state $(1/2, +)$.

Turning to the exit boundary condition, let the adjacent cell be designated as N. Let $\mathbf{U}_{N+1/2,-} = \mathbf{U}_N + \frac{1}{2}\Delta \mathbf{U}_N$, $\mathbf{U}_{N+1/2,+} = \left[\rho_{N+1/2,-}, \, p_b, \, u_{N+1/2,-}\right]$. The GRP at the exit point $N+1/2$ is then set up with the initial data

$$\mathbf{U}(r, 0) = \begin{cases} \mathbf{U}_{N+1/2,-} + (r-1)(\Delta \mathbf{U}_N/\Delta r), & r < 1, \\ \mathbf{U}_{N+1/2,+}, & r > 1. \end{cases} \quad (6.17)$$

In analogy to the inlet point, the initial slopes are zero for the right state and $(\Delta \mathbf{U}_N/\Delta r)$ for the left state. The only specified boundary condition here is the value of "background pressure" p_b imposed on the pressure $p(1)$; the density and velocity are simply extended by continuity across the boundary.

For the finite-difference computations, we deliberately selected a coarse grid, where the interval $[0 \leq r \leq 1]$ was divided into 22 equal cells. The integration was conducted with a time step $\Delta t = 0.009$, up to final time $t = 18$. This was done for the two previously specified cases (A) and (B), both producing approximately the same value of $\mu_{\text{CFL}} = 0.55$. In choosing the final time, we took notice of the nozzle "time of flow" t_{of}, that is, the time taken by a fluid particle to traverse the nozzle from inlet to exit at the steady flow. In case (A) we calculated from the exact solution $t_{\text{of}} = 1.31$, whereas in case (B) we found $t_{\text{of}} = 1.74$ (apparently because of the lower velocity downstream of the shock). Thus, the final computation time is $t > 10 \times t_{\text{of}}$, implying that a steady flow regime has probably been established in the nozzle by that time. Indeed, continuing the finite-difference computation for longer final time has produced flow profiles that were virtually identical to those obtained at $t = 18$.

Turning to the results for case (A), we first consider the Godunov scheme computation (Figure 6.31). Here the overall agreement with the exact solution is fairly good, except for the pressure in the interval $0 \leq r \leq 0.25$. In that interval, where the steady flow is subsonic, the computed pressure is overvalued whereas the Mach number is nearly correct. In the language of steady flow, this implies that the computed stagnation pressure is overvalued. The only explanation we can propose for this is the simple statement that the large error is due to performing a first-order computation with a coarse grid. (Indeed, with a grid of 66 cells and $\Delta t = 0.003$, the error decreased substantially, and the pressure profile became monotonically decreasing.)

When the computation of case (A) was repeated using the GRP scheme (Figure 6.32), a strikingly good agreement with the exact solution was obtained. This is a clear demonstration of the improved accuracy obtained by upgrading the scheme level of accuracy from first to second order. These results are particularly interesting, because in the converging part of the nozzle the flow is resolved by only 6 cells (more precisely 5.5), whereas the cross-sectional area ratio is about 4.9. We also note the discontinuity in flow gradient at the throat, which is well reproduced by the GRP solution.[9]

For case (B) (Figures 6.33 and 6.34), the most conspicuous feature here is that upstream of the shock wave the computed solutions are identical to the respective case (A) solutions. In the diverging part of the nozzle, the flow between the throat and the shock is resolved by a coarse grid of 12 cells. Good agreement is obtained by the Godunov scheme (Figure 6.33) and even better agreement is obtained by the GRP computation (Figure 6.34). The pressure profiles obtained by either scheme comply well with the exit boundary condition $p_b = 0.4$.

Notice that the shock in case (B) is captured with perfect "sharpness" (although not perfect accuracy). Obviously, this feature is related to the fact that here we are dealing with a steady flow and hence a *standing shock*. The Godunov or GRP schemes treat standing shocks particularly well, since at the cell interface separating the preshock and postshock states, the solution of the corresponding Riemann problem would be the exact shock wave if the states at the adjacent cells were accurately computed. This is not the case here, because the numerical error in smooth regions of the flow is generally nonvanishing. However, the error associated with the (second-order-accurate) GRP method is smaller than the respective error of the (first-order-accurate) Godunov method. Hence the former method replicates the exact solution better than the latter.

[9] It is noted that the flow variables at the cell whose midpoint coincides with the throat ($r = 0.25$) cannot approximate the values of the exact profiles with second-order accuracy, since the gradients of the exact solution are discontinuous at that point.

Part II

Numerical Implementation

7

From the GRP Algorithm to Scientific Computing

The GRP algorithm was developed in Part I and tested against a variety of analytical examples. The next challenge is to adapt it to the needs of "Scientific Computing," that is, to implement it in the simulation of problems of physical significance. This is a formidable task for any numerical algorithm. The problems encountered in applications usually are multidimensional and involve complex geometries and additional physical phenomena. In this chapter we first describe in more detail some of these problems, as a general background for the following chapters. We then proceed to the main goal of the chapter, namely, the upgrading of the GRP algorithm (which is essentially one dimensional) to a two-dimensional, second-order scheme. To this end we recall (Section 7.2) Strang's method of "operator splitting" and then (Section 7.3) apply it to the (planar) fluid-dynamical case. Although the method can be further extended to the three-dimensional case, we confine our presentation to the two-dimensional setup, which serves in our numerical examples (Chapters 8 and 10).

7.1 General Discussion

In Part I we studied the mathematical basis of the GRP method. We also considered a variety of numerical examples for which analytic (or at least asymptotic) solutions were available. This provided us with some measure of the accuracy of the computational results, and helped identifiy potential difficulties (such as the resolution of contact discontinuities). In the absence of any rigorous results concerning the convergence of the profiles to the exact solutions (at least in the case of systems–which is our main objective here), such test cases play a very important role in the construction of any numerical algorithm. To estimate the improvement achieved by this second-order scheme, we have systematically compared it to its first-order counterpart, the Godunov scheme (of which the GRP method is a natural extension, as noted in Sections 3.1 and 5.2). Although both schemes were quite successful in capturing the general features of wave patterns, the sharpness with which various discontinuities were resolved was substantially improved, thus justifying the transition from the first-order to the second-order scheme. However, the entire treatment of Part I (Section 3.3 excepted) was confined to the *one-dimensional* (or quasi-1-D) setting. Indeed, the very foundation of the Godunov (hence also the GRP) approach is the solution of the Riemann problem, which is of a one-dimensional nature. In contrast, the vast majority of physical or fluid-dynamical problems of interest deal with two- or even three-dimensional settings. Thus, the first step to be taken is the conversion of the GRP algorithm of Section 5.2 into a multidimensional scheme. The fundamental idea is that of "operator splitting," which has already been used in Section 3.3. It allows the construction of a multidimensional scheme by using alternately one-dimensional solvers ("sweeps") in directions parallel to the coordinate axes. Strang [105] has found a way of constructing such a scheme, which, in addition, is second-order accurate when this is true for its one-dimensional "factors." This is therefore exactly what we need to implement a second-order-accurate multidimensional scheme based on the (1-D) GRP solver. Section 7.2 is devoted to a discussion of Strang's method, in a rather general setting of time-dependent partial differential equations. In Section 7.3 we specialize to the case of two-dimensional compressible flow in the Cartesian plane. In principle, the method can be further extended to three-dimensional problems. However, since our examples (Chapters 8 and 10) are limited to the two-dimensional case we prefer to confine our discussion to this case. The flow equations are "natural candidates" for the operator-splitting method. Indeed, the spatial part decomposes as a sum of derivatives in the *x* and *y* directions ("divergence form"), so that the one-dimensional solvers bring us right back into the framework of Section 5.2. Invoking Strang's general construction we therefore obtain a second-order two-dimensional scheme. It enables us to solve flow problems in rectangular domains, with sides that are parallel to the coordinate axes. However, the vast majority of compressible flow problems of physical interest cannot be represented in the rectangular framework. Indeed, these problems often involve complex (stationary or moving) boundaries. If we imagine a Cartesian (rectangular) computational grid, the cells intersected by such a boundary assume irregular shapes. The resulting problem is (at least) twofold, involving, first, the geometric handling of those irregular cells and, second, the adaptation of the two-dimensional scheme in this case. This topic will be taken up in Chapter 8.

Problems of physical interest involve, in many cases, various significant mechanisms beyond the basic conservation laws of mass, momentum, and energy. Since this monograph deals only with nonlinear hyperbolic systems, which possess the property of "finite propagation speed" of waves, diffusive effects such as heat conduction or viscosity are excluded. Nonetheless, we deal in Chapter 9 with "reacting flows." The fluid in this case is subject to a chemical reaction that changes its composition, thus also changing its equation of state. Our model is perhaps the most elementary one, in which the fluid is a mixture of two species, "burnt" and "unburnt," and an exothermic reaction converts the unburnt component to the burnt one. This is a basic model for combustion, sufficient to describe "detonations" but not "deflagrations"; the latter involve a decrease of pressure and density in the burnt products and require effects such as heat conduction for their study. The basic building block in the numerical resolution of the reacting-flow system will again be a 1-D GRP solver, which deals with the *coupled* equations of compressible flow and chemical reaction. Indeed, the implementation of the 1-D GRP solver, either through spatial splitting or the addition of unknowns and equations, is a unifying theme in the handling of applications in this monograph.

Our final Chapter 10 deals with the (numerical) comparison of quasi-1-D and fully 2-D computations for a flow in a nozzle. The quasi-1-D approach was presented in Section 4.2 as a possible approximation to a flow that is nearly uniform on each cross section. Using the full capability of 2-D calculations, as developed in Chapter 8 (including the handling of rigid curvilinear boundaries by the MBT method; Section 8.3), we can try and validate this assumption. The test case studied in Chapter 10 illustrates the limitations of the quasi-1-D hypothesis.

7.2 Strang's Operator-Splitting Method

We consider here a solution ψ to an evolution equation of the type

$$\frac{\partial}{\partial t}\psi = c[\psi]. \tag{7.1}$$

Here $\psi(x, y, t)$ is a smooth function in the spatial coordinates $(x, y) \in \mathbb{R}^2$ and time $t \in [0, \infty)$. The operator $c[\psi]$ is a general nonlinear differential operator that may depend on x, y, t as well as on ψ, ψ_x, ψ_y, ψ_{xx}, ... (various spatial derivatives of ψ), but of course not on ψ_t. This dependence is in general nonlinear but smooth (as a function of all its variables). We now assume further that

$$c[\psi] = a[\psi] + b[\psi], \tag{7.2}$$

where $a[\psi]$, $b[\psi]$ are operators of the same type as $c[\psi]$. No further restrictions are imposed on a, b. In particular, a, b need not commute ($a[b[\psi]] \neq b[a[\psi]]$ in general). We now describe a method due to Strang [105], allowing us to obtain a second-order-accurate approximation to (7.1), when such approximations are already available for the "simpler" equations $\psi_t = a[\psi]$, $\psi_t = b[\psi]$.

We first address the notion of "order of accuracy" (see Definition 2.21) for the case at hand. To this end, we use the common notation

$$\partial^\alpha \psi = \left(\frac{\partial}{\partial x}\right)^{\alpha_1} \left(\frac{\partial}{\partial y}\right)^{\alpha_2} \psi,$$

where $\alpha = (\alpha_1, \alpha_2)$ is a "double index" of nonnegative integers, $\alpha_1, \alpha_2 = 0, 1, 2, \ldots$, and in particular $\psi = \partial^{(0,0)}\psi$. The operators a, b, c are functions of (x, y, t) and $\{\partial^\alpha \psi; \alpha \in J\}$, where J is a finite set of double indices, common to all three of them. For simplicity we designate this dependence as $a(\partial^\alpha \psi, x, y, t)$, suppressing J, where there is no risk of confusion. We indicate by z_α the argument corresponding to $\partial^\alpha \psi$ in the functional expressions.

Example 7.1 In Equation (3.49) we have $a(\psi) = -\frac{\partial}{\partial x} f(\psi)$, $b(\psi) = -\frac{\partial}{\partial y} g(\psi)$.

Differentiating Equation (7.1) with respect to t and using the chain rule we get

$$\psi_{tt} = \left(\frac{\partial}{\partial t}\right)^2 \psi = \frac{\partial}{\partial t} c\,(\partial^\alpha \psi, x, y, t) = c_t\,(\partial^\alpha \psi, x, y, t) + \sum_{\alpha \in J} \frac{\partial c}{\partial z_\alpha} \cdot \frac{\partial}{\partial t} \partial^\alpha \psi$$

$$= c_t\,(\partial^\alpha \psi, x, y, t) + \sum_{\alpha \in J} \frac{\partial c}{\partial z_\alpha} \cdot \partial^\alpha \left[c\,(\partial^\alpha \psi, x, y, t)\right], \quad (7.3)$$

where in the last step we have again used Equation (7.1). Note that $\frac{\partial c}{\partial z_\alpha}$ is evaluated at $(\partial^\alpha \psi, x, y, t)$.

We now fix a time step $k = \Delta t > 0$ and consider a discrete scheme C_k acting on functions of (x, y, t). The scheme is assumed to be "consistent" with Equation (7.1) in the sense that if $\psi(x, y, t)$ is a smooth solution then $C_k \psi(x, y, t)$ approximates $\psi(x, y, t + k)$. More precisely, as in Definition 2.21, we say that C_k is *second-order accurate* if

$$\psi(x, y, t + k) - C_k \psi(x, y, t) = O(k^3) \quad \text{as} \quad k \to 0. \tag{7.4}$$

---

[1] This equality means that, for any fixed t, a suitable norm [in (x, y)] of the left-hand side tends to zero as k^3, when $k \to 0$. Smoothness is needed if a pointwise norm is used. Compare with Footnote 12 in Section 5.2.

Taylor's theorem yields, in conjunction with (7.1) and (7.3),

$$\psi(x, y, t + k) = \psi(x, y, t) + k\psi_t(x, y, t) + \frac{k^2}{2}\psi_{tt}(x, y, t) + O(k^3)$$

$$= \psi(x, y, t) + kc\,(\partial^\alpha \psi, x, y, t) + \frac{k^2}{2}\left\{c_t\,(\partial^\alpha \psi, x, y, t)\right.$$

$$\left. + \sum_{\alpha \in J} \frac{\partial c}{\partial z_\alpha} \cdot \partial^\alpha \left[c\,(\partial^\alpha \psi, x, y, t)\right]\right\} + O(k^3). \tag{7.5}$$

We now compare (7.4) and (7.5). Like the operator c, the discrete approximation C_k may also depend explicitly on time (in addition to $\partial^\alpha \psi$, x, y) and we make this explicit by writing $C_k = C_k(t)$. The condition (7.4) for second-order accuracy can now be written as

$$C_k(t)\psi(x, y, t) = \psi(x, y, t) + kc\,(\partial^\alpha \psi, x, y, t) + \frac{k^2}{2}\left\{c_t\,(\partial^\alpha \psi, x, y, t)\right.$$

$$\left. + \sum_{\alpha \in J} \frac{\partial c}{\partial z_\alpha} \cdot \partial^\alpha \left[c\,(\partial^\alpha \psi, x, y, t)\right]\right\} + O(k^3). \tag{7.6}$$

Let us now suppose that $A_k(t)$ [resp. $B_k(t)$] is a second-order-accurate scheme for $\psi_t = a\,(\partial^\alpha \psi, x, y, t)$ [resp. $\psi_t = b\,(\partial^\alpha \psi, x, y, t)$]. Then Equation (7.6) is satisfied with $C_k(t)$ replaced by $A_k(t)$ [resp. $B_k(t)$] and c replaced by a [resp. b].

Before stating Strang's result we clarify the meaning of the time dependence of $C_k(t)$. It is related to the *explicit* dependence of $c\,(\partial^\alpha \psi, x, y, t)$ on the last variable t. Thus Equation (7.6) yields

$$C_k\left(t + \frac{k}{2}\right)\psi(x, y, t) = \psi(x, y, t) + kc\left(\partial^\alpha \psi(x, y, t), x, y, t + \frac{k}{2}\right)$$

$$+ \frac{k^2}{2}\left\{c_t\left(\partial^\alpha \psi(x, y, t), x, y, t + \frac{k}{2}\right) + \sum_{\alpha \in J} \frac{\partial c}{\partial z_\alpha}\right.$$

$$\left. \times \partial^\alpha \left[c\left(\partial^\alpha \psi(x, y, t), x, y, t + \frac{k}{2}\right)\right]\right\} + O(k^3).$$

$$\tag{7.7}$$

Observe that the time step here is k; it is used as the index labeling A_k, B_k, C_k. Also, the arguments in $\frac{\partial c}{\partial z_\alpha}$ are $\left(\partial^\alpha \psi(x, y, t), x, y, t + \frac{k}{2}\right)$. We can now state the following theorem by Strang [105].

Theorem 7.2 *Let* A_k, B_k *be, respectively, second-order-accurate approxima-tions for* $\psi_t = a[\psi]$, $\psi_t = b[\psi]$. *Then the product*

$$C_k(t) = B_{\frac{k}{2}}\left(t + \frac{k}{2}\right) A_k(t) B_{\frac{k}{2}}(t) \tag{7.8}$$

is a second-order approximation to $\psi_t = c[\psi]$, *where* c *satisfies (7.2).*

Proof The proof is just a computation aimed at verifying (7.6). We apply (7.6) first to $A_k \tilde\psi$, with $\tilde\psi(x, y, t) = B_{\frac{k}{2}}(t)\psi(x, y, t)$, and then to $B_{\frac{k}{2}}(t)\psi$. For no-tational simplicity, we omit the arguments (x, y, t) of ψ, $\tilde\psi$, $\partial^\alpha \psi$, and $\partial^\alpha \tilde\psi$. Then

$$A_k(t)B_{\frac{k}{2}}(t)\psi = \tilde\psi + ka\left(\partial^\alpha \tilde\psi, x, y, t\right) + \frac{k^2}{2}\left\{ a_t\left(\partial^\alpha \tilde\psi, x, y, t\right) + \sum_{\alpha \in J} \frac{\partial a}{\partial z_\alpha} \right.$$

$$\left. \times \partial^\alpha \left[a\left(\partial^\alpha \tilde\psi, x, y, t\right)\right]\right\} + O(k^3) = \psi + \frac{k}{2}b\left(\partial^\alpha \psi, x, y, t\right)$$

$$+ \frac{k^2}{8}\left\{ b_t\left(\partial^\alpha \psi, x, y, t\right) + \sum_{\alpha \in J} \frac{\partial b}{\partial z_\alpha} \cdot \partial^\alpha \left[b\left(\partial^\alpha \psi, x, y, t\right)\right]\right\}$$

$$+ k\left\{ a\left(\partial^\alpha \psi, x, y, t\right) + \sum_{\alpha \in J} \frac{\partial a}{\partial z_\alpha} \cdot \left(\partial^\alpha \tilde\psi - \partial^\alpha \psi\right)\right\}$$

$$+ \frac{k^2}{2}\left\{ a_t\left(\partial^\alpha \psi, x, y, t\right) + \sum_{\alpha \in J} \frac{\partial a}{\partial z_\alpha} \cdot \partial^\alpha \left[a\left(\partial^\alpha \psi, x, y, t\right)\right]\right\}$$

$$+ O(k^3). \tag{7.9}$$

Observe that in the first equality the arguments of $\frac{\partial a}{\partial z_\alpha}$ (like those of a) are $\left(\partial^\alpha \tilde\psi, x, y, t\right)$, whereas in the second equality the arguments, for both $\frac{\partial a}{\partial z_\alpha}$ and $\frac{\partial b}{\partial z_\alpha}$, are $(\partial^\alpha \psi, x, y, t)$. We have, by $\tilde\psi = B_{\frac{k}{2}}(t)\psi$, $\partial^\alpha \left(\tilde\psi - \psi\right) = \frac{k}{2}\partial^\alpha \left[b\left(\partial^\alpha \psi, x, y, t\right)\right] + O(k^2)$, and inserting this in (7.9) and collecting terms we get

$$A_k(t)B_{\frac{k}{2}}(t)\psi = \psi + k\left\{ \frac{1}{2}b\left(\partial^\alpha \psi, x, y, t\right) + a\left(\partial^\alpha \psi, x, y, t\right)\right\}$$

$$+ \frac{k^2}{2}\left\{ \frac{1}{4}b_t\left(\partial^\alpha \psi, x, y, t\right) + \sum_{\alpha \in J} \left(\frac{1}{4}\frac{\partial b}{\partial z_\alpha} + \frac{\partial a}{\partial z_\alpha}\right)\right.$$

$$\times \partial^\alpha \left[b\left(\partial^\alpha \psi, x, y, t\right)\right] + a_t\left(\partial^\alpha \psi, x, y, t\right)$$

$$\left. + \sum_{\alpha \in J} \frac{\partial a}{\partial z_\alpha} \cdot \partial^\alpha \left[a\left(\partial^\alpha \psi, x, y, t\right)\right]\right\} + O(k^3). \tag{7.10}$$

Finally, applying $B_{\frac{k}{2}}\left(t + \frac{k}{2}\right)$ to (7.10) we obtain

$$C_k(t)\psi = B_{\frac{k}{2}}\left(t + \frac{k}{2}\right) A_k(t) B_{\frac{k}{2}}(t)\psi$$

$$= \psi + k \left\{ \frac{1}{2} b\left(\partial^\alpha \psi, x, y, t\right) + a\left(\partial^\alpha \psi, x, y, t\right)\right\} + \frac{k^2}{2} \left\{ \frac{1}{4} b_t\left(\partial^\alpha \psi, x, y, t\right)\right.$$

$$+ \sum_{\alpha \in J}\left(\frac{1}{4}\frac{\partial b}{\partial z_\alpha} + \frac{\partial a}{\partial z_\alpha}\right) \partial^\alpha\left[b\left(\partial^\alpha \psi, x, y, t\right)\right] + a_t\left(\partial^\alpha \psi, x, y, t\right)$$

$$+ \sum_{\alpha \in J}\frac{\partial a}{\partial z_\alpha} \cdot \partial^\alpha\left[a\left(\partial^\alpha \psi, x, y, t\right)\right] \right\} + \frac{k}{2} b\left(\partial^\alpha \psi + k\partial^\alpha\left[\frac{1}{2} b\left(\partial^\alpha \psi, x, y, t\right)\right.\right.$$

$$\left. + a\left(\partial^\alpha \psi, x, y, t\right)\right], x, y, t + \frac{k}{2}\right) + \frac{k^2}{8}\left\{b_t\left(\partial^\alpha \psi, x, y, t + \frac{k}{2}\right)\right.$$

$$\left.\left. + \sum_{\alpha \in J}\frac{\partial b}{\partial z_\alpha} \cdot \partial^\alpha\left[b\left(\partial^\alpha \psi, x, y, t + \frac{k}{2}\right)\right]\right\} + O(k^3)\right.$$

$$= \psi + k\left\{a\left(\partial^\alpha \psi, x, y, t\right) + b\left(\partial^\alpha \psi, x, y, t\right)\right\}$$

$$+ \frac{k^2}{2}\left\{a_t\left(\partial^\alpha \psi, x, y, t\right) + b_t\left(\partial^\alpha \psi, x, y, t\right) + \sum_{\alpha \in J}\left(\frac{\partial a}{\partial z_\alpha} + \frac{\partial b}{\partial z_\alpha}\right)\right.$$

$$\left. \times \partial^\alpha\left[a\left(\partial^\alpha \psi, x, y, t\right) + b\left(\partial^\alpha \psi, x, y, t\right)\right]\right\} + O(k^3), \tag{7.11}$$

which confirms the claim of the theorem in view of (7.6) (with $c = a + b$).   □

In typical applications to fluid dynamics, as well as to practically all equations of physical origin, the operators a, b, c do not depend explicitly on time; they model phenomena that are invariant under a translation $t \to t + \beta$, $\beta = constant$ of the time scale. We shall henceforth assume that this is the case, so that the approximation operators A_k, B_k, C_k are also independent of t. The statement of Theorem 7.2 can then be simplified to say that if A_k, B_k are second-order accurate, then so is $C_k = B_{\frac{k}{2}} A_k B_{\frac{k}{2}}$. In fact, we can go a little further in this case.

Corollary 7.3 *If A_k, B_k, C_k are independent of t, then the approximation*

$$\widetilde{C}_{2k} = B_{\frac{k}{2}} A_k B_k A_k B_{\frac{k}{2}} \tag{7.12}$$

is second-order accurate.

Proof Clearly $C_k^2 = B_{\frac{k}{2}} A_k B_{\frac{k}{2}} B_{\frac{k}{2}} A_k B_{\frac{k}{2}}$ is second-order accurate. Indeed, applying C_k to Equation (7.4) we have $\psi(x, y, t + 2k) - C_k^2 \psi(x, y, t) = O(k^3)$. The point is that we want to replace the middle term $B_{\frac{k}{2}} B_{\frac{k}{2}}$ by B_k. But the second-order accuracy of B_k yields, as for C_k^2, $B_k \psi(x, y, t) - B_{\frac{k}{2}}^2 \psi(x, y, t) = O(k^3)$, which implies that $\left(C_k^2 - \widetilde{C}_{2k}\right) \psi = O(k^3)$. $\square$

Remark 7.4 The same method of "grouping together" the middle terms $B_{\frac{k}{2}} B_{\frac{k}{2}}$ can be repeated, thus leading to the following approximating scheme for the solution $\psi(x, y, t + nk)$:

$$\widetilde{C}_{nk} \, \psi(x, y, t) = B_{\frac{k}{2}} \, (A_k B_k)^{n-1} A_k B_{\frac{k}{2}} \, \psi(x, y, t). \tag{7.13}$$

This means that, apart from initial and final applications of $B_{\frac{k}{2}}$, the scheme consists of alternating applications of A_k and B_k, thus reducing its overall complexity. In fact, this is the scheme used in Section 3.3, where A_k, B_k, C_k correspond, respectively, to $H_{\Delta t}^x$, $H_{\Delta t}^y$, $H_{\Delta t}$ [see Equation (3.52)].

Remark 7.5 Observe that $\widetilde{C}_{nk}$ is slightly "nonsymmetric" with respect to the alternating operators A_k, B_k. In fact, there are a large variety of schemes, constructed from A_k, B_k, that retain second-order accuracy. An example that can easily be verified by the reader is given by the symmetrized form

$$\widetilde{\widetilde{C}}_k = \frac{1}{2} \left(A_k B_k + B_k A_k\right). \tag{7.14}$$

Note that the approximation $\left(\widetilde{\widetilde{C}}_k\right)^n$ requires $4n$ applications of either A_k or B_k, roughly twice as many as needed for $\widetilde{C}_{nk}$.

We refer to Strang [105], Gottlieb [58], and Teng [109] for more detailed discussions of such schemes.

Definition 7.6 ("Split scheme") An "operator-splitting method" is any method used to construct an approximating scheme to $\psi_t = c[\psi] = a[\psi] + b[\psi]$ using schemes associated with the simpler problems $\psi_t = a[\psi]$, $\psi_t = b[\psi]$. A scheme constructed by this method is referred to as a "split scheme."

The terms "fractional step method" or "time-splitting method" are also used by some authors.

In the treatment of 2-D fluid dynamical problems (using conservation laws) a natural "splitting" is obtained by taking a, b to be the one-dimensional (x or y) terms. This has already been done in Section 3.3 [see Example 7.1

and Equation (3.51). In our 2-D extension of the GRP method we always use this "dimensional" splitting. The 1-D approximating schemes are those presented in Section 5.2. A more detailed discussion of this methodology is given in Section 7.3. The split scheme used in our treatment is always of the type $\widetilde{C}_{nk}$ [see (7.13)]. It has the advantage of being purely "multiplicative", that is, it does not involve any addition of operators such as in (7.14). Thus, its (linearized) stability condition can be inferred from the corresponding conditions for the x and y schemes (compare Remark 3.17 in Section 3.3). It turns out to be less restrictive than that of scheme (7.14). We refer to Strang [105] for a further discussion of this topic.

The most important issue related to the construction of an approximating scheme is naturally that of its convergence to the exact solution in a suitable norm (see, e.g., Definition 2.25 in Section 2.2). In the case of a split scheme, its convergence is not obvious even when the two components A_k, B_k converge to the respective exact solutions. Take, for example, the case of the scalar equation (3.49), and let the split scheme $H_{\Delta t}$ be given by (3.52) [namely, (7.13)], where each of the components $H_{\Delta t}^x$, $H_{\Delta t}^y$ is the Godunov scheme. In view of Theorem 3.6 in Section 3.1, each component converges to the respective one-dimensional equation in (3.51). The fact that the split scheme $H_{\Delta t}$ (which is clearly only first-order accurate) indeed converges to the exact (entropy) solution to (3.49) follows from the work of Crandall and Majda [32, 33]. Since not much is known concerning the convergence of 1-D second-order-accurate schemes, we can expect less for a scheme like (7.13), even in the scalar case.

We conclude this general treatment by pointing out the connection between split schemes and "product formulas." Indeed, the scheme (7.13) consists of an "alternating" product of operators approximating (for "short time") the solutions of $\psi_t = a[\psi]$, $\psi_t = b[\psi]$. One can then ask what happens if these approximations are replaced by the *exact* solutions of the two equations. As an example, consider the simple case of a system of ordinary (constant coefficient) differential equations,

$$\frac{d}{dt}\psi = (A + B)\psi, \qquad \psi(0) = \psi_0, \qquad (7.15)$$

where $\psi(t) \in \mathbb{R}^n$ and A, B are real $n \times n$ matrices. As is well known, the solution is given by $\psi(t) = \exp[t(A + B)]\psi_0$. In contrast, the alternating product of the solution operators for A, B, using $k = \frac{t}{n}$, is

$$C_n\psi_0 = \left[\exp\left(\frac{t}{n}A\right)\exp\left(\frac{t}{n}B\right)\right]^n\psi_0.$$

If $AB \neq BA$, then $C_n\psi_0 \neq \psi(t)$. However, the well-known theorem by Lie states that $C_n\psi_0 \to \psi(t)$ as $n \to \infty$, for every fixed t. There are also theorems generalizing this "product formula." We refer to Chorin et al. [27] for a general survey of such results.

The structure of the split scheme (7.13) can be described in terms of two steps. First, the full solution operator is represented as (a limit of) a product of "short-term" solution operators of the two "partial" equations. Second, the short-term operators are replaced by suitable approximations. This description illustrates the difficulty involved in proving the convergence of the scheme.

7.3 Two-Dimensional Flow in Cartesian Coordinates

We consider here a split scheme for the numerical integration of the Euler equations of compressible, inviscid flow in the (x, y) plane. The "natural" dimensional splitting is used, whereby the equations are integrated alternately in the x and y directions. This has already been done in Section 3.3, in the case of a scalar conservation law.

The Euler equations governing the flow are derived from the three basic conservation laws (mass, momentum and energy), as in Section 4.2. We have two components (u, v) of velocity, so that the planar system (4.47) is now replaced by

$$\frac{\partial}{\partial t}\mathbf{U} + \frac{\partial}{\partial x}\mathbf{H}(\mathbf{U}) + \frac{\partial}{\partial y}\mathbf{L}(\mathbf{U}) = \mathbf{0},$$

$$\mathbf{U} = \begin{bmatrix} \rho \\ \rho u \\ \rho v \\ \rho E \end{bmatrix}, \qquad \mathbf{H}(\mathbf{U}) = \begin{bmatrix} \rho u \\ p + \rho u^2 \\ \rho u v \\ (\rho E + p)u \end{bmatrix}, \qquad \mathbf{L}(\mathbf{U}) = \begin{bmatrix} \rho v \\ \rho u v \\ p + \rho v^2 \\ (\rho E + p)v \end{bmatrix},$$

$$\mathbf{U}(x, y, 0) = \mathbf{U}_0(x, y). \tag{7.16}$$

The flow variables ρ (density), p (pressure), and E (total energy per unit mass) are all functions of x, y, t, as are the two velocity components. The total energy is given by $E = e + \frac{1}{2}(u^2 + v^2)$, where e is the specific internal energy. An equation of state $p = p(\rho, e)$ is given. In the majority of numerical applications (certainly those used as "test cases") the equation of state is assumed to be that of a perfect (γ-law) gas [see Equation (4.104)].

The Euler equations (7.16) cover a very rich panoply of fluid dynamical phenomena. Wave interactions lead to complex patterns involving triple points, Mach stems, and more (see Courant and Friedrichs [30]). We recall that this complexity was manifested even in the case of a scalar equation [see Equation (3.66)] with relatively simple (sector-wise constant) initial data.

The notion of a weak solution can be defined in the case of Equations (7.16), in analogy with the quasi-1-D case [see Equation (4.73)]. This leads to "jump conditions" across "surfaces of discontinuity," which are now moving curves in the (x, y) plane. These jump conditions are identical to the Rankine–Hugoniot condition [Equation (4.76)] when expressed in terms of the normal (to the surface) components of the velocity. From the mathematical viewpoint, very little is available in terms of rigorous results for solutions of the Euler system. This is true even for "Riemann-type" IVPs, namely, when the initial data are sector-wise constant (as in the scalar case of Section 3.3). Fortunately, numerical simulations have been very successful in reproducing laboratory experiments in a variety of nontrivial cases (e.g., Hillier [61], Falcovitz, Alfandary, and Ben-Dor [38], Sasoh, Maemura, Hirataka, Falcovitz, and Takayama [98], Igra, Falcovitz, Meguro, Takayama, and Heilig [66], Igra, Falcovitz, Reichenbach, and Heilig [67]).

We now turn to the split scheme for (7.16). As in the general treatment of Section 7.2, we construct separate schemes for the two equations,

$$\frac{\partial}{\partial t}\mathbf{U} + \frac{\partial}{\partial x}\mathbf{H}(\mathbf{U}) = 0, \tag{7.17}$$

$$\frac{\partial}{\partial t}\mathbf{U} + \frac{\partial}{\partial y}\mathbf{L}(\mathbf{U}) = 0, \tag{7.18}$$

where $\mathbf{U}(x, y, 0) = \mathbf{U}_0(x, y)$ is given. Observe that in the "x system" (7.17) the third equation (for v) can be written as

$$v_t + uv_x = 0, \tag{7.19}$$

by making use of the first equation $\rho_t + (\rho u)_x = 0$. Thus, the y component of the velocity is "passively" convected in the x direction, with speed u. The system can be viewed as a planar compressible system to which an "unknown" v is added and advected along the particle paths $\frac{dx}{dt} = u$. We shall encounter a similar extension in Chapter 9, where the added unknown is a "mass fraction" of unburnt gas. Even though such an extension remains very close to (4.47), the (three-component) system of planar flow, there are certain points that need to be clarified. We do this next, before resuming the description of the two-dimensional split scheme.

The Linear GRP for a Planar System with Advection

The system (7.17) conforms to the general abstract setting (4.1), so that the analytical framework of Section 4.1 is applicable to its study. However, its analysis can be practically reduced to that of the planar system (4.47). This is done by separating the equation for ρv from the other three equations. To do this we need to modify the energy equation, since $E = e + \frac{1}{2}(u^2 + v^2)$ depends on v. Note that by using the equation for ρv and its alternative form (7.19) we get

$$\frac{\partial}{\partial t}(\rho v^2) + \frac{\partial}{\partial x}(\rho v^2 u) = v\left[\frac{\partial}{\partial t}(\rho v) + \frac{\partial}{\partial x}(\rho v u)\right]$$

$$+ \rho v\left[\frac{\partial v}{\partial t} + u\frac{\partial v}{\partial x}\right] = 0, \qquad (7.20)$$

which implies that in the energy equation we can substitute $E^x = e + \frac{1}{2}u^2$ for E. Thus, defining

$$\mathbf{U}^x = (\rho, \ \rho u, \ \rho E^x)^T, \qquad \mathbf{H}^x(\mathbf{U}^x) = (\rho u, \ p + \rho u^2, \ (\rho E^x + p)u)^T, \quad (7.21)$$

we can rewrite the system (7.17) as

$$\frac{\partial}{\partial t}\mathbf{U}^x + \frac{\partial}{\partial x}\mathbf{H}^x(\mathbf{U}^x) = \mathbf{0}. \qquad (7.22)$$

$$\frac{\partial}{\partial t}(\rho v) + \frac{\partial}{\partial x}(\rho u v) = 0. \qquad (7.23)$$

The system (7.22) is independent of v and is in fact identical to the planar system (4.47). Its eigenvalues are [see (4.64)]

$$\lambda_1 = u - c, \qquad \lambda_2 = u, \qquad \lambda_3 = u + c, \qquad (7.24)$$

and the procedure of solving its Riemann problem is given in Construction 4.46. The solution to the linear GRP is outlined in Summary 5.29. Equation (7.23) is tied to the system (7.22) through the velocity field u. When written in the form (7.19) it yields the eigenvalue $\lambda_4 = u$, whereby the equation serves as its characteristic relation, as in the abstract setting of Equation (4.6).

Remark 7.7 Observe the full analogy between Equations (4.59) and (7.19). Repeating the proof of Claim 4.30 we see that the λ_4 family is also linearly degenerate. Since $\lambda_2 = \lambda_4$ the full system (7.17) is "doubly linearly degenerate."

In accordance with the discussion following Equation (4.77), the jumps of v can take place only across a contact discontinuity. The solution for v in the context of the linear GRP [to the system (7.17)] is thus reduced to the scalar case, as in Construction 3.10. We summarize these considerations in the following corollary. Note that for simplicity we omit the dependence on y (that serves as a parameter in the treatment of Equation (7.22)).

Corollary 7.8 [RP and the linear GRP for the system (7.17)] *Let the initial data* $\mathbf{U}(x, 0)$ *be given by*

$$\mathbf{U}(x, 0) = \begin{cases} \mathbf{U}_\mathrm{L} + x\mathbf{U}'_\mathrm{L}, & x < 0, \\ \mathbf{U}_\mathrm{R} + x\mathbf{U}'_\mathrm{R}, & x > 0, \end{cases} \tag{7.25}$$

where $\mathbf{U}_\mathrm{L}$, $\mathbf{U}_\mathrm{R}$, $\mathbf{U}'_\mathrm{L}$, $\mathbf{U}_\mathrm{R}$ *are constant (four-component) vectors.*

Let $\mathbf{U}(x, t)$ *be the solution to (7.17) and let [see Equations (5.3) and (5.4)]*

$$\mathbf{U}_0 = \lim_{t \to 0+} \mathbf{U}(0, t), \qquad \left(\frac{\partial}{\partial t} \mathbf{U} \right)_0 = \lim_{t \to 0+} \frac{\partial}{\partial t} \mathbf{U}(0, t).$$

Then:

(I) *The values* $\mathbf{U}_0^x$, $\left(\frac{\partial}{\partial t} \mathbf{U}^x \right)_0$ *are derived from the planar system (7.22), namely, Construction 4.46 (for* $\mathbf{U}_0^x$*) and Summary 5.29 [for* $\left(\frac{\partial}{\partial t} \mathbf{U}^x \right)_0$ *].*

(II) *Let* u^* *be the initial speed of the contact discontinuity. We have, by (7.19),*

$$\left(v_0, \left(\frac{\partial v}{\partial t} \right)_0 \right) = \begin{cases} (v_\mathrm{L}, -u_\mathrm{L} v'_\mathrm{L}), & if \quad u^* > 0, \\ (v_\mathrm{R}, -u_\mathrm{R} v'_\mathrm{R}), & if \quad u^* < 0. \end{cases} \tag{7.26}$$

Note that if $u^* = 0$ then v is discontinuous across $x = 0$, but the flux $\rho v u^*$ vanishes.

The Split Scheme for (7.16)

The numerical simulation is always performed on a rectangular grid $\left\{ x_{j+1/2}, \ y_{k+1/2} \right\}_{j,k}$. For simplicity we assume a fixed mesh size, $x_{j+1/2} - x_{j-1/2} = \Delta x$, $y_{k+1/2} - y_{k-1/2} = \Delta y$. As in the 1-D case (Section 5.2) we take the grid lines $x = x_{j+1/2}$, $y = y_{k+1/2}$ to be the lines of jump discontinuities (for flow variables and their slopes). The computational cell $(x_{j-1/2}, x_{j+1/2}) \times (y_{k-1/2}, y_{k+1/2})$ is labeled as "cell (j, k)." It is centered at $(x_j, y_k) = \frac{1}{2}(x_{j-1/2} + x_{j+1/2}, \ y_{k-1/2} + y_{k+1/2})$. At time t_n, the average values $\mathbf{U}_{j,k}^n$ are given, as well as the constant slopes $(\mathbf{U}_x)_{j,k}^n$, $(\mathbf{U}_y)_{j,k}^n$. Given a time step Δt, we let $A_{\Delta t}^x$, $B_{\Delta t}^y$ be the GRP schemes for (7.17), (7.18) respectively. As already noted, the scheme $A_{\Delta t}^x$ approximates the system (7.22), (7.23) while $B_{\Delta t}^y$ approximates the analogous system in the y direction. Each of them can be selected to be of the E_1 (Definition 5.37) or the E_∞ type (Definition 5.41), or any other intermediate scheme. In each direction the scheme uses the cell averages and the slopes in that direction as the initial data for a 1-D computation.

Let us discuss the numerical procedure in some more detail. The implementation of $A_{\Delta t}^x$ is carried out row by row, fixing k and "sweeping" over j, and

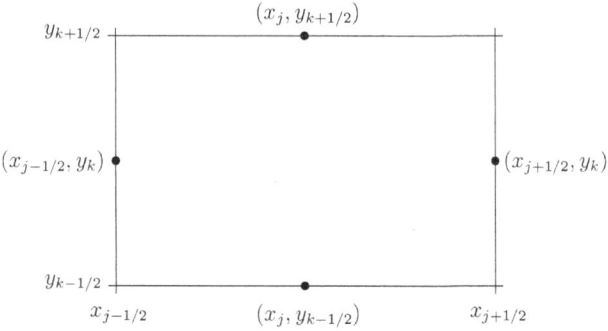

Figure 7.1. Setup for the Euler split scheme.

then repeating the process with k replaced by $k + 1$. Thus, in cell (j, k) (see Figure 7.1) the only numerical fluxes needed at this stage are $\mathbf{H}^{n+1/2}_{j-1/2,k}$, $\mathbf{H}^{n+1/2}_{j+1/2,k}$, corresponding to the two vertical sides. The fluxes corresponding to the three components $\mathbf{H}^x(\mathbf{U}^x)$ [see (7.21)] are computed, in view of Corollary 7.8, by the 1-D GRP algorithm [Equation (5.110)]. Next, the values $v^{n+1/2}_{j\pm1/2,k}$ are obtained as in (7.26), and the numerical fluxes for $\rho u v$ are evaluated by

$$(\rho u v)^{n+1/2}_{j\pm1/2,k} = (\rho u)^{n+1/2}_{j\pm1/2,k}\, v^{n+1/2}_{j\pm1/2,k}.$$

Equation (7.17) is now discretized as

$$\widetilde{\mathbf{U}}^{n+1}_{j,k} = \mathbf{U}^n_{j,k} - \frac{\Delta t}{\Delta x}\left(\mathbf{H}^{n+1/2}_{j+1/2,k} - \mathbf{H}^{n+1/2}_{j-1/2,k}\right). \tag{7.27}$$

Additionally, the new x slopes $\left(\widetilde{\mathbf{U}}_x\right)^{n+1}_{j,k}$ are obtained by the 1-D procedure, including the appropriate application of the slope limiter [see Equations (5.108) and (5.109)].

The "intermediate" cell averages $\widetilde{\mathbf{U}}^{n+1}_{j,k}$ serve as initial data for the application of $B^y_{\Delta t}$. Observe that the slopes $\left(\widetilde{\mathbf{U}}_y\right)^{n}_{j,k}$ must be subjected to the monotonicity requirement with respect to the new averages $\widetilde{\mathbf{U}}^{n+1}_{j,k}$. Doing this we obtain the slopes $\left(\widetilde{\mathbf{U}}_y\right)^{n+1}_{j,k}$ and the y scheme $B^y_{\Delta t}$ can be invoked to get the updated averages $\mathbf{U}^{n+1}_{j,k}$ and slopes $\left(\mathbf{U}_x\right)^{n+1}_{j,k}$, $\left(\mathbf{U}_y\right)^{n+1}_{j,k}$.

The full scheme approximating (7.16) is given by

$$C_{\Delta t} = B^y_{\Delta t}\, A^x_{\Delta t}. \tag{7.28}$$

This scheme is of the type (7.13). As noted in Remark 7.4 we must apply $A^x_{\Delta t}\, B^y_{\frac{\Delta t}{2}}$ initially at $t = 0$ and $B^y_{\frac{\Delta t}{2}}$ at the end of the integration ($T = n\Delta t$).

The Split Scheme and Conservation Form

The full (2-D) system is in conservation form. This means that it can be recast in the form of a balance equation, analogous to Equation (2.3). Integrating over a space–time box $[x_1, x_2] \times [y_1, y_2] \times [t_1, t_2]$ we get

$$
\int_{y_1}^{y_2} \int_{x_1}^{x_2} [\mathbf{U}(x, y, t_2) - \mathbf{U}(x, y, t_1)] \, dx \, dy = - \int_{t_1}^{t_2} \int_{y_1}^{y_2} [\mathbf{H}(\mathbf{U}(x_2, y, t))
$$

$$
- \; \mathbf{H}(\mathbf{U}(x_1, y, t))] \, dy \, dt - \int_{t_1}^{t_2} \int_{x_1}^{x_2} [\mathbf{L}(\mathbf{U}(x, y_2, t)) - \mathbf{L}(\mathbf{U}(x, y_1, t))] \, dx \, dt.
$$

$$(7.29)$$

In designing the numerical scheme for the one-dimensional cases we have always been guided by the principle of "conservative differencing," namely, compliance with the balance equation [see Equations (3.1) and (5.88)]. Imposing the same principle on the two-dimensional equation (7.29), with

$$
[x_1, x_2] \times [y_1, y_2] = [x_{j-1/2}, x_{j+1/2}] \times [y_{k-1/2}, y_{k+1/2}] \text{ and } [t_1, t_2] = [t_n, t_{n+1}],
$$

we should seek numerical fluxes $\mathbf{H}_{j\pm1/2,k}^{n+1/2}$, $\mathbf{L}_{j,k\pm1/2}^{n+1/2}$ (referring to the setup in Figure 7.1) such that

$$
\mathbf{U}_{j,k}^{n+1} - \mathbf{U}_{j,k}^{n} = -\frac{\Delta t}{\Delta x} \left(\mathbf{H}_{j+1/2,k}^{n+1/2} - \mathbf{H}_{j-1/2,k}^{n+1/2} \right) - \frac{\Delta t}{\Delta y} \left(\mathbf{L}_{j,k+1/2}^{n+1/2} - \mathbf{L}_{j,k-1/2}^{n+1/2} \right).
$$

$$(7.30)$$

This discrete scheme seems to be consistent with (7.28), where the x step $A_{\Delta t}^x$ is given by (7.27), with a similar expression for $B_{\Delta t}^y$. However, in the case of (7.28), the step $B_{\Delta t}^y$ uses the *intermediate* values $\widetilde{\mathbf{U}}_{j,k}^{n+1}$, $(\widetilde{\mathbf{U}}_y)_{j,k}^{n+1}$ as the *initial data* in the one-dimensional second step leading to $\mathbf{U}_{j,k}^{n+1}$. Thus, the one-dimensional steps $A_{\Delta t}^x$, $B_{\Delta t}^y$ use *different* initial data [in cell (j, k)] while carrying out the computation in the time interval $[t_n, t_{n+1}]$. This is not what is "naturally" expected of the numerical fluxes in (7.30). In general when these fluxes are all derived from the same discrete data (at time $t = t_n$) we say that the scheme is of the "finite-volume" class (see Kröner [72]). This means that the values $\mathbf{U}_{j,k}^n$ can be regarded as *volume averages*, rather than point values, and the scheme (7.30) serves as a discrete analog to the integral ("balance") equation (7.29) rather than the differential system (7.16). Thus, for (7.30) to be a "finite-volume" scheme we need to compute $\mathbf{H}_{j\pm1/2,k}^{n+1/2}$, $\mathbf{L}_{j,k\pm1/2}^{n+1/2}$ on the basis of the data $\mathbf{U}_{j,k}^n$, $(\mathbf{U}_x)_{j,k}^n$,

$(\mathbf{U}_y)^n_{j,k}$. The general discussion of Section 7.2 has shown that in this case the resulting finite-difference approximation is only first-order accurate. We therefore lose the great advantage built into the second-order-accuracy approach, namely, the considerable improvement in wave resolution.

Finally, it should be noticed that even though the GRP split scheme (7.28), (7.30) as described here is not "finite volume," it certainly remains "conservative"; taking a computational domain large enough so that the numerical fluxes vanish on its boundary, we obtain, from (7.30),

$$\sum_{j,k} \mathbf{U}^{n+1}_{j,k} = \sum_{j,k} \mathbf{U}^n_{j,k}. \tag{7.31}$$

Remark 7.9 A basic issue that arises in the use of split (rectangular) schemes is their compatibility with problems that are radially symmetric or, more generally, involve large variations in directions that are not aligned with the coordinate axes. For example, this is the case when one is trying to compute the flow in a "radial shock tube" where, initially, there are two different constant states (of the same gas) separated by a circle. In this case the solution retains the radial symmetry and can be accurately computed by the quasi-1-D scheme. However, it is possible to resolve the same problem using a split scheme like (7.28) on a Cartesian mesh. This poses a problem right at the initial stage: How does one assign values (density, pressure, etc.) to the boundary mesh cells, divided between the two states? This can be done in a variety of ways (e.g., averaging by area ratio or assigning the values of one state to the complete cell). The 2-D calculation is certainly sensitive to the choice of initial data. However, once a "suitable" choice is made, this calculation fits surprisingly well to the accurate quasi-1-D result. We refer the reader to Ben-Artzi, Falcovitz, and Feldman [13] and Birman, Har'el, Falcovitz, Ben-Artzi, and Feldman [17] for detailed studies of such test cases. The second reference actually employs a 3-D split (GRP) scheme to study the problem considered in Section 6.3 of this monograph.

8

Geometric Extensions

This chapter addresses one of the most central issues of computational fluid dynamics, namely, the simulation of flows under complex geometric settings. The diversity of these issues is briefly outlined in Section 8.1, which points out the role played by the present extensions: the (1-D) "singularity tracking" and the (2-D) "moving boundary tracking" (MBT) schemes. Section 8.2 deals with the first extension, and Section 8.3 is devoted to an outline of the second one. In the former we present the scheme methodology and refer to GRP papers for examples. In the latter, the basic principles of the method are presented, and we refer to [39] for more algorithmic details. Finally, an illustrative example of the MBT method shows how an oval disk is "kicked-off" by a shock wave.

8.1 Grids That Move in Time

In Part I of this monograph we dealt with finite-difference approximations to the quasi-1-D hydrodynamic conservation laws, where the underlying grid was *fixed* and equally spaced in the majority of cases. In our two-dimensional numerical extension (Section 7.3) we restricted the treatment to a *Cartesian* (rectangular) grid. Naturally, finite-difference approximations assume their simplest form on such grids, and the motivation for seeking geometric extensions comes primarily from *physical applications*.

In the 1-D setting, a geometric extension is typically needed when it is desired to "track" special singularities of the flow. Such a "singularity tracking" extension is considered in Section 8.2, where selected flow singularities (shock, material interface, contact, and gradient discontinuity) are tracked as special (moving) grid points. More generally, in the context of the GRP methodology one can treat grid points having an arbitrary law of motion, thus allowing for a "dynamically changing" grid. This has already been done in the ALE

251

computation of the shock tube problem (Section 6.1), as well as in the expanding sphere problem (Section 6.4). In these cases the motion of grid points was designed to maintain computational cells of uniform size, thus contributing to the overall numerical accuracy. It should be noted that although this capability arises naturally from the GRP analysis, which provides the full range of directional derivatives (see Remark 5.31), it is not the case for most conservation law schemes, which are based on a fixed (Eulerian) or material (Lagrangian) grid.

In the multidimensional case there exists a rich variety of fluid dynamical problems involving time-changing geometry caused by "moving boundaries" (e.g., pistons in various machines). Furthermore, even fixed boundaries, (i.e., rigid wall surfaces) require complex grid geometries if they are not aligned with (Cartesian) coordinate directions. The numerical simulation of fluid flows in such "extended geometries," involving time-changing grids and "front tracking," has become a highly demanding technological endeavor. We refer to Richtmyer and Morton [96, Chapter 13], Chern et al. [24], Glimm et al. [50, 51, 52], Tabak [106], and references therein for such treatments. In the context of the present monograph, however, our aim is to focus on some basic principles of geometric extensions, and our scope of presentation will be far narrower.

A full multidimensional realization of the 1-D "tracking" idea is extremely complex and probably impossible to achieve in general geometric settings. Our two-dimensional extension, presented in Section 8.3, is thus restricted to tracking of regular curves, which move so that the relative fluid velocity is tangential to them (in other words, no fluid crosses the moving curve). The motion of the curve is imposed as a known boundary condition; it may also be stationary, enabling the treatment of arbitrary rigid boundaries. For a detailed description of the MBT method, with particular emphasis on its geometrical aspects, we refer to Falcovitz, Alfandary, and Hanoch [39].

As was already noted, the (1-D) extension presented in Section 8.2 draws on the analytic features of the GRP solution. However, the treatment of moving boundaries in Section 8.3 is practically independent of the particular finite-difference method.

8.2 Singularity Tracking

Consider quasi-one-dimensional flow in a duct of varying cross section, which is governed by the Euler equations (4.45). Its GRP numerical integration was described in Section 5.2, using a fixed (Eulerian) spatial grid, or a grid moving with the fluid (Lagrangian; see Remark 5.40). We now seek to extend the scheme to the case where the grid points are allowed to move arbitrarily with respect to the Eulerian coordinate r. In so doing we draw on the GRP analytic solution of directional derivatives as described in Remark 5.31.

Let $\Sigma_t = \{r \,|\, a(t) \le r \le b(t)\}$ be a moving segment of the flow for which endpoints $a(t)$, $b(t)$ trace out smooth trajectories. Using the equality

$$\frac{d}{dt} \int_{\Sigma_t} A(r) \mathbf{U}(r, t) \, dr = \int_{\Sigma_t} A(r) \frac{\partial}{\partial t} \mathbf{U}(r, t) \, dr + b'(t) A(b(t)) \mathbf{U}(b(t), t) \tag{8.1}$$
$$- a'(t) A(a(t)) \mathbf{U}(a(t), t),$$

and expressing $A(r) \frac{\partial}{\partial t} \mathbf{U}(r, t)$ in terms of spatial derivatives by (4.45), we get

$$\frac{d}{dt} \int_{\Sigma_t} A(r) \mathbf{U}(r, t) \, dr = -\Big[(\mathbf{F}(\mathbf{U}) - \Lambda \mathbf{U}) \, A(r)\Big]_{a(t),t}^{b(t),t} - \int_{\Sigma_t} A(r) \frac{\partial}{\partial r} \mathbf{G}(\mathbf{U}) \, dr, \tag{8.2}$$

where $\Lambda(a(t), t) = a'(t)$, $\Lambda(b(t), t) = b'(t)$.

Identifying $\Sigma_t = [r_{j-1/2}(t), r_{j+1/2}(t)]$ with a single cell of the grid (which varies in time), we let $\Lambda_{j\pm1/2}^{n+1/2} = r'_{j\pm1/2}(t_{n+1/2})$ denote the mean velocity of the cell endpoints. We can now set up a finite-difference approximation to (8.2) as follows:

$$\mathbf{U}_j^{n+1} = \frac{(\Delta v)_j^n}{(\Delta v)_j^{n+1}} \mathbf{U}_j^n - \frac{\Delta t}{(\Delta v)_j^{n+1}} \left\{ \Big[(\mathbf{F}(\mathbf{U}) - \Lambda \mathbf{U}) \, A\Big]_{j+1/2}^{n+1/2} \right.$$
$$- \Big[(\mathbf{F}(\mathbf{U}) - \Lambda \mathbf{U}) \, A\Big]_{j-1/2}^{n+1/2} - \frac{1}{2}\Big[\mathbf{G}(\mathbf{U})_{j+1/2}^{n+1/2} - \mathbf{G}(\mathbf{U})_{j-1/2}^{n+1/2}\Big]$$
$$\left. \times \left(A_{j+1/2}^{n+1/2} + A_{j-1/2}^{n+1/2} \right) \right\}, \tag{8.3}$$

where

$$(\Delta v)_j^n = \int_{r_{j-1/2}(t_n)}^{r_{j+1/2}(t_n)} A(r) \, dr$$

is the volume of cell j at time $t = t_n$, and $A_{j+1/2}^{n+1/2} = A(r_{j+1/2}(t_{n+1/2}))$. Let $D_{j+1/2}^n$ denote the directional derivative along the trajectory $r_{j+1/2}(t)$ [i.e., $D_{j+1/2}^n(Q) = \frac{\partial Q}{\partial t} + r'_{j+1/2}(t)\frac{\partial Q}{\partial r}$], evaluated at $(r, t) = (r_{j+1/2}(t_n), t_n)$. The values at half time step $t = t_{n+1/2}$ are then evaluated as follows: Let the half-time-step state be approximated as in the linear GRP by

$$\mathbf{U}_{j+1/2}^{n+1/2} = \mathbf{U}_{j+1/2}^n + \frac{\Delta t}{2} D_{j+1/2}^n(\mathbf{U}). \tag{8.4}$$

The half-time-step fluxes in (8.3) (again evaluated to within second-order accuracy; see Remark 3.9) are then expressed as

$$\mathbf{F}_{j\pm1/2}^{n+1/2} = \mathbf{F}\big(\mathbf{U}_{j\pm1/2}^{n+1/2}\big),$$
$$\mathbf{G}_{j\pm1/2}^{n+1/2} = \mathbf{G}\big(\mathbf{U}_{j\pm1/2}^{n+1/2}\big).$$

(8.5)

Assume further that the half-time-step speed $\Lambda_{j+1/2}^{n+1/2}$ is known (i.e., as when the motion of the tracked point is externally imposed; other cases will be considered later). The new coordinates of moving grid points are thus given by

$$r_{j+1/2}^{n+1} = r_{j+1/2}^{n} + \Delta t\, \Lambda_{j+1/2}^{n+1/2},$$

(8.6)

and hence the cell volumes $(\Delta v)_j^{n+1}$ are also known. This completes the evaluation of all variables needed in (8.3).

In this integration scheme, not only may the grid spacings $(r_{j+1/2}^{n} - r_{j-1/2}^{n})$ change "dynamically," but their (spatial) order may change too. This will be the case, for example, when there is a fixed "underlying" grid while some moving grid points are assigned to a contact discontinuity, a shock, or a $C_\pm$ characteristic curve. There are two aspects to the algorithm required in such cases. One is organizational, which consists in updating the information related to the (spatial) order of grid points and cells. The other is a "cell-merging" procedure, intended to avoid computation by (8.3) of a "small" cell, which would bring the time step down to unacceptable values. Thus, when a moving grid point approaches a fixed grid point, producing a cell of size smaller than some preassigned value, that cell is "merged" with the adjacent one, resulting in a bigger combined cell. The new values computed by (8.3) for the "merged cell" are then assigned to the two "constituent" cells.

We now resume the discussion of the grid point speed $\Lambda_{j+1/2}^{n+1/2}$. We proceed according to the type of singularity at that point, as follows:

(a) The simplest case is when $r_{j+1/2}(t)$ is an imposed smooth function (independent of the solution to the fluid-dynamical equations). We then have $\Lambda_{j+1/2}^{n+1/2} = r_{j+1/2}'(t_{n+1/2})$. The moving grid points in some test cases of Chapter 6 were treated this way (see Sections 6.1 and 6.4).

(b) When $r_{j+1/2}(t)$ is assigned to a contact discontinuity, $\Lambda_{j+1/2}^{n+1/2} = u_{j+1/2}^{*} + \frac{\Delta t}{2}\big(\frac{\partial u}{\partial t}\big)_{j+1/2}^{*}$. (See Theorem 5.7 and Claims 5.17 and 5.18. Here the grid point $j + 1/2$ corresponds to the point $r = 0$ in the treatment of Section 5.1.)

(c) When a $C_\pm$ characteristic curve is tracked, the point speed is $\Lambda_{j+1/2}^{n+1/2} = (u \pm c)_{j+1/2}^{n+1/2}$. These values, in turn, are obtained by adding to $(u \pm c)_{j+1/2}^{n}$ the term $\frac{\Delta t}{2} D_{j+1/2}^{n}(u \pm c)$, where $D_{j+1/2}^{n}$ is now the directional derivative along $C_\pm$ (see Remarks 5.4 and 5.31).

(d) When $r_{j+1/2}(t)$ is a shock trajectory, we proceed as follows: Let $\mathbf{U}_{j+1/2}^{n+1/2}$ denote the half-time-step state *behind* the shock, evaluated as in (8.4) by taking the directional derivative on the postshock side. Assuming a 3-shock, we let the half-time-step state ahead be

$$\left(\mathbf{U}_{j+1/2}^{n+1/2}\right)_+ = \left(\mathbf{U}_{j+1/2}^{n}\right)_+ + \frac{\Delta t}{2} D_{j+1/2}^{n}(\mathbf{U})_+,$$

where the time derivative ahead of the shock is evaluated from (4.45), using the linear data in cell $j+1$ at time $t = t_n$ [see also Equation (5.57)]. Then, to within second-order accuracy, the speed of the shock is given by the jump condition [as in (4.89)]

$$\Lambda_{j+1/2}^{n+1/2} = u_+ + \tau_+ \frac{p - p_+}{u - u_+},$$

where u and p are taken from $\mathbf{U}_{j+1/2}^{n+1/2}$, and u_+, τ_+, p_+ are taken from $\left(\mathbf{U}_{j+1/2}^{n+1/2}\right)_+$.

As previously mentioned, the moving grid points in Sections 6.1 and 6.4 were treated as in case (a) here. Option (b) is used for tracking either a contact discontinuity or a material interface (see Ben-Artzi and Birman [4] for the latter case). The tracking of characteristic curves [option (c)] is typically used to better resolve discontinuities in flow gradients across the head or tail characteristics of a rarefaction wave. And finally, option (d) is used to track shock waves. We refer to Falcovitz and Birman [41], where diverse test cases employing grid scheme of options (a)–(d) (and combinations thereof) are presented.

8.3 Moving Boundary Tracking (MBT)

In Section 7.3 we presented a basic "numerical extension" to the GRP method, whereby a two-dimensional computation was obtained by operator splitting on a Cartesian grid. Here we consider a further expansion of the scope of fluid-dynamical problems, where a "geometric extension" of the grid is introduced, aimed at treating curved moving (or stationary) boundary lines. Generally speaking, there are two different approaches to the construction of such grids. In the first approach the grid is "body-fitted" to the boundary line, whereas in the second the boundary line is treated as a separate geometric entity, cutting across a fixed underlying mesh. The MBT method (Falcovitz et al. [39]) belongs to the second category, and before presenting it in detail we briefly review the cardinal features of each approach.

Body-fitted grid generation is analogous to covering a floor in a room bounded by curved walls with tiles, using tiles of various shapes. The resulting grid usually requires an irregular (nonrectangular) connectivity (as for grids used by finite-element methods). When the boundary moves, the grid deforms "elastically" with it, so as to maintain the shape and size of all cells within an acceptable range. For example, a grid of this type was used in the simulation of the gas dynamics within an internal combustion engine by Wierse [119]. In this case, entire cell layers had to be periodically added or deleted to avoid excessive cell stretching or compression (caused by reciprocating piston motion). We observe that although the implementation of the boundary condition at the moving surface is simple in this type of grid, the treatment of the entire "dynamically" changing mesh is quite complex. It involves algorithms that typically require specific "tuning" for each problem, resulting in a nonrobust computational scheme.

Consider now the second approach of "moving boundary tracking" where the flow is approximated on a fixed Eulerian grid. Here the finite-difference computation of the flow at interior cells is unaffected by the moving boundary. The entire scheme, therefore, consists only of the algorithms needed to integrate the conservation laws in the cells intersected by the moving boundary line. In our presentation of the MBT method we show that its basic principles can be formulated at a modest level of complexity. This scheme is also robust, requiring no heuristic "tuning" tailored to particular applications. It is noted, however, that like any other algorithm of this family, the MBT is "geometrically intensive." By this we mean that it involves a good deal of geometrical analysis concerning the "monitoring" of a moving boundary relative to an underlying grid.

Let us first clarify the meaning of "tracking" in regards to the present MBT scheme. Given the flow profiles at time level $t = t_n$ [which includes the boundary curve location $\Lambda(t_n)$], the new location $\Lambda(t_{n+1})$ is explicitly determined and is totally unrelated to the details of the flow field evolving from $t = t_n$ to $t = t_{n+1}$. In particular, this includes the special case of a rigid wall [in which $\Lambda(t_n) \equiv \Lambda$ is a fixed curve]. However, tracking an "internal boundary" such as a shock front, the motion of which constitutes an integral part of the solution, lies outside the scope of the method presented here.

As for the case of a material interface, it is strictly speaking analogous to a shock front in that its motion is derived as part of the overall solution. However, in various computational approaches, notably the CEL (Coupled Eulerian–Lagrangian) method due to Noh [93], two adjacent fluids are described by different frames, one being Lagrangian while the other is Eulerian. In this case the material interface is the boundary of the "Lagrangian" side. Its motion between $t = t_n$ and $t = t_{n+1}$ is then fully determined (to first-order accuracy) by the Lagrangian velocity field at $t = t_n$ and is imposed on the Eulerian side as an external condition. It is thus admissible as a "moving boundary" (for the Eulerian calculation) in the sense of the MBT scheme. In particular, a rigid body of fixed mass immersed in a compressible fluid and moving under the action of the pressure exerted by the fluid constitutes a reduced case of the previously cited CEL method. In the following, we shall consider a simple problem of this type, where a rigid elliptic disk lying on flat ground is "kicked up" by an incident shock wave.

8.3.1 Basic Setup

Here we introduce the fundamental concepts of the MBT scheme, assuming a simplified geometric setting. Some additional aspects related to the general algorithm are briefly discussed in the following subsection. We start our description of the scheme by a presentation of the geometric setup and then follow with an outline of the finite-difference scheme.

Referring to the schematic display in Figure 8.1, we see that the boundary line is specified as a single *closed polygon*, or as the union of several closed polygons, with the fluid occupying the domain exterior to them. Every such polygon is non-self-intersecting. Additionally, the polygons are free to move about the (x, y) plane, including entering or exiting the (rectangular) computational domain, as long as they do not intersect each other.

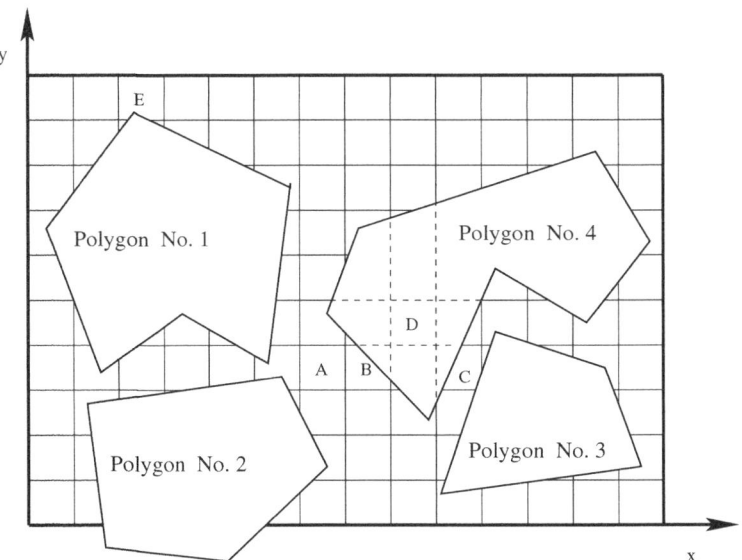

Figure 8.1. Boundary polygons and underlying Cartesian grid.

Several types of cells are formed by the intersection between boundary polygons and the grid. Cell A is uncut by any polygon and is completely filled with fluid. We then refer to it as a "regular cell." Cell D, however, is completely covered by a polygon and contains no fluid, so it is labeled a "covered cell." The three cells B, C, and E are intersected by the boundary line and are consequently referred to as "boundary cells" (the term "cut cells" is also used in the literature). We now observe that a boundary cell may be intersected by more than one polygon (as is the case of cell C), and a side of a boundary cell may also be intersected by two different polygons (see cell C again), or by two different segments of the same polygon (see cell E). These observations point out that the calculation of fractional areas and side lengths in boundary cells is algorithmically difficult. This is further complicated by the fact that the configuration is not static.

Our finite-difference scheme should provide an approximation to the flow evolving over a time interval $[t_n, t_{n+1}]$, taking into account the geometric effects of the moving boundary in boundary cells. This is achieved by extending the balance equation (7.29) to a control volume that consists of a single grid cell intersected by the moving boundary line. The resulting balance equation leads to a finite-difference approximation of the hydrodynamic conservation laws in *boundary cells*, extending the split scheme (7.30).

Before considering the balance equation in this case let us focus on the boundary cell geometry, referring to the cell $[x_{j-1/2}, x_{j+1/2}] \times [y_{k-1/2}, y_{k+1/2}]$ shown in Figure 8.2. The control volume boundary is of a "mixed" type, meaning that it consists of whole or partial sides of the regular cell, complemented by segments of the (moving) boundary line. This structure varies over the time interval $[t_n, t_{n+1}]$. In Figure 8.2, we show the initial ($t = t_n$), intermediate ($t_n < t < t_{n+1}$), and final ($t = t_{n+1}$) configurations. It is assumed that the vertices of the polygonal line move at constant (given) velocities over the time interval. This in

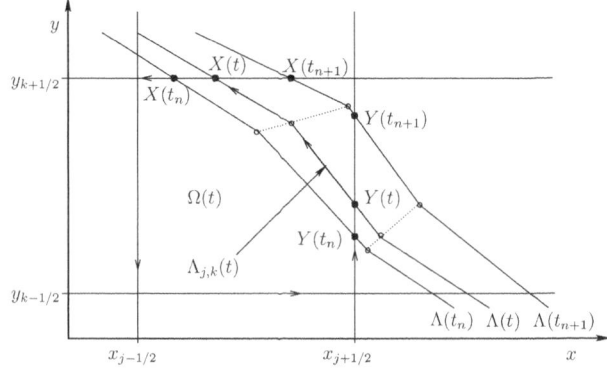

Figure 8.2. Cell (j, k) intersected by a moving boundary.

turn determines the position of the line $\Lambda(t)$ at any given time $[t_n \leq t \leq t_{n+1}]$. In particular, it follows that the location $Y(t)$, where $\Lambda(t)$ intersects the side $x = x_{j+1/2}$ of the cell, can be expressed as a function of t. Assume for simplicity (as is the case in Figure 8.2) that $Y(t)$ remains on the same side during the entire time interval. We take $Y_{n+1/2} = \frac{1}{\Delta t} \int_{t_n}^{t_{n+1}} Y(t)\,dt$ and regard it as an average location of the intersection of the moving line with that side. Similarly, referring again to Figure 8.2, we take the average location on the side $y = y_{k+1/2}$ as

$$X_{n+1/2} = \frac{1}{\Delta t} \int_{t_n}^{t_{n+1}} X(t)\,dt.$$

Let the exposed (fluid-filled) area of this boundary cell at time t be denoted by $\Omega(t)$, and let the (closed) boundary line of that cell be $\partial\Omega(t)$. We then denote by $\Lambda_{j,k}(t) = \partial\Omega(t) \cap \Lambda(t)$ the "moving boundary" part of $\partial\Omega(t)$, that is, that part of $\Lambda(t)$ which is located within the cell at time t. Following the common mathematical practice, we take the counterclockwise orientation around $\partial\Omega(t)$, as shown by the arrows in Figure 8.2. In particular, this determines the direction of the partial segment $\Lambda_{j,k}(t)$, along which we take the length coordinate s. The other ("static") part of $\partial\Omega(t)$ is designated as $\partial\Omega_{\text{stat}}(t)$, so that

$$\partial\Omega(t) = \partial\Omega_{\text{stat}}(t) \cup \Lambda_{j,k}(t). \tag{8.7}$$

Referring to the procedure leading to the balance equation (7.29), we repeat its derivation in the present "geometrically extended" setting, by integrating the conservation law (7.16) over the area of $\Omega(t)$:

$$\iint_{\Omega(t)} \left(\mathbf{U}_t + \mathbf{H}(\mathbf{U})_x + \mathbf{L}(\mathbf{U})_y \right) dx\,dy = \mathbf{0} \tag{8.8}$$

By Green's theorem we get

$$\iint_{\Omega(t)} \mathbf{U}_t\,dx\,dy + \oint_{\partial\Omega(t)} \mathbf{H}(\mathbf{U})\,dy - \mathbf{L}(\mathbf{U})\,dx = \mathbf{0}. \tag{8.9}$$

The integral of $\mathbf{U}_t$ over $\Omega(t)$ is now expressed as

$$\iint_{\Omega(t)} \mathbf{U}_t\,dx\,dy = \frac{d}{dt} \iint_{\Omega(t)} \mathbf{U}\,dx\,dy - \int_{\Lambda_{j,k}(t)} \mathbf{U}\left[\underline{V}(s) \cdot \underline{n}(s)\right] ds, \tag{8.10}$$

where the line integral over $\Lambda_{j,k}(t)$ "compensates" for the change in the area $|\Omega(t)|$ resulting from the motion of the boundary. Here $\underline{V}(s)$ is the given velocity of the moving boundary, and $\underline{n}(s)$ is the outward-pointing unit normal. The product $\underline{V}(s) \cdot \underline{n}(s)$ is thus the normal component of the boundary velocity. By assumption, this component is equal to the normal component of the fluid velocity (u, v), so that

$$\left[\underline{V}(s) \cdot \underline{n}(s)\right] ds = (u, v) \cdot (dy, -dx) = u\,dy - v\,dx. \qquad (8.11)$$

Inserting (8.10) and (8.11) into (8.9), and noting (8.7), we get

$$\frac{d}{dt} \iint_{\Omega(t)} \mathbf{U}\,dx\,dy + \int_{\partial\Omega_{\text{stat}}(t)} \mathbf{H}(\mathbf{U})\,dy - \mathbf{L}(\mathbf{U})\,dx + \int_{\Lambda_{j,k}(t)} \left(\mathbf{H}(\mathbf{U}) - u\mathbf{U}\right) dy$$

$$+ \int_{\Lambda_{j,k}(t)} \left(-\mathbf{L}(\mathbf{U}) + v\mathbf{U}\right) dx = \mathbf{0}. \qquad (8.12)$$

We can rewrite (8.12) as[1]

$$\frac{d}{dt} \iint_{\Omega(t)} \mathbf{U}\,dx\,dy + \int_{\partial\Omega_{\text{stat}}(t)} \mathbf{H}(\mathbf{U})\,dy - \mathbf{L}(\mathbf{U})\,dx$$

$$+ \int_{\Lambda_{j,k}(t)} \mathbf{M}(\mathbf{U})\,dy - \mathbf{N}(\mathbf{U})\,dx = \mathbf{0}, \qquad (8.13)$$

where

$$\mathbf{M}(\mathbf{U}) = \mathbf{H}(\mathbf{U}) - u\mathbf{U} = \begin{bmatrix} 0 \\ p \\ 0 \\ up \end{bmatrix}, \qquad \mathbf{N}(\mathbf{U}) = \mathbf{L}(\mathbf{U}) - v\mathbf{U} = \begin{bmatrix} 0 \\ 0 \\ p \\ vp \end{bmatrix}. \qquad (8.14)$$

The balance equation (8.13) is now split along the (x, y) directions as follows:

$$\frac{d}{dt} \iint_{\Omega(t)} \mathbf{U}\,dx\,dy + \int_{\partial\Omega_{\text{stat}}(t)} \mathbf{H}(\mathbf{U})\,dy + \int_{\Lambda_{j,k}(t)} \mathbf{M}(\mathbf{U})\,dy = \mathbf{0}, \qquad (8.15a)$$

$$\frac{d}{dt} \iint_{\Omega(t)} \mathbf{U}\,dx\,dy - \int_{\partial\Omega_{\text{stat}}(t)} \mathbf{L}(\mathbf{U})\,dx - \int_{\Lambda_{j,k}(t)} \mathbf{N}(\mathbf{U})\,dx = \mathbf{0}. \qquad (8.15b)$$

Note that this form is analogous to the split system (7.17), (7.18) in the sense that only the x-flux components $\mathbf{H}$, $\mathbf{M}$ are retained in Equation (8.15a), and likewise only the y-flux components $\mathbf{L}$, $\mathbf{N}$ in Equation (8.15b).

---

[1] Observe that the contribution of the moving boundary to the balance relation (8.13) involves only "pressure terms," which produce the force exerted by $\Lambda_{j,k}(t)$ on the fluid in $\Omega(t)$, and the work done by this force.

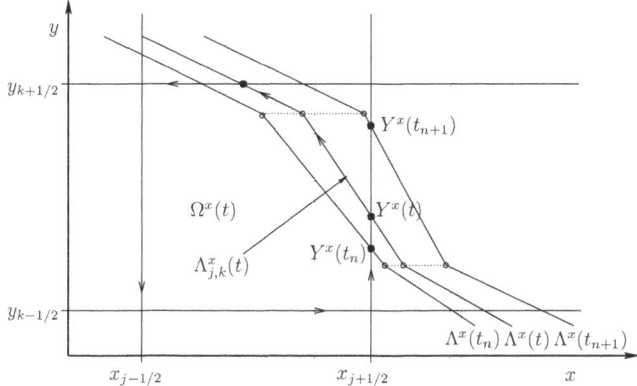

Figure 8.3. Cell (j, k) intersected by a boundary moving in the x direction.

When solving Equations (8.15a), (8.15b) separately, we are obliged to also "split" the geometry. By this we mean that in the process of solving (8.15a) we take into account only the x motion of the boundary line, and similarly for (8.15b). Thus, taking only the x component of the velocity, the boundary line $\Lambda(t)$ moves horizontally, and we denote its position at time t by $\Lambda^x(t)$. The situation is now completely analogous to the previous two-dimensional setup. In particular, the boundary $\partial\Omega(t)$ is now replaced by $\partial\Omega^x(t)$, where $\partial\Omega^x(t) = \partial\Omega^x_{\mathrm{stat}}(t) \cup \Lambda^x_{j,k}(t)$ (see Figure 8.3). Similarly, when solving (8.15b) we take into account only the y component of the motion of $\Lambda(t)$.[2] Note that our assumption that the velocity of each vertex of $\Lambda(t)$ is constant throughout the time interval implies that the final position of the boundary line, when the x and y integration steps have been completed, is identical to $\Lambda(t_{n+1})$. We now integrate (8.15a), replacing $\Lambda(t)$ by $\Lambda^x(t)$, thus obtaining

$$\iint\limits_{\Omega^x(t_{n+1})} \mathbf{U}\,dx\,dy - \iint\limits_{\Omega^x(t_n)} \mathbf{U}\,dx\,dy$$

$$= -\int\limits_{t_n}^{t_{n+1}} \int\limits_{\partial\Omega^x_{\mathrm{stat}}(t)} \mathbf{H(U)}\,dy\,dt - \int\limits_{t_n}^{t_{n+1}} \int\limits_{\Lambda^x_{j,k}(t)} \mathbf{M(U)}\,dy\,dt. \qquad (8.16)$$

Denoting by $V^n_{j,k}$, $V^{n+1}_{j,k}$, $\mathbf{U}^n_{j,k}$, $\mathbf{U}^{n+1}_{j,k}$ the values of the boundary cell area and average $\mathbf{U}$ at the two time levels, we recast (8.16) in the form of a finite-difference

---

[2] It is understood that the geometric setup at the beginning of each integration step is identical to that prevailing at the end of the preceding step. Thus, for a y step beginning at $t = t_n$, following an x step that ended at $t = t_{n+1}$, $\Omega^y(t_n) \equiv \Omega^x(t_{n+1})$, $\Lambda^y(t_n) \equiv \Lambda^x(t_{n+1})$, etc.

approximation,

$$V_{j,k}^{n+1} \mathbf{U}_{j,k}^{n+1} - V_{j,k}^n \mathbf{U}_{j,k}^n = -\Delta t \left[A_{j+1/2,k}^{n+1/2} \, \mathbf{H}_{j+1/2,k}^{n+1/2} - A_{j-1/2,k}^{n+1/2} \, \mathbf{H}_{j-1/2,k}^{n+1/2} \right]$$
$$- \Delta t \left[B_{j,k}^{n+1/2} \, \mathbf{M}_{j,k}^{n+1/2} \right],$$

$$\mathbf{H}_{j\pm1/2,k}^{n+1/2} = \mathbf{H} \left(\mathbf{U}_{j\pm1/2,k}^{n+1/2} \right), \qquad \mathbf{M}_{j,k}^{n+1/2} = \mathbf{M} \left(\mathbf{U}_b^{n+1/2} \right), \quad (8.17)$$

where $A_{j\pm1/2,k}^{n+1/2}$ are the effective lengths for the flux through the sides $x = x_{j\pm1/2}$, and $B_{j,k}^{n+1/2}$ is the mean projection of $\Lambda_{j,k}^x(t)$ on the side $x = x_{j+1/2}$. For the geometric setup in Figure 8.3, they are determined as follows: Let $Y^x(t)$ be the point of intersection of $\Lambda^x(t)$ with the side $x = x_{j+1/2}$, and let

$$\left(Y^x \right)^{n+1/2} = \frac{1}{\Delta t} \int_{t_n}^{t_{n+1}} Y^x(t) \, dt$$

be its mean value. We then let

$$A_{j+1/2,k}^{n+1/2} = \left(Y^x \right)^{n+1/2} - y_{k-1/2}. \tag{8.18}$$

The side $x = x_{j-1/2}$ is exposed throughout the time interval in this case, so that $A_{j-1/2,k}^{n+1/2} = y_{k+1/2} - y_{k-1/2}$. It is noted that according to these definitions, $A_{j\pm1/2,k}^{n+1/2}$ is always positive (or zero).

Turning to the mean boundary projected length $B_{j,k}^{n+1/2}$, we see that it follows from (8.16) that it is generally given by

$$B_{j,k}^{n+1/2} = \frac{1}{\Delta t} \int_{t_n}^{t_{n+1}} \int_{\Lambda_{j,k}^x(t)} dy \, dt. \tag{8.19}$$

Here, the sign of $B_{j,k}^{n+1/2}$ is determined by the orientation of the line integral along $\Lambda_{j,k}^x(t)$; it may therefore be either negative or positive. In the particular setup of Figure 8.3, $B_{j,k}^{n+1/2}$ is equal to the mean "covered" part of side $x = x_{j+1/2}$ and is thus given by

$$B_{j,k}^{n+1/2} = y_{k+1/2} - \left(Y^x \right)^{n+1/2}. \tag{8.20}$$

Now, consider the fluid states $\mathbf{U}_{j\pm1/2,k}^{n+1/2}$, $\mathbf{U}_b^{n+1/2}$. The states $\mathbf{U}_{j\pm1/2,k}^{n+1/2}$ are determined by the solution to the GRP at the cell boundaries ($j \pm 1/2, k$), as described in Section 7.3, with the added assumption that the slopes in the boundary cell are taken as zero. As for the value of $U_b^{n+1/2}$ at the moving boundary, it requires careful attention and we evaluate it as follows.

At any given point on $\Lambda_{j,k}^x(t)$ we consider a "one-dimensional" interaction problem normal to the boundary. Denoting by $u^\perp$ the normal component of the fluid velocity, we obtain the state $U^\perp = \left[\rho, \rho u^\perp, \rho E\right]^T$, in which values of ρ, $u^\perp$, E are obtained from $U_{j,k}^n$. Let now $V^\perp$ be the (imposed) normal velocity of the boundary $\Lambda_{j,k}^x(t)$ at that point. We have thus obtained an interaction problem as follows. The state $U^\perp$ is taken to be a right state. The state $U_b^{n+1/2}$ is viewed as being connected to $U^\perp$ by a 3-wave, thus lying on the interaction curve $I_3^r(U^\perp)$ (see Section 6.2, Figure 6.6). Requiring that the normal velocity in $U_b^{n+1/2}$ be equal to $V^\perp$ fully determines $U_b^{n+1/2}$ (the tangential velocity in this state is set equal to that of $U_{j,k}^n$). In fact, this description was confined to a fixed boundary point at time t, whereas the value of $U_b^{n+1/2}$ requires averaging over $\Lambda_{j,k}^x(t)$ and over the time interval. Note that as a result of the motion of $\Lambda_{j,k}^x(t)$ the normal velocities $u^\perp$, $V^\perp$ vary in time. In contrast, the values $U_{j,k}^n$ are held fixed throughout the time interval.

The second equation (8.15b) in the y direction is now treated in exactly the same fashion. The sequencing of the x and y integration steps is done as in (7.28).

The integration time step in (8.17) is naturally limited by the CFL condition. However, in dealing with a boundary cell, we always impose additional restrictions on the time step, which are summarized in the following remark.

Remark 8.1 (Time step restrictions for boundary cells) There are two additional restrictions on the time step for cells intersected by a moving boundary. One is because boundary cells are reduced in size; the other is related to the velocity of the boundary.

(1) The CFL condition is $\Delta t < \mu_{\mathrm{CFL}} \frac{\Delta x}{S_{\mathrm{max}}}$, where S_{max} is the maximal wave speed and $\mu_{\mathrm{CFL}} < 1$. Here we take Δx to be the (reduced) "x length" of the boundary cell. This length is estimated by $\Delta x = 2V_{j,k}^n/(A_{j-1/2,k}^n + A_{j+1/2,k}^n)$, where $A_{j\pm1/2,k}^n$ are the exposed side lengths at $t = t_n$.

(2) It is required that a covered cell will not become fully exposed over a single time step (or vice versa), since the MBT scheme cannot handle such a transition. In other words, when a cell changes status from covered to regular (or vice versa), it must go (for at least one time level) through the intermediate status of a boundary cell. This additional restriction is obtained when S_{max} in the aforementioned inequality is replaced by the maximal speed of the boundary line. It is noted, however, that in typical physical problems the boundary speed is lower than S_{max}, so that this additional

restriction is usually satisfied without having to be explicitly imposed. In fact, computational experience has shown that cells normally change status over *several* time steps.

We conclude this presentation with the following remark on uniform flow.

Remark 8.2 (The case of uniform flow) To illustrate the difficulty involved in dealing with boundary cells, we make the following observation concerning the MBT algorithm. Suppose that one or several polygons (as in Figure 8.1) are placed in a uniform flow field, moving with the fluid at the same (uniform) velocity. Then the MBT algorithm guarantees that (as expected of the exact solution) flow variables in boundary cells retain their constant values for all times. Even this seemingly elementary property turns out to be nontrivial in the context of algorithms aimed at treating flows with moving boundaries or interfaces.

8.3.2 *Survey of the Full MBT Algorithm*

In the preceding subsection the MBT method was presented in the context of a simple geometric setup. Here we briefly outline the main issues associated with a general geometric setting. In particular we focus our discussion on the algorithms for the average side lengths $A_{j\pm1/2,k}^{n+1/2}$, $B_{j,k}^{n+1/2}$ needed to handle the full range of situations arising from the intersection of moving boundary polygons and the underlying grid. In formulating these algorithms, great care was taken to retain the uniform flow property (Remark 8.2). Our presentation here is rather brief, and we generally refer to Falcovitz et al. [39] for a more complete description of the MBT algorithms.

The geometric quantities to be evaluated are $V_{j,k}^{n}$, $V_{j,k}^{n+1}$, $A_{j\pm1/2,k}^{n+1/2}$, and $B_{j,k}^{n+1/2}$ [see (8.17)], as well as $A_{j\pm1/2,k}^{n}$, which serve to determine the time step Δt (see Remark 8.1). The "static" values (at $t = t_n$ or $t = t_{n+1}$) are relatively simple to evaluate, and we need not expound on the algorithms involved. In determining the average exposed side lengths, the situation is complicated since the boundary polygons move, and a changing "intersection geometry" has to be considered. Using the notation $\Omega^x(t)$ introduced in Subsection 8.3.1, and noting the identity

$$\oint_{\partial\Omega^x(t)} dy = \int_{\partial\Omega_{\text{stat}}^x(t)} dy + \int_{\Lambda_{j,k}^x(t)} dy = 0, \tag{8.21}$$

we get the relation

$$A_{j+1/2,k}^{n+1/2} - A_{j-1/2,k}^{n+1/2} + B_{j,k}^{n+1/2} = 0. \tag{8.22}$$

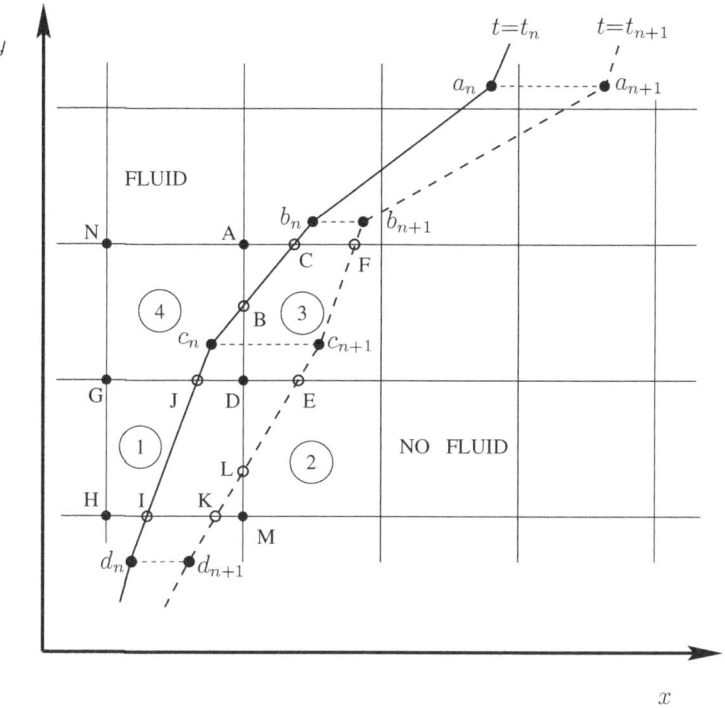

Figure 8.4. Boundary moving one time step in the x direction.

Observe that whereas the side lengths $A_{j\pm1/2,k}^{n+1/2}$ are taken as positive, the sign of $B_{j,k}^{n+1/2}$ may be either positive or negative. Consequently, only $A_{j\pm1/2,k}^{n+1/2}$ need to be determined by the time-averaging procedure.

In the following we illustrate the diversity of situations involved in evaluating $A_{j\pm1/2,k}^{n+1/2}$, by considering some of the "geometric scenarios" shown in Figure 8.4. Consider the side AD, which is partially exposed (AB) at $t = t_n$ and fully exposed at $t = t_{n+1}$. We note that its fully exposed status is reached at some intermediate time $t_n < t_i < t_{n+1}$, which is determined by the "sliding" motion of the intersection point from J to E. All this has to be taken into account in evaluating the time-averaged length of that side. A similar scenario takes place for side DM, which starts as fully covered and becomes partially exposed (DL). Here too, the same intermediate time t_i is required in evaluating the average side length. Obviously, an algorithm that would handle such situations on a case-by-case basis would be unreasonably complex. Instead, we make use of the fact that at this stage the boundary moves purely horizontally (in the x direction). Therefore the projection of each individual side, such as $b_n c_n$ of the polygonal boundary, on the cell side (AD) is invariant, and it is relatively simple to determine its exposed portion (AB). Repeating this procedure for all boundary sides in regards to their projection on the cell side (AD), we obtain $A_{j+1/2,k}^{n+1/2}$ as a sum of the individual contributions.

Inspection of Equation (8.17), seems to indicate that all needed geometric values have been determined. However, referring again to the scenario of Figure 8.4, we consider the case of boundary cell 1. The numerical fluxes across side GH (which is fully exposed at $t = t_n$) are determined by the GRP solver in the usual way (with the slopes in cell 1 set to zero). By contrast, the side DM is completely covered at $t = t_n$, so that a GRP cannot be solved for that side. In this situation we apply a "merging" procedure. More specifically, we modify the geometric setup pertinent to this boundary cell. Whereas $\Omega^x(t_n)$ remains as before (i.e., the polygon $HIJG$), the polygon $\Omega^x(t_{n+1})$ is taken to be the "merged" polygon $HKEG$, which comprises not only the exposed portion of boundary cell 1 but also that of boundary cell 2 (DLE). Thus, the fluxes through side DL are no longer needed. This scenario arises whenever a newly exposed boundary cell is formed. Similarly, when an existing boundary cell becomes fully covered during the time interval $[t_n, t_{n+1}]$, the same "merging" is applied in reverse. Such a case can be envisioned by considering an interchange of the old and new positions of the boundary line in Figure 8.4, so that cell 2 becomes fully covered and cell 1 remains a boundary cell of reduced area.

Another reason for merging of two adjacent boundary cells is to avoid an excessively small time step (see Remark 8.1). Take for example the situation of cells 3 and 4 in Figure 8.4. Here the time step would be limited in proportion to the size of cell 3 at $t = t_n$ (ABC). If that restriction is deemed excessive, we can follow a merging procedure similar to that just used for cells 1 and 2. For $\Omega^x(t_n)$, $\Omega^x(t_{n+1})$ we take the polygons $NGJc_nC$, $NGEc_{n+1}F$, respectively, thus avoiding the restriction of Δt by the size of ABC. In all cases where a "pair merging" of two adjacent cells is performed, and when neither cell becomes fully covered, the values calculated by (8.17) at time $t = t_{n+1}$ are assigned to both cells.

8.3.3 An Example: Shock Lifting of an Elliptic Disk

Here we consider a demonstrative case using the MBT method to calculate the (2-D) motion of a solid disk immersed in a compressible (inviscid) fluid. Specifically, we take an elliptic disk that lies at rest on the ground; it is "kicked-off" by an incident planar shock wave (see Figure 8.5, where the ">" inside the disk indicates the shock-facing end). The motion of the disk is calculated step by step with the integration of the fluid-dynamical equations (8.17), using the

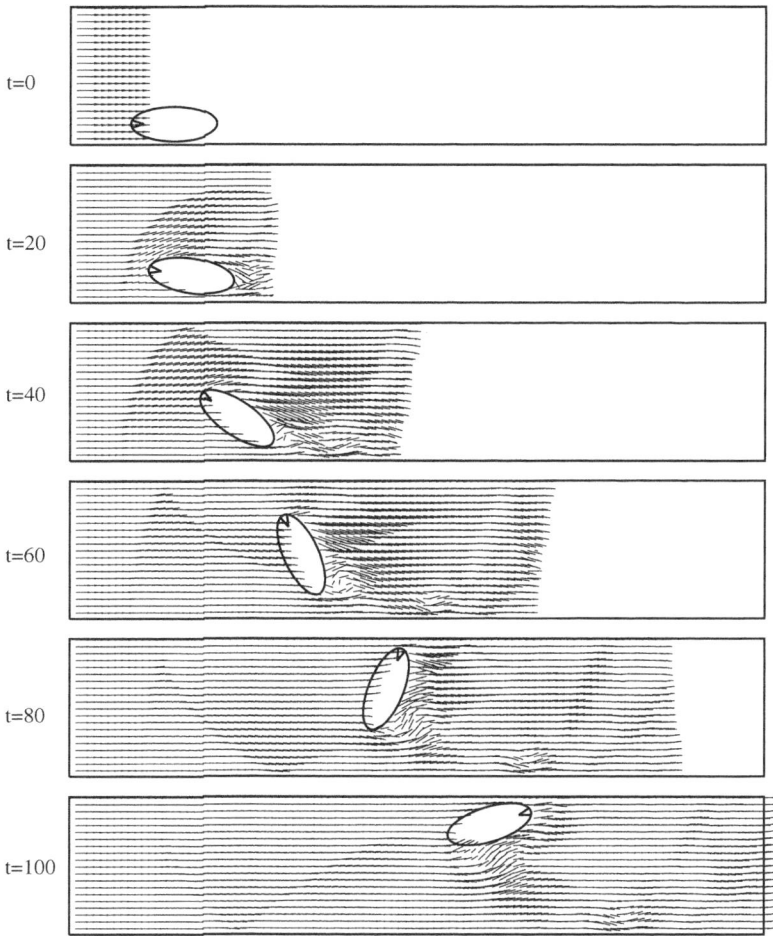

Figure 8.5. Kickoff of an elliptic disk by an $M_s = 3$ shock in air.

classical equations governing the translation and rotation of a solid (rigid) disk
in two space dimensions. The smooth disk is approximated as a polygon, and
the total force and moment (about the center of the disk) are simply calculated
by summing up the contributions of the fluid pressure on every side. From the
force and moment we determine the velocity of the center point (x_c, y_c) and the
rotational velocity about that point, assuming that both are constant throughout
the time interval $[t_n, t_{n+1}]$. The displacement $(\Delta x_q, \Delta y_q)$ of each polygon vertex
q from $t = t_n$ to $t = t_{n+1}$ is then obtained as a superposition of the translational
displacement of (x_c, y_c) and the (finite-angle) rotation about that point. Finally,
we calculate the vertex velocities $(u_q, v_q) = (x_q/\Delta t, \ \Delta y_q/\Delta t)$ and assume that

(in compliance with the MBT scheme) they are also constant throughout the time interval $[t_n, t_{n+1}]$.

The data for the specific example are as follows: The computational domain is an x, y rectangle of dimensions 100×20, which is divided into a grid of 200×40 square cells. The fluid is a perfect gas with $\gamma = 1.4$ and initial state $[\rho_0, p_0, u_0, v_0] = [1.3, 0.1, 0, 0]$. The planar right-facing shock wave (3-shock) is initially at $x = 8$, and its intensity is specified by the shock Mach number $M_s = 3$. Thus, the postshock state is given by $[\rho_1, p_1, u_1, v_1] = [5.0143, 1.0333, 0.72926, 0]$. The boundary conditions are a rigid wall at the top ($y = 20$) and bottom ($y = 0$) planes and the postshock state at the left plane ($x = 0$). At the right plane ($x = 100$) we set up a "nonreflecting" boundary condition intended to let 3-waves propagate "out" of the computational rectangle, as follows: Let j, k indicate a cell adjacent to that boundary. We first set the slopes $\mathbf{U}_{j,k}'^{n}$ equal to zero. Then we define the data $\mathbf{U}_L, \mathbf{U}_R$ for the Riemann problem at the boundary point $x_{j+1/2}$ by $\mathbf{U}_L = \mathbf{U}_{j,k}^n$ and by "continuation" of the flow, $\mathbf{U}_R = \mathbf{U}_L$.

The disk is taken as an ellipse having major and minor axes $a = 6.25$ and $b = 2.5$, centered at $(x, y) = (15, 3)$ (hence its bottom is 0.5 above the "ground plane" $x = 0$, and its leftmost tip is located at $x = 8.75$, i.e., 0.75 ahead of the initial shock front). The solid disk has a uniform mass density $\rho_c = 10\rho_0$, so that its total mass and moment of inertia are $m_c = \pi a b \rho_c$ and $I_c = \frac{1}{4} m_c (a^2 + b^2)$, respectively. The ellipse is approximated by a sixty-sided polygon, whose vertices (x_q, y_q) are given by

$$(x_q, y_q) = \left(15 + a \cos \left(\frac{2\pi q}{60} \right), 3 + b \sin \left(\frac{2\pi q}{60} \right) \right), \qquad q = 1, 2, \ldots, 60.$$

In Figure 8.5 we see that the asymmetric shock diffraction (due the ground reflection) causes initial liftoff and rotational motion of the disk. This effect seems to persist through the entire computation, producing a "tumbling" motion as shown. The ">" marker helps indicate that by time $t = 100$ the disk has rotated by an angle of nearly π.

9

A Physical Extension: Reacting Flow

In this chapter we consider the system of equations governing compressible reacting flow. The fluid is a homogeneous mixture of two species. The evolution of the flow under the mechanical conservation laws of mass, momentum and energy is coupled to the (continuous or abrupt) conversion of the "unburnt" species to the "burnt" one. We take the simplest model of such a reaction, namely, an irreversible exothermic process. The equation of state of the fluid depends on its chemical composition. The resulting (augmented) system is still nonlinear hyperbolic (in the sense of Chapter 4) and is amenable to the GRP methodology. The basic hypotheses are presented in Section 9.1, leading to the derivation of the characteristic relations and jump conditions. In Section 9.2 we describe the classical Chapman–Jouguet model of deflagrations and detonations, and the Zeldovich–von Neumann–Döring (Z–N–D) solution is presented in Section 9.3. In Section 9.4 we study the generalized Riemann problem for the system of reacting flow. The treatment here is close to that of the basic GRP case (Section 5.1), but there are significant differences because of the reaction equation. In Section 9.5 we outline briefly the resulting GRP numerical scheme and study a physical problem of ozone decomposition.

So far our physical model of compressible flow was based on the three conservation laws: those of mass, momentum and energy [see Equations (4.44)]. In the absence of external forces the internal "driving force" of the fluid motion is the pressure gradient. It is stipulated that, as a basic hypothesis, two independent thermodynamic variables (commonly taken as the density ρ and the internal energy e) serve to determine all others (such as the pressure p, the entropy S, or the temperature T) by a suitable equation of state. The first law of thermodynamics [see (4.49)] expresses the dependence of S on (e, ρ) by

$$T\,dS = de + p\,d\tau, \qquad \tau = 1/\rho \tag{9.1}$$

(see Footnote 7 in Section 4.2).

There exists a great number of fluid-dynamical situations where various physical effects (or assumptions) are added to this basic system of compressible flow, without changing the character of the governing system of equations. In other words, the (extended or modified) system still belongs to the mathematical class of "systems of conservation laws," in the sense of Chapter 4. We mention two well-known cases included in this class:

(i) The equations of magnetohydrodynamics in plasma physics (and astrophysics). In this case the (compressible) fluid is a medium with electromagnetic properties and Maxwell's equations are added to the basic conservation laws (see Courant and Hilbert [31, Chapter 6] and Richtmyer and Morton [96, Chapter 12]).

(ii) Shallow-water equations. In this case, the equations of *incompressible* flow in a "shallow" (three-dimensional) layer are simplified by the assumption that the velocity field does not depend on the "depth" coordinate and the pressure is hydrostatic, depending linearly on depth (see Courant and Friedrichs [30, Chapter I]). This model is widely used in meteorology (see Alcrudo and Garcia-Navarro [1] and Garcia-Navarro, Hubbard, and Priestley [47]). In fact, when the equations are written using spherical coordinates, they do not conform exactly to the class of conservation laws, as the fluxes depend explicitly on the (spatial) coordinates.

These (nonlinear hyperbolic) systems possess a complete set of real eigenvalues (with corresponding eigenvectors). They all share the fundamental property of "finite propagation speed," thus enabling the construction of "local numerical fluxes" (see Remark 3.4). The Godunov scheme, as well as the full GRP methodology, can be adapted to the numerical resolution of such systems. In practice, however, there has only been a very limited amount of work in these directions.

In this chapter we study an important physical extension, commonly known as "reacting flow." It combines the three conservation laws with chemical reactions. The latter are responsible for the gradual or abrupt change of the chemical composition of the fluid, leading to corresponding changes in the equation of state. As an example one can think of the decomposition of a mixture of gases (or even isotopes of the same gas) under sufficiently high pressure. Another prototypical example is the process of (exothermic) combustion, in which the fluid undergoes changes caused by reactive interactions of its constituent species. It is because of this example that the whole topic is labeled "reacting flow." In particular, it covers the phenomena of "deflagrations" and "detonations," which will be discussed in Sections 9.2 and 9.3. We refer to Landau and Lifshitz [74, Chapter 14] for a broader discussion of issues related to combustion. The model that we discuss here is perhaps the simplest one for "continuous" reaction. It assumes that the fluid consists of a (homogeneous) mixture of two species, "burnt" and "unburnt." Their ratio varies in time (and space) as a result of the chemical process.

9.1 The Equations of Compressible Reacting Flow

Let $0 \leq z \leq 1$ be the mass fraction of unburnt gas. Thus $z = 1$ (resp. $z = 0$) represents a completely unburnt (resp. burnt) mixture. In general, the mixture is assumed to be homogeneous and possesses a single set of thermodynamic variables ρ, e, p, etc. It undergoes an irreversible chemical reaction, whereby the fraction of the unburnt component always decreases. This rule is expressed by the "reaction law" (see Courant and Friedrichs [30, Chapter III, Section 93]),

$$\frac{D}{Dt} z = -k(e, \rho, z),\qquad(9.2)$$

where we have used the "total" (or "Lagrangian") time derivative $\frac{D}{Dt} = \frac{\partial}{\partial t} + u \frac{\partial}{\partial x}$ [see Equations (4.50) and (4.82)]. The "reaction rate" function k is assumed to be non negative and is indeed the rate at which the reaction (i.e., decrease of z) takes place along the trajectory of a fluid element.

As the mixture composition certainly affects the thermodynamic variables, we have to modify our previous assumptions. We take (e, ρ, z) as the three basic (independent) variables, and these are used to determine all other thermodynamic quantities, such as the entropy $S = S(e, \rho, z)$ or the pressure $p = p(e, \rho, z)$. Some caution must be exercised when applying the basic thermodynamical relations in this extended setting. Thus, by fixing z, the mixture retains a fixed ratio of its components and is therefore subject to the first law of thermodynamics (9.1). Since no change of z takes place, this law can be written as

$$T\,dS_{z=\mathrm{const}} \equiv T\left(\frac{\partial S}{\partial e} de + \frac{\partial S}{\partial \rho} d\rho\right) = de + p\,d\tau.\qquad(9.3)$$

In particular $\frac{\partial S}{\partial e} = \frac{1}{T} > 0$. Solving for e we obtain a dependence $e = e(\rho, S, z)$ and substituting this in the equation of state $p = p(e, \rho, z)$ we get $p = p(\rho, S, z)$.

In accordance with (4.61) the speed of sound c is now defined by

$$c^2 = \left(\frac{\partial p}{\partial \rho}\right)_{S,z}. \tag{9.4}$$

Following these preliminaries, we can now write the system of equations that governs *planar* reacting flow, namely, the compressible planar flow [as in Equations (4.47)] of a reacting fluid mixture, subject to the reaction law (9.2):

$$\frac{\partial}{\partial t}\mathbf{U} + \frac{\partial}{\partial x}\mathbf{F}(\mathbf{U}) = \mathbf{K}(\mathbf{U}), \tag{9.5}$$

$$\mathbf{U} = \begin{bmatrix} \rho \\ \rho u \\ \rho\left(e + \frac{1}{2}u^2\right) \\ \rho z \end{bmatrix}, \quad \mathbf{F}(\mathbf{U}) = \begin{bmatrix} \rho u \\ \rho u^2 + p \\ \rho u\left(e + \frac{1}{2}u^2\right) + pu \\ \rho z u \end{bmatrix}, \quad \mathbf{K}(\mathbf{U}) = \begin{bmatrix} 0 \\ 0 \\ 0 \\ -k\rho \end{bmatrix}. \tag{9.6}$$

Observe that the fourth (reaction) equation is written in the quasi-conservative form

$$\frac{\partial}{\partial t}(\rho z) + \frac{\partial}{\partial x}(\rho z u) = -k\rho, \tag{9.7}$$

which follows from (9.2) in view of the first (conservation of mass) equation.

The system (9.5) is not homogeneous; it contains a nonvanishing "source" term $\mathbf{K}(\mathbf{U})$. The situation encountered here is analogous to that of the quasi-one-dimensional system (4.48). In both cases, dropping the undifferentiated (source) term leads to a system of conservation laws of the class studied in the general setting of Section 4.1 [Equation (4.1)]. In the quasi-one-dimensional case, this reduction led to the planar system of compressible flow (4.47). In the case of the system (9.5) the reduction yields

$$\frac{\partial}{\partial t}\mathbf{U}^{\text{red}} + \frac{\partial}{\partial x}\mathbf{F}(\mathbf{U}^{\text{red}}) = \mathbf{0}, \quad \mathbf{U}^{\text{red}} = \left(\rho, \rho u, \rho\left(e + \frac{1}{2}u^2\right), \rho z^{\text{red}}\right)^T. \tag{9.8}$$

This system is completely analogous to (7.17) [or, equivalently, (7.22) and (7.23)]; instead of the transversal velocity component v we have here the "reduced" mass fraction z^{red}. These variables satisfy, respectively, Equations (7.19) and [in the case of (9.8)] $z_t^{\text{red}} + uz_x^{\text{red}} = 0$. Both are convected "passively" along the particle paths $\frac{dx}{dt} = u$. The eigenvalues of (9.8), hence also of (9.5), are

$$\lambda_1 = u - c, \qquad \lambda_2 = u, \qquad \lambda_3 = u + c, \qquad \lambda_4 = u, \tag{9.9}$$

[compare (7.24)]. The system is therefore "doubly linearly degenerate" (see Remark 7.7). In particular, the discontinuities of z^{red} can take place only across contact discontinuities.

Let us now turn back to the reacting flow system (9.5). In this case Equation (9.2) implies that the mass fraction (of the unburnt component) decreases along particle paths. We shall see that as a result the entropy S also varies along these paths, in contrast to the adiabatic law (4.59), which is valid in regions of smooth (nonreacting) flow. Note first that the derivation of Equation (4.58) was based solely on the conservation laws (4.45), namely, the first three equations in (9.5). We conclude that in the reacting case too,

$$\frac{De}{Dt} + p\frac{D\tau}{Dt} = 0, \tag{9.10}$$

from which we infer, by (9.3),

$$\frac{\partial S}{\partial e}\frac{De}{Dt} + \frac{\partial S}{\partial \rho}\frac{D\rho}{Dt} = 0 \tag{9.11}$$

(meaning that, for a fixed z, there is no change of entropy along smooth particle paths). The full variation of entropy is now given by

$$\frac{DS}{Dt} = \frac{\partial}{\partial z}S(e, \rho, z)\frac{Dz}{Dt} = -k\frac{\partial}{\partial z}S = -f(e, \rho, z), \tag{9.12}$$

where the right-hand side $-f$ is a thermodynamic function depending on the equation of state $S = S(e, \rho, z)$ and the reaction rate k. It expresses the change in entropy of a mixed element of the fluid as it undergoes chemical reaction.

As in the case of nonreacting flow, it is useful to formulate the system (9.5) in the Lagrangian framework. Since the flow is planar, the Lagrangian coordinate ξ [see (4.79)] is defined here by

$$\xi = \int_{x_0}^{x} \rho(s, 0)ds. \tag{9.13}$$

Following the derivation of (4.83) and noting (9.2) we get

$$\tfrac{\partial}{\partial t}\mathbf{V} + \tfrac{\partial}{\partial \xi}\Phi(\mathbf{V}) = \Psi(\mathbf{V}),$$

$$\tag{9.14}$$

$$\mathbf{V} = \begin{bmatrix} \tau \\ u \\ E \\ z \end{bmatrix}, \quad \Phi(\mathbf{V}) = \begin{bmatrix} -u \\ p \\ pu \\ 0 \end{bmatrix}, \quad \Psi(\mathbf{V}) = \begin{bmatrix} 0 \\ 0 \\ 0 \\ -k \end{bmatrix},$$

where $E = e + \frac{1}{2}u^2$. Note that the definition of the Lagrangian coordinate implies that $\frac{\partial}{\partial t}$ here is identical to the total derivative $\frac{D}{Dt}$. In regions of smooth flow Equation (9.12) can be exploited in order to simplify (9.14), replacing E by S, giving

$$\tfrac{\partial}{\partial t}\widetilde{\mathbf{V}} + \tfrac{\partial}{\partial \xi}\widetilde{\mathbf{\Phi}}(\widetilde{\mathbf{V}}) = \widetilde{\mathbf{\Psi}}(\widetilde{\mathbf{V}}),$$

$$\widetilde{\mathbf{V}} = \begin{bmatrix} \tau \\ u \\ S \\ z \end{bmatrix}, \quad \widetilde{\mathbf{\Phi}}(\widetilde{\mathbf{V}}) = \begin{bmatrix} -u \\ p \\ 0 \\ 0 \end{bmatrix}, \quad \widetilde{\mathbf{\Psi}}(\widetilde{\mathbf{V}}) = \begin{bmatrix} 0 \\ 0 \\ -f \\ -k \end{bmatrix}. \tag{9.15}$$

The Characteristic Relations

The last two equations in (9.15) serve as the characteristic equations associated with the double eigenvalue $\lambda_2 = \lambda_4 = u$ [compare (4.67)(ii)]. The other two equations can be directly extracted from (9.15) as follows.

If we use $p = p(\rho, S, z)$ as in (9.4) the momentum equation yields

$$u_t + c^2 \rho_\xi + p_S S_\xi + p_z z_\xi = 0,$$

and adding to it suitable multiples of the other equations we get

$$u_t \pm \rho c u_\xi \pm \frac{c}{\rho}(\rho_t \pm \rho c \rho_\xi) \pm \frac{p_S}{\rho c}(S_t \pm \rho c S_\xi) \pm \frac{p_z}{\rho c}(z_t \pm \rho c z_\xi)$$

$$= \mp \frac{1}{\rho c}(p_S f + p_z k). \tag{9.16}$$

The left-hand side of (9.16) is seen to be equal to $\frac{du}{dt} \pm \frac{1}{g}\frac{dp}{dt}$ along the characteristic directions $C_\pm : \frac{d\xi}{dt} = \pm g = \pm \rho c$, where $g = \rho c$ is the Lagrangian speed of sound. As for the right-hand side in (9.16), we define the thermodynamic function

$$\Lambda(e, \rho, z) = p_S f + p_z k = k \left[\frac{\partial}{\partial S} p(\rho, S, z) \frac{\partial}{\partial z} S(e, \rho, z) + \frac{\partial}{\partial z} p(\rho, S, z) \right]$$

$$= k(e, \rho, z) \frac{\partial}{\partial z} p(e, \rho, z), \tag{9.17}$$

so that (9.16) can be written as

$$g\, du \pm dp = \mp \Lambda \qquad \text{along} \quad C_\pm : \frac{d\xi}{dt} = \pm g \tag{9.18}$$

[compare (4.85)]. The role of the function Λ is further highlighted in the relation

$$\frac{Dp}{Dt} = c^2 \frac{D\rho}{Dt} - \Lambda, \tag{9.19}$$

where, as in (9.10)–(9.12), the total derivatives $\frac{D}{Dt}$ are evaluated along particle paths.

To prove (9.19) we use the relations (9.4) and (9.17), in conjunction with equations (9.2) and (9.12), to obtain, with $p = p(\rho, S, z)$,

$$\frac{Dp}{Dt} = c^2 \frac{D\rho}{Dt} + p_S \frac{DS}{Dt} + p_z \frac{Dz}{Dt} = c^2 \frac{D\rho}{Dt} - \Lambda.$$

Comparing (9.19) to the adiabatic relation (4.67), we see that $-\Lambda$ expresses the rate of change of the pressure resulting from the chemical reaction along a particle trajectory (beyond the adiabatic effect of compression).

The set of characteristic relations can be written in the Eulerian framework, along the directions given by the eigenvalues $\lambda_1, \ldots, \lambda_4$ [see (9.9)]. We have, as in (4.67),

$$g\, du \pm dp = \mp \Lambda dt, \qquad \text{along} \quad \frac{dx}{dt} = u \pm c, \qquad (9.20)$$

$$dS = -f\, dt, \quad dz = -k\, dt, \qquad \text{along} \quad \frac{dx}{dt} = u. \qquad (9.21)$$

Discontinuities and Centered Rarefaction Waves

In dealing with the discontinuities of solutions to the system (9.5) we first recall that jumps in z can take place only across contact discontinuities [see the discussion following Equation (9.8)]. In particular, z is continuous across shocks associated with the λ_1, λ_3 families, and consequently there is no change in the mixture composition across such shocks. In the study of the jump (Rankine–Hugoniot) conditions we may therefore assume a fixed value of $z = z_0$. The first three equations in (9.5) are then identical to the planar system[1] (4.47) [with $p = p(e, \rho, z_0)$ etc.]. The corresponding jump relations are then given by (4.76), or, in a more explicit form, by (4.86). The whole treatment of the shock conditions, leading up to Summary 4.36, can now be repeated verbatim. It should be kept in mind, however, that the Hugoniot curve $H(\tau, p)$ [see (4.92)] depends on the parameter z_0 (which determines the exact composition of the fluid and hence its equation of state). When the jumps in pressure and density across a shock front are known, the velocity jump is determined by (4.95) [resp. (4.96)] for a 3-shock [resp. 1-shock].

In the case of a contact discontinuity the pressure and velocity are continuous, whereas the mass fraction z and the density may undergo jumps.

---

[1] As already observed in the paragraph following (4.76), the presence of source terms such as $\mathbf{K}(\mathbf{U})$ does not affect the Rankine–Hugoniot conditions.

We now turn to the case of a centered rarefaction wave (CRW). As observed in Section 4.2 this type of nonlinear wave is associated with the λ_1, λ_3 characteristic families. As in the quasi-1-D case (see the discussion preceding Claim 4.42) we cannot expect (generally speaking) a "self-similar" CRW, owing to the presence of a nonvanishing $\mathbf{K}(\mathbf{U})$ in (9.5). This is still the case even if the CRW is assumed to connect two constant states. However, the reduced system (9.8) belongs to the general class of systems studied in Section 4.1 [recall the analogy with (7.17)]. As observed in the paragraph following (9.8), it decomposes into a convection equation for z^{red} and a system identical to that of planar flow. The former implies that z^{red} is constant throughout the CRW (connecting two constant states). It follows that all other flow variables can be obtained from the Riemann invariants, as in (4.98). Note that the speed of sound depends on the (constant) value of z^{red}. For later reference we summarize the above considerations as follows (see Claim 4.42).

Summary 9.1 *A CRW solution of the reduced system (9.8), connecting two constant states, is self-similar, isentropic, and carries a constant value of z^{red}. The values of the other flow variables are obtained in terms of the Riemann invariants of isentropic planar flow (4.98).*

As in the case of quasi-1-D flow, the "reduced" CRW is the "limit" (at the singularity) for the full CRW associated with (9.5). Heuristically, it means that the effect of the chemical reaction (through changes in z) is vanishing as we observe the solution closer and closer to the initial interaction (at the singularity).

9.2 The Chapman–Jouguet (C–J) Model

Before dealing with the Riemann and generalized Riemann problems for the full reacting flow system (9.5), we consider a well-known simplified model, the Chapman–Jouguet (C–J) model (see Courant and Friedrichs [30, Chapter III, Section 84]). The basic hypothesis of a continuous "homogeneous coexistence" of the burnt and unburnt species (represented by the mass fraction z) is suppressed. Instead, it is assumed that the two species are separated by a sharp discontinuity (shock front).[2] Each species has its own equation of state, corresponding to the extreme cases of $z = 0$ (burnt) and $z = 1$ (unburnt) of the continuous model. The reaction equation (9.7) is dropped, and the system (9.5) is replaced by (9.8). However, the two species are governed by two different

---

[2] Formally, this model can be obtained by letting $k \to +\infty$ in (9.7), for pressures (or rather temperatures) beyond a "critical" value. Thus, when the unburnt gas is "ignited", it is sharply converted into the fully burnt species.

equations of state. We designate these equations by $p = p^{(1)}(e, \rho)$ (unburnt) and $p = p^{(0)}(e, \rho)$ (burnt). Solving for e we have $e = e^{(1)}(\rho, p)$ and $e = e^{(0)}(\rho, p)$ for the two species. Across a shock front separating the two species, the jump relations (4.77) are still valid, but the internal energies on the two sides are given by the respective equations of state. Let us take a fixed unburnt right state (τ_+, p_+) as the preshock state. As in Summary 4.36, the set of all (left) burnt states (τ, p) satisfying the Rankine–Hugoniot jump condition [i.e., connected to (τ_+, p_+) by a 3-shock] is represented by the Hugoniot curve

$$H^{(0)}(\tau, p) = \frac{(p + p_+)}{2}(\tau - \tau_+) + e^{(0)}(\tau, p) - e^{(1)}(\tau_+, p_+) = 0. \quad (9.22)$$

Observe that, as $\tau \to \tau_+$, the pressure p approaches a value p_l such that

$$e^{(0)}(\tau_+, p_l) = e^{(1)}(\tau_+, p_+). \quad (9.23)$$

Both $e^{(0)}$ and $e^{(1)}$ are increasing functions of pressure. We now add to it the physically plausible assumption that *for any pressure value p, $e^{(1)}(\tau_+, p) >$ $e^{(0)}(\tau_+, p)$.* This means that the unburnt species contains "chemical energy," which is released in the reaction process. To achieve the inequality (9.23), we must therefore have $p_l > p_+$.

Similarly, taking $p = p_+$ in (9.22) we obtain $\tau = \tau_l$ such that

$$p_+(\tau_l - \tau_+) + e^{(0)}(\tau_l, p_+) - e^{(1)}(\tau_+, p_+) = 0. \quad (9.24)$$

As before $e^{(1)}(\tau, p_+) > e^{(0)}(\tau, p_+)$ for any value of τ. However, $e^{(0)}$ and $e^{(1)}$ are decreasing functions of τ (since they increase with increasing density), so that (9.24) yields $\tau_l > \tau_+$ [otherwise the left-hand side in (9.24) is negative]. It follows that the Hugoniot curve (9.22) is as shown in Figure 9.1. In particular, it does not pass through (τ_+, p_+), and the portion between (τ_l, p_+) and (τ_+, p_l) is omitted. The latter fact is due to the relation (4.89)(ii),

$$-M^2 = \frac{p - p_+}{\tau - \tau_+}, \quad (9.25)$$

which is satisfied by all admissible left states (τ, p). The omitted portion corresponds to

$$\frac{p - p_+}{\tau - \tau_+} > 0.$$

In the nonreacting case, the upper and lower parts of the Hugoniot curve (see Figure 4.11) correspond to interchanged roles of right and left states, as in Remark 4.37. In the present case, by contrast, the roles played by the two

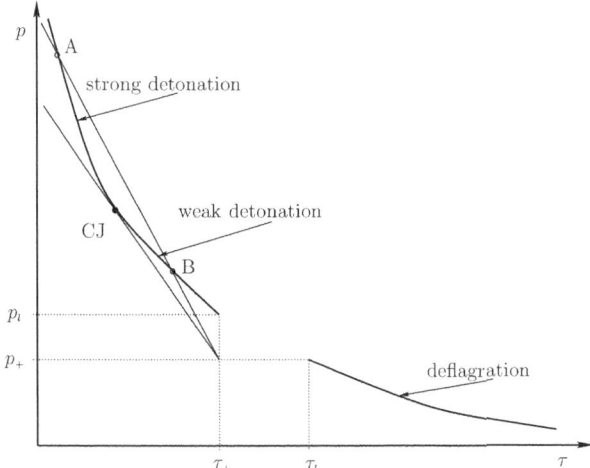

Figure 9.1. The C–J Hugoniot curve $H^{(0)}(\tau, p) = 0$: Burnt states connected to an unburnt state (τ_+, p_+) by a detonation or a deflagration front.

parts of the Hugoniot curve are very different. The lower part $(\tau > \tau_l)$ is called the "deflagration branch" of the curve. It represents burnt states for which both the pressure and the density are lower than those of the unburnt state. This is in contrast to the nonreacting case where, in view of entropy considerations, postshock values (of pressure and density) are higher than the preshock ones (Claim 4.35). Here too, the deflagration states are excluded on the basis of entropy considerations. In particular, it can be shown that for such states the shock is subsonic with respect to the unburnt (preshock) state, thus violating (4.78). We shall not deal any further with this branch and refer to Courant and Friedrichs [30, Chapter III, Section 88] for a detailed discussion.

The upper part $(p > p_l)$ of the Hugoniot curve is the "detonation branch." The pressure and the density of any state on this branch are higher than the respective values of the unburnt state. In this respect this branch "agrees" with the admissible (postshock) states of nonreacting flow. It therefore serves as a basic model for the physical phenomenon of detonation, where the chemical transition is fully completed over an "infinitesimally thin" layer, producing higher values of pressure and density in the burnt substance (while preserving mass, momentum and energy). Moreover, under plausible physical hypotheses on the equation of state it can be shown that the detonation branch is convex (compare Summary 4.36). It follows that a straight line [in the (τ, p) plane] emanating from the unburnt state (τ_+, p_+) meets the detonation branch at exactly two points (or none at all), such as A and B in Figure 9.1. There is one exception: The line can be tangent, touching the curve at the point marked as CJ in Figure 9.1. A straight line like this is called a "Rayleigh line." By (9.25) its slope is equal to $-M^2$. Recall that [see (4.88) with $A = 1$] M is the speed of the shock in the Lagrangian frame. If σ_3 is the speed in the Eulerian frame and u_+ is the velocity (of the unburnt state), then the mass flux across the shock is $M = \rho_+(\sigma_3 - u_+) > 0$. Thus, the slope of the Rayleigh line in the (τ, p) plane determines the shock speed σ_3 [the state (τ_+, p_+, u_+) is held fixed]. In particular we have the following conclusion, which will be needed in the next section.

Corollary 9.2 *The shock speeds corresponding to the two detonation states on the same Rayleigh line (such as A and B in Figure 9.1) are identical. The shock speed at the tangency point CJ is minimal.*

Denoting by (τ_A, p_A) and (τ_B, p_B) the states A and B, respectively, and substituting both of them in (9.22) we get, by subtraction,

$$\frac{p_A + p_+}{2}(\tau_A - \tau_+) + e^{(0)}(\tau_A, p_A) - \left[\frac{p_B + p_+}{2}(\tau_B - \tau_+) + e^{(0)}(\tau_B, p_B)\right] = 0.$$

Using (9.25) (with the same M) in this equation we readily obtain

$$\frac{p_A + p_B}{2}(\tau_A - \tau_B) + e^{(0)}(\tau_A, p_A) - e^{(0)}(\tau_B, p_B) = 0. \tag{9.26}$$

This equation has a very simple interpretation. Comparing it to Equation (4.92) we see that it means that the state A lies on the Hugoniot curve passing through B (in the burnt gas, where $e = e^{(0)}(\tau, p)$ is the internal energy) and vice versa. Let u_A, u_B be the velocities at the states A, B respectively. Since state A lies on the admissible part of the Hugoniot curve through B (see P_3^+ in Figure 4.11 and Claim 4.35) the entropy condition (4.78) is satisfied, namely,

$$u_A + c_A > \sigma_3. \tag{9.27}$$

However, the state B lies on the inadmissible (lower) part of the Hugoniot curve through A and the entropy condition (4.78) is violated,

$$u_B + c_B < \sigma_3. \tag{9.28}$$

At the extreme case CJ (Figure 9.1), both points A and B coalesce at CJ, so that

$$u_{CJ} + c_{CJ} = \sigma_3. \tag{9.29}$$

Detonations corresponding to points above CJ, namely, for which the pressure jump is higher than $p_{CJ} - p_+$, are called *strong detonations*. Similarly, detonations corresponding to points below CJ are called *weak detonations* (see Figure 9.1). The detonation corresponding to the point CJ is called a *C–J detonation*. We can summarize the above considerations as follows:

Corollary 9.3 *The detonation front is*

(a) subsonic with respect to the state behind it for a strong detonation;
(b) supersonic with respect to the state behind it for a weak detonation; and
(c) sonic with respect to the state behind it for a C–J detonation.

In particular, weak detonations are in violation of the entropy condition (4.78) and are therefore inadmissible.

This corollary is known as "Jouguet's Rule." We refer to Courant and Friedrichs [30, Chapter III, Section 88] for a different derivation of this rule (based on Weyl's method). Thus, in addition to the full branch of deflagrations, we are also excluding the weak detonations from our admissible class of jump discontinuities. We should keep in mind, however, that these exclusions stem from the fact that we have used the simplified C–J model. Physically, the excluded waves can actually occur when other physical factors (such as heat conduction) come into play. We refer to Landau and Lifshitz [74, Chapter 14] for a further discussion of this topic. An interpretation, based on the Z–N–D solution, of these exclusions will be given in the next section.

We note the special case of a C–J detonation. The shock speed satisfies (9.29), meaning that the shock trajectory coincides with the characteristic line $\frac{dx}{dt} = u_{CJ} + c_{CJ}$. This is a "limiting case" for the entropy condition (4.78), where the characteristic line does not "run into" the shock front (as t grows) but is instead tangent to it. If we express the (detonation) Hugoniot curve (9.22) as $p = p(\tau)$, we have at the CJ point $\frac{dp}{d\tau} = \frac{p - p_+}{\tau - \tau_+}$. However, by (9.22), we have

$$\frac{(p_{CJ} + p_+)}{2} d\tau + \frac{(\tau_{CJ} - \tau_+)}{2} dp + de^{(0)}(\tau_{CJ}, p_{CJ}) = 0,$$

which, in view of (9.1), yields (at the CJ state)

$$T_{CJ} \, dS + \frac{1}{2} [(\tau_{CJ} - \tau_+) \, dp - (p_{CJ} - p_+) \, d\tau] = 0;$$

hence $dS = 0$ at the point CJ and the Hugoniot curve there is tangent to the isentropic curve $S(\tau, p) = S_{CJ}$. The states on this curve, with $p < p_{CJ}$, are those connected to the CJ state by a CRW in the burnt gas (Claim 4.42). Since the (C–J) shock trajectory itself is a characteristic line, it can serve as the head characteristic of such a CRW. We have therefore obtained the following conclusions, where we assume, as before, that the C–J detonation is a 3-shock.

Corollary 9.4 *Let* (τ, p) *be a state of the burnt gas, of entropy* $S(\tau, p)$ *equal to* S_{CJ} *and such that* $p < p_{CJ}$. *Then it can be connected to the (unburnt) state* (τ_+, p_+) *by a C–J detonation followed by a centered rarefaction wave (associated with the same eigenvalue* $\lambda = u + c$*).*

The Riemann problem in the C–J model deals with the flow evolving from initial data that consist of a constant unburnt state on the right and a constant burnt state on the left. The solution procedure is similar to the one discussed in Construction 4.46. However, in the present case the wave Γ_3 (which is the detonation separating the burnt and unburnt states) is either a strong detonation (when $p > p_{CJ}$) or the combined wave discussed in Corollary 9.4. We shall not pursue this topic any further and refer to Courant and Friedrichs [30, Chapter III, Section 90] and Teng, Chorin, and Liu [110] for a detailed discussion.

9.3 The Z–N–D (Zeldovich–von Neumann–Döring) Solution

We now turn back to the full reacting flow system (9.5).

Fix $0 \leq z \leq 1$ and let (τ_+, p_+) be a fixed state in the unburnt gas. We define

$$H^{(z)}(\tau, p) = \frac{(p + p_+)}{2}(\tau - \tau_+) + e(\tau, p, z) - e^{(1)}(\tau_+, p_+), \qquad (9.30)$$

where $e^{(1)}(\tau_+, p_+) = e(\tau_+, p_+, 1)$ as in (9.22). If we view the state (τ_+, p_+) as a right preshock state, the Hugoniot curve

$$H^{(z)}(\tau, p) = 0 \qquad (9.31)$$

is the locus of (left) postshock states with ratio $(1-z)/z$ of burnt to unburnt gas, which are connected to (τ_+, p_+) by a 3-shock (satisfying the Rankine–Hugoniot jump condition). Thus the chemical composition of the fluid changes discontinuously across the shock; it is fully unburnt ahead of it but contains only a fraction z of the unburnt species behind it. The curve (9.22) of the C–J model corresponds to $z = 0$, whereas the one with $z = 1$ is the Hugoniot curve in the unburnt species. These two, along with an intermediate one ($0 < z < 1$) are shown in Figure 9.2. In view of the discussion in the preceding section, we show only the admissible (detonation) branches of the curves.

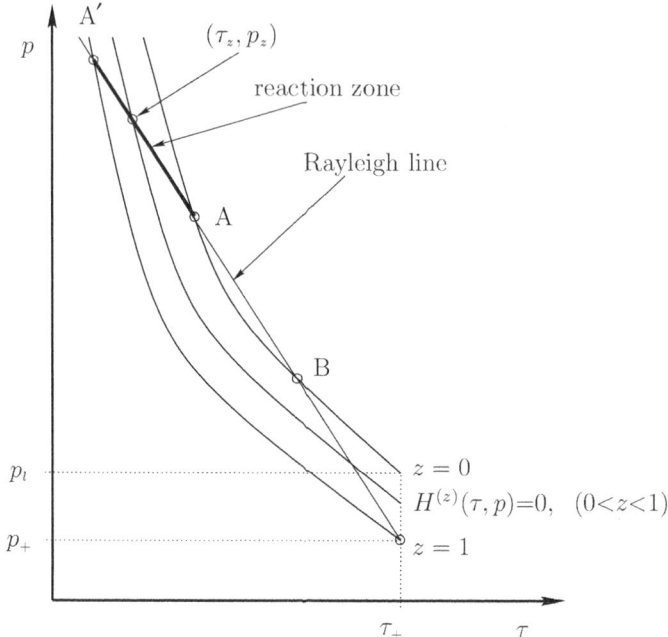

Figure 9.2. The Z–N–D solution. The burnt state A (Figure 9.1) is obtained through a shock front (A′) in the unburnt gas, followed by a reaction zone (A′ → A) lying on the Rayleigh line.

The equation of the Rayleigh line through A and B (Figure 9.2) is given by (9.25). It intersects the curve $H^{(1)} = 0$ at A′, and we denote by (τ_z, p_z) its point of intersection with $H^{(z)} = 0$, $0 \le z \le 1$. The speed $\sigma_3 = M\tau_+ + u_+$ is common to all shocks with postshock states (τ_z, p_z), $0 \le z \le 1$ [and preshock state (τ_+, p_+)].

Given τ_z, all other postshock values are uniquely determined, and we denote them as u_z, S_z, etc. The smaller the unburnt fraction z, the smaller the pressure jump $p_z - p_+$. The largest jump is attained at A′ [i.e., (τ_1, p_1)]. In this case the shock separates two fully unburnt states. For simplicity we shall henceforth refer to this shock as an "A′-shock" (Figure 9.2). Note that the entropy condition (4.78) yields (because the shocks are "strong detonations"; see Corollary 9.3)

$$u_z + c_z > \sigma_3 > u_z, \qquad 0 \le z \le 1. \tag{9.32}$$

Consider the A′-shock starting at $x = 0$ and traveling into the uniform unburnt state (τ_+, p_+). It is located at $x = \sigma_3 t$ at time t so that $z(\sigma_3 t, t) = 1$. We seek a special solution to the system (9.5), in the region behind the shock ($x \le \sigma_3 t$), for which all flow variables depend only on the distance $\eta = x - \sigma_3 t$. Without the risk of confusion, we write these variables as functions of a single variable, $u = u(\eta)$, $p = p(\eta)$, etc. A solution of this type is clearly a "traveling wave", namely, the profile of each variable moves, unmodified, at speed σ_3. In particular, the profile $z(\eta)$ is expected to decrease from $z(0) = 1$ to 0 as η decreases to $-\infty$.

To construct the solution, we start with Equation (9.2). We take for u, e, ρ their values on the Rayleigh line u_z, e_z, ρ_z, which means that for a given $z = z(\eta)$ the state $(\rho(\eta), p(\eta), u(\eta))$ is the one corresponding to z on the Rayleigh line. Equation (9.2) then reduces to

$$z'(\eta)(-\sigma_3 + u_{z(\eta)}) = -k\left(e_{z(\eta)}, \rho_{z(\eta)}, z(\eta)\right). \qquad (9.33)$$

Conservation of mass [Equation (4.87)(i)] implies $u_z - \sigma_3 = -M\tau_z$; hence Equation (9.33) can be rewritten as

$$z'(\eta) = M^{-1}\rho_{z(\eta)}\, k\left(e_{z(\eta)}, \rho_{z(\eta)}, z(\eta)\right), \qquad (9.34)$$

which is an ordinary first-order differential equation for $z(\eta)$, $-\infty < \eta \le 0$, subject to the initial condition $z(0) = 1$. In particular, $z'(\eta) \ge 0$ and we make the following assumption:

Equation (9.34) admits a unique solution $z(\eta)$, $-\infty < \eta \le 0$.

This solution is monotonic and nondecreasing, and either $z(\eta_0) = 0$ (9.35)

for some $-\infty < \eta_0 < 0$, or $\displaystyle\lim_{\eta \to -\infty} z(\eta) = 0$.

Using this assumption we proceed to find a "traveling" solution to the first three equations of (9.5), which can now be written as

$$-\sigma_3 \widetilde{\mathbf{U}}'(\eta) + \tfrac{d}{d\eta}\widetilde{\mathbf{F}}(\widetilde{\mathbf{U}}(\eta)) = \mathbf{0}, \qquad -\infty < \eta \le 0,$$

$$\widetilde{\mathbf{U}}(\eta) = \begin{bmatrix} \rho(\eta) \\ \rho(\eta)u(\eta) \\ \rho(\eta)\left(e(\eta) + \tfrac{1}{2}u(\eta)^2\right) \end{bmatrix}, \qquad (9.36)$$

$$\widetilde{\mathbf{F}}(\widetilde{\mathbf{U}}(\eta)) = \begin{bmatrix} \rho(\eta)u(\eta) \\ \rho(\eta)u(\eta)^2 + p(\eta) \\ u(\eta)\left[\rho(\eta)\left(e(\eta) + \tfrac{1}{2}u(\eta)^2\right) + p(\eta)\right] \end{bmatrix}.$$

This system is supplemented by the initial condition

$$\widetilde{\mathbf{U}}(0) = \begin{bmatrix} \rho_1 \\ \rho_1 u_1 \\ \rho_1 \left(e_1 + \frac{1}{2}u_1^2\right) \end{bmatrix} \qquad \text{(the state at A'; Figure 9.2).}$$

Integrating (9.36) we obtain the equivalent form

$$-\sigma_3 \left[\widetilde{\mathbf{U}}(\eta) - \widetilde{\mathbf{U}}(0)\right] + \widetilde{\mathbf{F}}(\widetilde{\mathbf{U}}(\eta)) - \widetilde{\mathbf{F}}(\widetilde{\mathbf{U}}(0)) = \mathbf{0}. \tag{9.37}$$

The solution to (9.37) is readily available. Indeed,

$$\widetilde{\mathbf{U}}(\eta) = \widetilde{\mathbf{U}}_{z(\eta)}, \tag{9.38}$$

namely, the state on the Rayleigh line at $z = z(\eta)$. The verification of (9.37) is straightforward: It is the Rankine–Hugoniot jump condition between the states (τ_+, p_+) and $(\tau_{z(\eta)}, p_{z(\eta)})$. The speed of this shock is σ_3, common to all states $(\tau_{z(\eta)}, p_{z(\eta)})$ on the Rayleigh line.

Summary 9.5 (The Z–N–D traveling solution) *Assuming (9.35), there exists a solution $\widetilde{\mathbf{U}}(\eta) = \widetilde{\mathbf{U}}(x - \sigma_3 t)$, $-\infty < \eta \leq 0$, to the reacting-flow system (9.5). This solution has the following properties:*

(i) *$\widetilde{\mathbf{U}}(\eta)$ lies on the Rayleigh line (Figure 9.2) for all η.*
(ii) *$\widetilde{\mathbf{U}}(0)$ is the state A' of the unburnt gas.*
(iii) *The pressure $p(\eta)$, density $\rho(\eta)$, and mass fraction $z(\eta)$ are all monotonic functions of η, decreasing as $\eta \to -\infty$.*

Observe that property (iii) follows from the monotonicity of $z(\eta)$. As $z(\eta)$ decreases, the point $(\tau(\eta), p(\eta)) = (\tau_{z(\eta)}, p_{z(\eta)})$ moves down the Rayleigh line (Figure 9.2) from A' to A.

Definition 9.6 The solution described in Summary 9.5 is called the "Z–N–D profile." It is shown in Figure 9.3.

The Z–N–D profile represents the "reaction-mechanism" built into the reacting-flow system (9.5). It starts with a leading shock in the unburnt gas, raising the pressure and density to their peak values p_1, ρ_1 (A' in Figure 9.2). There is also a corresponding rise in temperature, which initiates the chemical reaction, that is, the "ignition" of the unburnt gas. The reactive process that follows gradually transforms the unburnt species to the burnt one, thereby decreasing pressure and density. This whole process takes place on the segment

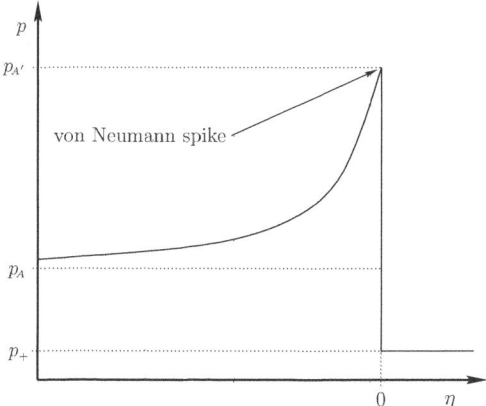

Figure 9.3. The Z–N–D profile, $p = p(\eta)$.

A'A of the Rayleigh line, which is therefore labeled as the "reaction zone" in Figure 9.2.

We also note that the state B of weak detonation (Figure 9.2) is not accessible by this mechanism; the gas is already completely burnt at the strong detonation state A.

Observe that if k in Equation (9.34) is increased, the reaction process is accelerated and $z(\eta)$ decreases to 0 at a higher rate. In other words, the reaction zone becomes narrower and the strong detonation A (in the fully burnt gas) is attained more quickly. The Z–N–D profile then comes closer to the C–J model (compare Footnote 2).

Note that the Z–N–D profile is *not* a solution to the Riemann problem associated with (9.5). Indeed, setting $t = 0$ we see that $\mathbf{U}(\eta)$ solves the IVP for (9.5) with initial data identical to the Z–N–D profile (which only reiterates the fact that this profile is a "traveling wave"). In the remaining sections we describe the GRP solution to the system (9.5). This solution does not rely, of course, on the Z–N–D profile. In fact, if we resolve numerically a "reacting" Riemann problem (Section 9.4), the computed solution is found to possess the main features of that profile.

Finally, we note that there exist no theoretical results concerning the global existence and uniqueness of solutions to the IVP associated with the system (9.5). In fact, such results do not exist even in the case of the planar nonreacting system (4.47), except for certain cases (isentropic flow or initial data "close" to constant). A detailed discussion of this topic is beyond the scope of this monograph. The local existence of a solution to (9.5), with initial data having a jump at $x = 0$, is shown in Ying and Wang [124]. In particular, it is shown there that the (local) wave pattern is identical to that of the homogeneous system (9.8), thus providing a basis to our GRP solution.

A simplified model to the reacting flow system has been proposed in Majda [88]. It is an extension of Burgers' equation, adding an unknown that plays the role of the mass fraction z, with a suitable "reaction mechanism." A "traveling wave" solution corresponding to the Z–N–D profile was given in Majda [88], and a proof of the existence and uniqueness of a solution to the general IVP was obtained in Levy [82] and Ying and Teng [123]. The paper by Levy [82] also provides a proof of the convergence to the C–J model (in this simple case) and a numerical GRP algorithm.

9.4 The Linear GRP for the Reacting-Flow System

The linear GRP for the system (9.5) is stated as in the nonreacting case (Definition 5.3). The initial data $\mathbf{U}(x, 0)$ are piecewise linear with a possible jump at $x = 0$,

$$\mathbf{U}(x, 0) = \begin{cases} \mathbf{U}_L + x\mathbf{U}_L', & x < 0, \\ \mathbf{U}_R + x\mathbf{U}_R', & x > 0, \end{cases} \tag{9.39}$$

where $\mathbf{U}_L$, $\mathbf{U}_R$, $\mathbf{U}_L'$, $\mathbf{U}_R'$ are constant (four-component) vectors. The situation is somewhat analogous to that of the system (7.17), where the added variable v is "passively" advected along particle paths. There is, nevertheless, a substantial difference between Equations (7.19) and (9.2). The unknown v can be effectively decoupled from the other equations, as in (7.22), (7.23). However, z varies along a particle path, thus affecting the other flow variables (via the equation of state). This strong coupling within the system (9.5) is evident in Equation (9.19) [whereas in the case of (7.17) the adiabatic relation (4.68) remains unchanged].

The Associated Riemann Problem

The "associated RP" to (9.5), (9.39) (see Definition 5.2) is the Riemann problem for the reduced system (9.8), subject to the initial data

$$\mathbf{U}^{\text{red}}(x, 0) = \begin{cases} \mathbf{U}_L, & x < 0, \\ \mathbf{U}_R, & x > 0. \end{cases} \tag{9.40}$$

Since the system and the initial data are homogeneous, the solution $\mathbf{U}^{\text{red}}(x, t)$ is "self-similar," depending only on $\frac{x}{t}$. Employing the same notation as in the nonreacting case [see the paragraph following Equation (5.2)], we write

$$\mathbf{U}^{\mathrm{red}}(x, t) = \mathbf{R}^{A} \left(\frac{x}{t}; \mathbf{U}_{L}, \mathbf{U}_{R} \right). \tag{9.41}$$

The system (9.8) is analogous to (7.17). The first three equations (conservation laws) are identical to the planar system (4.47), and the resulting wave structure is determined in the same way [see Figures 4.16 and 5.1(b)]. In particular, since z^{red} is invariant along particle paths,

$$z^{\mathrm{red}}(x, t) = \begin{cases} z_{L}, & x < u^{*}t, \\ z_{R}, & x > u^{*}t, \end{cases} \tag{9.42}$$

where u^{*} is the speed of the contact discontinuity. To obtain the solution $\mathbf{R}^{A} \left(\frac{x}{t}; \mathbf{U}_{L}, \mathbf{U}_{R} \right)$ we follow Construction 4.46. The main idea is that the interaction curves I_{3}^{r}, I_{1}^{l} [in the (u, p) plane] intersect at the point (u^{*}, p^{*}) (see Figure 4.18). These curves consist (each) of a "shock branch" and a "CRW branch," as in Summary 4.45. Note, however, that in the present case these curves depend on two different equations of state, namely, $p = p(e, \rho, z_{L})$ [resp. $p = p(e, \rho, z_{R})$] to the left (resp. right) of the contact discontinuity. The speeds of sound, needed in the isentropic branch of the interaction curves [the lower parts in Equations (4.101) and (4.102)] are evaluated as in (9.4), with S_{L}, z_{L} (resp. S_{R}, z_{R}) for I_{1}^{l} (resp. I_{3}^{r}). Also, the Hugoniot curves in the (τ, p) plane [hence the upper parts in (4.101) and (4.102)] depend on the values of z.[3]

Remark 9.7 (γ-law equation of state) A commonly used simple equation of state, which serves in virtually all numerical test cases of combustion, is that of a perfect (γ-law) gas. [See (4.104)–(4.106) for the nonreacting case.] In the present case, the additional dependence on z is introduced as follows:

$$p = (\gamma - 1)\rho(e - q_{0}z), \qquad \gamma > 1, \quad q_{0} > 0, \tag{9.43}$$

where γ is independent of z. The constant $q_{0} > 0$ is the specific chemical energy contained in the unburnt gas and released in the combustion process. For any fixed z, the constant $q_{0}z$ is therefore just a "translation" of the "zero level" of the internal energy. As such, it has no effect on the other thermodynamical formulas. In particular, the entropy S is a monotonic function of $p\tau^{-\gamma}$ [see

---

[3] Note, however, that this is different from the Hugoniot curve $H^{(z)}(\tau, p) = 0$ in Figure 9.2, which is related to an unburnt preshock state and does not pass through (τ_{+}, p_{+}). In the present Riemann problem, the mass fraction z^{red} and hence the composition of the fluid are identical on the two sides of the shock, implying that the Hugoniot curves pass through (τ_{L}, p_{L}), (τ_{R}, p_{R}), respectively.

(4.105)], which is therefore identical for two isentropic states with the same z. The speed of sound is given by (4.107), regardless of the value of z. The whole array of the Hugoniot equations in Summary 4.49 remains intact, since Equation (4.92) is independent of z. We conclude that the "Riemann solver" of Appendix C carries over to the present case.

Remark 9.8 A different procedure for the solution of the Riemann problem in the γ-law case is given in Chorin [26]. Even though z is assumed to satisfy Equation (9.2) (see Chorin [26, Equation 13f]), the solution to the Riemann problem is performed using a "C–J model" allowing jumps of z both on the contact discontinuity and on the shock front. This is not consistent with the quasi-conservation form (9.7), which permits jump discontinuities of z only across a contact discontinuity. The continuity of z across the shock front [in the case of the system (9.5)] is used in a substantial way in the following treatment of the linear GRP.

Structure of the Solution to the Linear GRP–The Main Theorem

Given the initial data (9.39), let $\mathbf{U}(x, t)$ be the solution to (9.5). The *linear GRP* consists in finding the limiting values

$$\left(\frac{\partial}{\partial t}\mathbf{U}\right)_0 = \lim_{t \to 0+} \frac{\partial}{\partial t}\mathbf{U}(0, t), \qquad (9.44)$$

as in Definition 5.3.

This problem is solved by a sequence of steps almost identical to those employed in the nonreacting case (Section 5.1). These steps can be summarized as follows:

(I) Using the associated RP [(9.8), (9.40)], determine the local wave configuration.

(II) Define the Lagrangian coordinate $\xi = \int_0^x \rho(s, 0)\,ds$ as in (4.79) [with $A(s) \equiv 1$]. The system (9.5) is then transformed to [compare (4.83)]

$$\tfrac{\partial}{\partial t}\mathbf{V} + \tfrac{\partial}{\partial \xi}\Phi(\mathbf{V}) = \tilde{\mathbf{K}}(\mathbf{V}),$$

$$\mathbf{V}(\xi, t) = \begin{bmatrix} \tau \\ u \\ e + \frac{1}{2}u^2 \\ z \end{bmatrix}, \quad \Phi(\mathbf{V}) = \begin{bmatrix} -u \\ p \\ pu \\ 0 \end{bmatrix}, \quad \tilde{\mathbf{K}}(\mathbf{V}) = \begin{bmatrix} 0 \\ 0 \\ 0 \\ -k \end{bmatrix}. \qquad (9.45)$$

The line $\xi = 0$ is the contact discontinuity, and the initial slopes are obtained from (5.7) and (5.8).

(III) Evaluate [see (5.10)]

$$\left(\frac{\partial u}{\partial t}\right)^* = \lim_{t \to 0+} \frac{\partial}{\partial t} u(\xi = 0, t), \qquad \left(\frac{\partial p}{\partial t}\right)^* = \lim_{t \to 0+} \frac{\partial}{\partial t} p(\xi = 0, t),$$

(9.46)

which are the limiting values of the directional derivatives along the contact discontinuity (across which u, p are continuous).

(IV) The line $x = 0$ is expressed in the (ξ, t) framework [see (5.71)] and the solution (9.44) is obtained by the chain rule (Propositions 5.26 and 5.27).

We now proceed to describe these steps in more detail, underlining the modifications resulting from the additional chemical equation. The wave structure (determined by the associated RP) is assumed to be that of Figure 5.1. We refer consistently to the notation introduced in Section 5.1, especially Table 5.5 (see Figure 5.2). Our presentation here follows Ben-Artzi [3].

The central point is that the derivatives (9.46) can be obtained by solving a pair of linear equations (compare Theorem 5.7). We state this theorem in full generality, along with its proof. Subsequently, we remark on the special cases of the "acoustic approximation" (Corollary 9.10) and the γ-law gas (Summary 9.14). The reader may wish to skip the proof and proceed directly to Corollary 9.10 and its simple proof.

Theorem 9.9 *The derivatives $\left(\frac{\partial u}{\partial t}\right)^*$, $\left(\frac{\partial p}{\partial t}\right)^*$ are determined by a pair of linear equations*

$$a_L \left(\frac{\partial u}{\partial t}\right)^* + b_L \left(\frac{\partial p}{\partial t}\right)^* = d_L, \qquad (9.47)_L$$

$$a_R \left(\frac{\partial u}{\partial t}\right)^* + b_R \left(\frac{\partial p}{\partial t}\right)^* = d_R. \qquad (9.47)_R$$

The coefficients a_L, b_L, d_L (resp. a_R, b_R, d_R) depend on $\mathbf{V}_L^*$, $\mathbf{V}_L$, $\mathbf{V}_L'$ (resp. $\mathbf{V}_R^*$, $\mathbf{V}_R$, $\mathbf{V}_R'$). They also depend on the equation of state (see Footnote 5, in Theorem 5.7).

Proof We start with a_R, b_R, d_R, employing the notational conventions introduced prior to the statement of Claim 5.18. The shock trajectory Γ_3 is parametrized as $\xi(\theta)$, $t(\theta)$, and limiting values ahead and behind the shock front are denoted by $Q_+(\theta)$, $Q(\theta)$ respectively. The unburnt mass fraction z is continuous,

$z(\theta) = z_+(\theta)$. The shock relation (4.102) yields (see Footnote 3)

$$u(\theta) = u_+(\theta) + \Phi\left(p(\theta); \rho_+(\theta), p_+(\theta), z(\theta)\right), \qquad (9.48)$$

where the function $\Phi = \Phi(p; \rho_I, p_I, z_I)$ is a function of four variables. Equation (9.48) is now differentiated with respect to θ, using the chain rule (5.52). The ensuing equation is identical to (5.53), apart from the term

$$\Phi_{z_1}\left(p(\theta); \rho_+(\theta), p_+(\theta), z(\theta)\right) \cdot \left[\frac{\partial z_+}{\partial t} + \sigma(\theta)\frac{\partial z_+}{\partial \xi}\right], \qquad (9.49)$$

which is added to the right-hand side [$\sigma(\theta) = \frac{p(\theta) - p_+(\theta)}{u(\theta) - u_+(\theta)}$ is the shock speed]. We get Equation $(9.47)_R$ by letting θ tend to 0, as in (5.58). To obtain only t derivatives behind the shock and ξ derivatives ahead of it, the substitutions (5.54)–(5.57) are invoked. Some minor modifications are needed, as follows:

(i) We set $A(0) = 1$ and $\lambda = 0$ in these formulas.
(ii) In view of (9.19) we have, ahead of the shock,

$$\left(\frac{\partial p}{\partial t}\right)_R = c_R^2\left(\frac{\partial \rho}{\partial t}\right)_R - \Lambda(e_R, \rho_R, z_R) = -g_R^2\left(\frac{\partial u}{\partial \xi}\right)_R - \Lambda(e_R, \rho_R, z_R). \quad (9.50)$$

(iii) By the same equation, behind the shock,

$$\left(\frac{\partial u}{\partial \xi}\right)_R^* = -(g_R^*)^{-2}\left[\left(\frac{\partial p}{\partial t}\right)^* + \Lambda(e_R^*, \rho_R^*, z_R)\right].$$

(iv) The initial derivatives (u_R' etc.) in (5.58) are Lagrangian. We use in Equation (9.51) the Eulerian initial derivatives of (9.39).

The limiting equation takes the form

$$\left[1 + \sigma(0)\Phi_p(p^*; \rho_R, p_R, z_R)\right]\left(\frac{\partial u}{\partial t}\right)^* + \left[-\sigma(0)(g_R^*)^{-2}\right.$$

$$\left. - \Phi_p(p^*; \rho_R, p_R, z_R)\right]\left(\frac{\partial p}{\partial t}\right)^* = \left[\sigma(0) - \rho_R^2\Phi_{\rho_I}(p^*; \rho_R, p_R, z_R)\right.$$

$$\left. - g_R^2\Phi_{\rho_I}(p^*; \rho_R, p_R, z_R)\right]\rho_R^{-1}u_R' + \left[-1 + \sigma(0)\Phi_{p_I}(p^*; \rho_R, p_R, z_R)\right]\rho_R^{-1}p_R'$$

$$+ \sigma(0)\Phi_{\rho_I}(p^*; \rho_R, p_R, z_R)\rho_R^{-1}\rho_R' - \Lambda(e_R, \rho_R, z_R)\Phi_{p_I}(p^*; \rho_R, p_R, z_R)$$

$$+ \Phi_{z_1}(p^*; \rho_R, p_R, z_R)\left[-k(e_R, \rho_R, z_R) + \sigma(0)\rho_R^{-1}z_R'\right]$$

$$+ \sigma(0)(g_R^*)^{-2}\Lambda(e_R^*, \rho_R^*, z_R) \qquad \left(\sigma(0) = \frac{p^* - p_R}{u^* - u_R}\right). \qquad (9.51)$$

Turning to the wave Γ_1, which is assumed to be a CRW, we adopt the notation used in Figure 5.4. The eigenvalues of the system (9.45), associated with the Γ_1, Γ_3 characteristic families, are $\pm g = \pm \rho c$ [see (4.85)]; hence the characteristic coordinates can be introduced as in (5.19)–(5.22) [with $A(0) = 1$]. The key ingredient in the treatment of the CRW resides in the fact that the directional derivative $a(\beta) = \frac{\partial u}{\partial \alpha}(0, \beta)$ is obtained by an explicit integration (see Proposition 9.12). The left-side initial data U_L, U'_L determine $a(1)$. This process amounts to a "propagation" (or "transport") of the directional derivative $\frac{\partial u}{\partial \alpha}$ from the head to the tail characteristic of the CRW. Assuming $a(\beta^*)$ is known, we repeat Equation (5.49) to get

$$\frac{\partial p}{\partial \alpha}(0, \beta^*) = -g_L^{-1}(\beta^*)^{-1/2}\left(\frac{\partial p}{\partial t}\right)^* - (\beta^*)^{1/2}\left(\frac{\partial u}{\partial t}\right)^*. \tag{9.52}$$

Now, the characteristic relation (9.20) (along $\frac{dx}{dt} = u - c$) yields

$$\frac{\partial p}{\partial \alpha}(0, \beta^*) = g_L \beta^* \frac{\partial u}{\partial \alpha}(0, \beta^*) - \Lambda(e_L^*, \rho_L^*, z_L)\frac{\partial t}{\partial \alpha}(0, \beta^*)$$

$$= g_L \beta^* a(\beta^*) + g_L^{-1}(\beta^*)^{-1/2}\Lambda(e_L^*, \rho_L^*, z_L). \tag{9.53}$$

Incorporating (9.53) into (9.52) and noting $\beta^* = g_L^{-1}g_L^*$ we get

$$\left(\frac{\partial u}{\partial t}\right)^* + (g_L^*)^{-1}\left(\frac{\partial p}{\partial t}\right)^* = -(g_L g_L^*)^{1/2}a(\beta^*) - (g_L^*)^{-1}\Lambda(e_L^*, \rho_L^*, z_L), \tag{9.54}$$

which establishes Equation $(9.47)_L$. $\qquad\qquad\square$

The Acoustic Approximation

The acoustic approximation was discussed in Section 5.1 (Proposition 5.9 and Remark 5.21). In this case the initial data at the singularity, U_L, U_R are assumed to be "close." The CRW then shrinks to a single characteristic curve, which entails a substantial simplification of the coefficients as follows.

Corollary 9.10 *In the acoustic approximation the coefficients in Equations* $(9.47)_{L,R}$ *are given by*

$$a_L = 1, \ b_L = (g_L^*)^{-1}, \ d_L = -(g_L^*)^{-1}\left\{c_L(g_L u'_L + p'_L) + \Lambda(e_L, \rho_L, z_L)\right\}, \tag{9.55}$$

$$a_R = -1, \ b_R = (g_R^*)^{-1}, \ d_R = -(g_R^*)^{-1}\left\{c_R(g_R u'_R - p'_R) + \Lambda(e_R, \rho_R, z_R)\right\},$$

where u'_L, p'_L, u'_R, p'_R *are the Eulerian initial slopes as in* (9.39). *In particular, the coefficients depend only on the left-side initial data and the associated Riemann solution.*

Proof We use the method of proof of Proposition 5.9. This means that in (9.52) we take $\beta^* = 1$ and equate the directional derivatives of p as evaluated on either side:

$$\left(\frac{\partial p}{\partial t}\right)^* - g_L^*\left(\frac{\partial p}{\partial \xi}\right)^* = \left(\frac{\partial p}{\partial t}\right)_L - g_L\left(\frac{\partial p}{\partial \xi}\right)_L. \qquad (9.56)$$

Using $\left(\frac{\partial p}{\partial \xi}\right)^* = -\left(\frac{\partial u}{\partial t}\right)^*$ and Equation (9.50) (with "L" instead of "R") in (9.56) we obtain (9.55) [note that $\left(\frac{\partial p}{\partial \xi}\right)_L = \rho_L^{-1} p'_L$ etc.], for a_L, b_L, d_L. The expressions for a_R, b_R, d_R follow by "reflection," $\xi \to -\xi$, $t \to t$, $u \to -u$, $p \to p$.   □

Resolution of the Centered Rarefaction Wave

We resume the study of the CRW, assuming the configuration and notation of Figure 5.4. In particular, the characteristic coordinates (α, β) play an essential role in this analysis. Our main goal is to obtain an ordinary differential equation for $a(\beta) = \frac{\partial u}{\partial \alpha}(0, \beta)$, $\beta^* \leq \beta \leq 1$, and the method is close to that employed in the proof of Proposition 5.12. The reader may skip it on first reading.

Recall [Equation (9.42)] that in the solution to the associated Riemann problem $z^{red} = z_L$ throughout the CRW; hence also $S^{red} = S_L$ ($S = $ entropy), as there is no change in chemical composition. Since the limiting values [at $(0, \beta)$] of the GRP solution are identical to those of the associated RP, we have

$$z(0, \beta) = z_L, \qquad S(0, \beta) = S_L, \qquad \beta^* \leq \beta \leq 1. \qquad (9.57)$$

The coordinates $\xi = \xi(\alpha, \beta)$, $t = t(\alpha, \beta)$ are given by (5.22), with $A(0) = 1$. In an attempt to imitate the proof of Proposition 5.12, we face the difficulty that z and hence also S are not constant along particle paths, thus invalidating the crucial equation (5.29) in the present case. Instead, we now use the three variables p, S, z to write

$$g = \rho c = G(p, S, z), \qquad (9.58)$$

where c is given in (9.4). Our first step is to obtain expressions for

$$S_\alpha(0, \beta) = \frac{\partial S}{\partial \alpha}(0, \beta) \quad \text{and} \quad z_\alpha(0, \beta) = \frac{\partial z}{\partial \alpha}(0, \beta)$$

in terms of the thermodynamic functions f [see (9.12)] and k [see (9.2)], which are expressed here as $f(p, S, z)$, $k(p, S, z)$.

Proposition 9.11 *The functions* $S_\alpha(0, \beta)$, $z_\alpha(0, \beta)$ *satisfy*

$$\frac{d}{d\beta}\left(\beta^{-1/2}S_\alpha(0, \beta)\right) = -g_L^{-1}\beta^{-2}f(p(0, \beta), S_L, z_L),\qquad (9.59)$$

$$\frac{d}{d\beta}\left(\beta^{-1/2}z_\alpha(0, \beta)\right) = -g_L^{-1}\beta^{-2}k(p(0, \beta), S_L, z_L),\qquad (9.60)$$

supplemented by the initial conditions

$$
\begin{aligned}
S_\alpha(0, 1) &= \rho_L^{-1}S_L' + g_L^{-1}f(p_L, S_L, z_L),\\
z_\alpha(0, 1) &= \rho_L^{-1}z_L' + g_L^{-1}k(p_L, S_L, z_L),
\end{aligned}
\qquad (9.61)
$$

where S_L', z_L' *are the initial Eulerian slopes (9.39).*

Proof We prove for S; the proof for z is identical. Following the convention in Section 5.1, we use both the notation $Q(\xi, t)$ and $Q(\alpha, \beta)$ to represent any flow variable Q throughout the CRW. Equation (9.12) is then written as

$$\frac{\partial}{\partial\xi}S(\xi, t)\frac{\partial\xi}{\partial\alpha}(\alpha, \beta) - f(p(\alpha, \beta), S(\alpha, \beta), z(\alpha, \beta))\frac{\partial t}{\partial\alpha}(\alpha, \beta) = \frac{\partial}{\partial\alpha}S(\alpha, \beta),$$

and letting $\alpha \to 0$ we get, by (5.22),

$$\frac{\partial S}{\partial\xi}(0, \beta)\beta^{1/2} + f(p(0, \beta), S_L, z_L)g_L^{-1}\beta^{-1/2} = S_\alpha(0, \beta).\qquad (9.62)$$

Note that, even though $\xi(0, \beta) \equiv 0$, the limiting value of $\frac{\partial S}{\partial\xi}(\alpha, \beta)$, as $\alpha \to 0$, depends on β. To evaluate it we use (9.12) once more to get, by differentiation,

$$
\begin{aligned}
\frac{\partial^2 S}{\partial t\partial\xi}(\xi, t) &= -\frac{\partial}{\partial\xi}f(p, S, z) = -f_p\frac{\partial p}{\partial\xi} - f_S\frac{\partial S}{\partial\xi} - f_z\frac{\partial z}{\partial\xi}\\
&= f_p\frac{\partial u}{\partial t} - f_S\frac{\partial S}{\partial\xi} - f_z\frac{\partial z}{\partial\xi}.
\end{aligned}
$$

Expressing $t = t(\xi, \beta)$ and multiplying the last equation by $\frac{\partial t}{\partial\beta}$ we get

$$\frac{\partial}{\partial\beta}\frac{\partial S}{\partial\xi}(\xi, \beta) = f_p\frac{\partial u}{\partial t}\frac{\partial t}{\partial\beta} - \left[f_S\frac{\partial S}{\partial\xi} + f_z\frac{\partial z}{\partial\xi}\right]\cdot\frac{\partial t}{\partial\beta}\qquad (9.63)$$

[equality as functions of (ξ, β)]. Fixing β, we now let $\xi \to 0$ (i.e., along a fixed C_- characteristic), getting

$$\frac{\partial}{\partial\beta}\frac{\partial S}{\partial\xi}(0, \beta) = f_p(p(0, \beta), S_L, z_L)\frac{\partial u}{\partial\beta}(0, \beta).\qquad (9.64)$$

The other terms in the right-hand side of (9.63) vanish (as $\xi \to 0$) since $\frac{\partial S}{\partial \xi}$, $\frac{\partial z}{\partial \xi}$ remain bounded in view of (9.62) and the similar equation satisfied by z.

Differentiating (9.62) with respect to β and using (9.64) we have

$$
\frac{d}{d\beta}(\beta^{-1/2}S_\alpha(0, \beta)) = f_p(p(0, \beta), S_L, z_L)\frac{\partial u}{\partial \beta}(0, \beta)
$$

$$
+ g_L^{-1}\frac{d}{d\beta}\left[\beta^{-1}f(p(0, \beta), S_L, z_L)\right]
$$

$$
= f_p(p(0, \beta), S_L, z_L)\left[\frac{\partial u}{\partial \beta}(0, \beta) + g(0, \beta)^{-1}\frac{\partial p}{\partial \beta}(0, \beta)\right]
$$

$$
- g_L^{-1}\beta^{-2}f(p(0, \beta), S_L, z_L), \tag{9.65}
$$

where $g(0, \beta) = g_L\beta$ (by the definition of β). The characteristic relation (9.20), applied at $(0, \beta)$, yields $\frac{\partial u}{\partial \beta}(0, \beta) + g(0, \beta)^{-1}\frac{\partial p}{\partial \beta}(0, \beta) = 0$, so that (9.59) follows from (9.65).

The initial condition (9.61) is obtained by taking $\beta = 1$ in (9.62) and noting that the flow is smooth in the sector between the head characteristic $\beta = 1$ and the negative ξ axis, so that $\frac{\partial S}{\partial \xi}(0, 1) = \lim_{\xi \to 0-} \frac{\partial S}{\partial \xi}(\xi, 0) = \rho_L^{-1}S_L'$. $\quad\square$

With $S_\alpha(0, \beta)$, $z_\alpha(0, \beta)$ at our disposal, we can study the transport of $a(\beta) = \frac{\partial u}{\partial \alpha}(0, \beta)$ as in Proposition 5.12.

Proposition 9.12 *Consider the CRW as in Figure 5.4. Then* $a(\beta) = \frac{\partial u}{\partial \alpha}(0, \beta)$, $\beta^* \le \beta \le 1$, *satisfies*

$$
a'(\beta) = -\frac{1}{2}g_L^{-2}\beta^{-1/2}\frac{d}{d\beta}\left[\beta^{-1}\Lambda(0, \beta)\right]
$$

$$
-\frac{1}{2}g_L^{-1}\beta^{-1}\left[G_S \cdot S_\alpha(0, \beta) + G_z \cdot z_\alpha(0, \beta)\right]\frac{d}{d\beta}u(0, \beta), \tag{9.66}
$$

where the function G is introduced in (9.58), and the derivatives $G_S = \frac{\partial G}{\partial S}$, $G_z = \frac{\partial G}{\partial z}$ are evaluated at $(p(0, \beta), S_L, z_L)$, and $\Lambda(\alpha, \beta) = \Lambda(e(\alpha, \beta), \rho(\alpha, \beta), z(\alpha, \beta))$ [see (9.17)].

Equation (9.66) is supplemented by the initial condition

$$
a(1) = \rho_L^{-1}(u_L' + g_L^{-1}p_L'), \tag{9.67}
$$

where u_L', p_L' are the Eulerian initial slopes [as in (9.39)].

Proof The characteristic relations (9.20) can be written as

$$g\frac{\partial u}{\partial \alpha} - \frac{\partial p}{\partial \alpha} = \Lambda\frac{\partial t}{\partial \alpha},$$

$$g\frac{\partial u}{\partial \beta} + \frac{\partial p}{\partial \beta} = -\Lambda\frac{\partial t}{\partial \beta}.$$

(9.68)

Differentiating the first equation with respect to β and the second with respect to α, we can eliminate the unknown p by addition. Recall that $g(0, \beta) = g_L\beta$, so that we have (at $\alpha = 0$)

$$2g_L\beta a'(\beta) + g_L a(\beta) + \frac{\partial g}{\partial \alpha}(0, \beta)\frac{d}{d\beta}u(0, \beta) = -g_L^{-1}\beta^{-1/2}\frac{d}{d\beta}\Lambda(0, \beta),$$

(9.69)

where we have used the expression (5.22) for $t(\alpha, \beta)$.

To evaluate $\frac{\partial g}{\partial \alpha}(0, \beta)$ we use (9.58) and the characteristic relation (9.68) to get

$$\frac{\partial g}{\partial \alpha}(0, \beta) = G_p \cdot \frac{\partial p}{\partial \alpha}(0, \beta) + G_S \cdot S_\alpha(0, \beta) + G_z \cdot z_\alpha(0, \beta)$$

$$= G_p\left[g_L\beta a(\beta) + \Lambda(0, \beta)g_L^{-1}\beta^{-1/2}\right] + G_S \cdot S_\alpha(0, \beta) + G_z \cdot z_\alpha(0, \beta),$$

where G_p, G_S, G_z are evaluated at $(p(0, \beta), S_L, z_L)$. Incorporating the last equation in (9.69) we get

$$2g_L\beta a'(\beta) + g_L\left[1 + \beta G_p \cdot \frac{d}{d\beta}u(0, \beta)\right]a(\beta) + \left[\Lambda(0, \beta)g_L^{-1}\beta^{-1/2}G_p\right.$$

$$\left. + G_S \cdot S_\alpha(0, \beta) + G_z \cdot z_\alpha(0, \beta)\right]\frac{d}{d\beta}u(0, \beta) = -g_L^{-1}\beta^{-1/2}\frac{d}{d\beta}\Lambda(0, \beta).$$

(9.70)

Differentiating the identity $g(0, \beta) = G(p(0, \beta), S_L, z_L)$ and using (9.68) we have

$$g_L = G_p\frac{d}{d\beta}p(0, \beta) = -G_p \cdot g_L\beta\frac{d}{d\beta}u(0, \beta),$$

so that the coefficient of $a(\beta)$ in (9.70) vanishes identically. We also have

$$\Lambda(0, \beta)\beta^{-1/2}G_p \cdot \frac{d}{d\beta}u(0, \beta) + \beta^{-1/2}\frac{d}{d\beta}\Lambda(0, \beta)$$

$$= -\Lambda(0, \beta)\beta^{-3/2} + \beta^{-1/2}\frac{d}{d\beta}\Lambda(0, \beta) = \beta^{1/2}\frac{d}{d\beta}\left(\beta^{-1}\Lambda(0, \beta)\right).$$

Inserting this relation in (9.70) we obtain (9.66). Equation (9.67) follows from the chain rule and (5.22) and (9.45), with $\beta = 1$,

$$\frac{\partial u}{\partial \alpha}(0, 1) = \left(\frac{\partial u}{\partial \xi}\right)_L \frac{\partial \xi}{\partial \alpha}(0, 1) - \left(\frac{\partial p}{\partial \xi}\right)_L \frac{\partial t}{\partial \alpha}(0, 1)$$

$$= \rho_L^{-1}\left[u_L' + g_L^{-1}p_L'\right].$$

□

Remark 9.13 (Solution of the associated RP) As observed earlier, the solution $\mathbf{U}^{\text{red}}$ to the associated RP is isentropic throughout the CRW, with $S = S_L$, and a constant mass fraction $z = z_L$ (hence also a fixed equation of state). Since $g(0, \beta) = g_L\beta$, the thermodynamic variables $p(0, \beta)$, $\rho(0, \beta)$ are evaluated from the equation of state [as functions of g on the isentropic curve $g = G(p, S_L, z_L)$]. The characteristic relation $g(0, \beta)\frac{d}{d\beta}u(0, \beta) = -\frac{d}{d\beta}p(0, \beta)$ yields $u(0, \beta)$. See Equations (5.39)–(5.41) for a γ-law gas.

Conclusion of the Linear GRP

When $a(\beta^*)$ is known, we have the coefficients in $(9.47)_L$ [see (9.54)] and together with $(9.47)_R$ we can solve for the time derivatives $\left(\frac{\partial u}{\partial t}\right)^*$, $\left(\frac{\partial p}{\partial t}\right)^*$ (at the singularity) in the direction of the contact discontinuity. The time derivatives of ρ, S, z (on either side of the contact discontinuity) are now evaluated by (9.19), (9.12), and (9.2), respectively. Next, as in Section 5.1 [see Equation (5.64)] we need the ξ derivatives on the two sides of the contact discontinuity. By (9.45) we have $\left(\frac{\partial p}{\partial \xi}\right)^* = -\left(\frac{\partial u}{\partial t}\right)^*$, and as in (9.50) we have

$$\left(\frac{\partial u}{\partial \xi}\right)_R^* = -\left(g_R^*\right)^{-2}\left[\left(\frac{\partial p}{\partial t}\right)^* + \Lambda(e_R^*, \rho_R^*, z_R^*)\right],$$

with an analogous expression for $\left(\frac{\partial u}{\partial \xi}\right)_L^*$. The expressions for $\left(\frac{\partial p}{\partial \xi}\right)_L^*$, $\left(\frac{\partial p}{\partial \xi}\right)_R^*$ are more difficult to obtain (compare Claim 5.22) and we refer to Ben-Artzi [3] for details. These derivatives are easier to get in the acoustic case (see Claim 5.23).

The solution to the linear GRP in the Eulerian framework [see (9.44)] follows from the Lagrangian solution just described. The line $x = 0$ is represented in the (ξ, t) plane as the curve $\xi = \xi(t)$, $\xi(0) = 0$, satisfying the equation $\xi'(t) = -\rho(\xi, t)u(\xi, t)$ [see Equation (5.71)], so that its initial slope is $-\rho_0 u_0$. The limiting values ρ_0, u_0 are obtained from the associated solution $\mathbf{R}^A(0; \mathbf{U}_L, \mathbf{U}_R)$ [see Equation (9.41)]. We distinguish two cases:

(a) The curve $\xi(t)$ is not contained in a CRW (nonsonic case). The solution is obtained by an application of the chain rule, as in Proposition 5.26.

(b) The curve $\xi(t)$ is contained in a CRW (sonic case). It is then expressed in terms of the characteristic coordinates, as $(\alpha(t), \beta(t))$, and the solution is obtained as in Proposition 5.27.

It turns out that Propositions 5.26 and 5.27 carry over to the present case without modification (see Ben-Artzi [3]). Hence, the Eulerian solution to the linear GRP (9.44) is deduced from the Lagrangian one by exactly same relations as in the nonreacting case.

The γ-Law Case

Certain simplifications take place when the equation of state is assumed to be of the form (9.43).

As already observed in Remark 9.7, the Hugoniot curve does not depend on z. In particular, the function Φ in (9.48) does not depend on z and is thus identical to the function Φ in the nonreacting case [Equation (5.61)]. Also, as noted in Remark 9.7, we have $c^2 = \frac{\gamma p}{\rho}$, so that the function $G(p, S, z)$ of (9.58) is independent of z. We record the ensuing simplifications.

Summary 9.14 (The γ-law equation of state) *When the equation of state is given by (9.43), the key equations $(9.47)_L$, $(9.47)_R$, are explicitly determined as follows:*

(a) In Equation (9.51), which is the explicit form of $(9.47)_R$, the function Φ is given by (5.61). The term containing Φ_{z_1} is therefore eliminated.
(b) The entropy S can be defined by

$$S = \frac{1}{\gamma - 1} \frac{p}{\rho^\gamma} = \rho^{1-\gamma}(e - q_0 z) \tag{9.71}$$

and the function G of (9.58) is

$$G(p, S, z) = \sqrt{\gamma p \rho} = \left(\frac{\gamma}{(\gamma - 1)^{1/\gamma}} \right)^{\frac{1}{2}} S^{-\frac{1}{2\gamma}} p^{\frac{\gamma+1}{2\gamma}}. \tag{9.72}$$

In particular, the term involving G_z in (9.66) is eliminated and $a(\beta^*)$, and hence the coefficients in (9.54) [which is the explicit form of $(9.47)_L$], can be explicitly determined (when Λ is known).

Note that even though the functions Φ, G are independent of z, the effect of the reaction equation still exists owing to the presence of Λ in Equations (9.51) and (9.66) and f in (9.59) [which affects $S_\alpha(0, \beta)$ and hence also $a(\beta)$].

9.5 The GRP Scheme for Reacting Flow

The resolution of the linear GRP leads directly to the GRP numerical methodology, as presented in Section 5.2. The procedure is straightforward. At time $t = t_n$ the flow variables (including the mass fraction z) are linearly distributed in the computational mesh cells, with jumps at the grid points (cell boundaries). Using the solution to the linear GRP, we can evaluate the numerical fluxes, new cell averages, and new slopes, as in Equations (5.108)–(5.110). In fact, this framework covers a full array of possible schemes. They range from the very basic (acoustic) scheme (Definition 5.37) to the most accurate one (Definition 5.41). The latter is based on the exact solution to the linear GRP (at cell boundaries), whereas the former approximates this solution (retaining second-order accuracy).

Note that in every GRP scheme all four equations of (9.5) are simultaneously resolved. In other words, the fourth (reaction) equation is *fully coupled* to the first three conservation laws. Alternatively, the two parts can be "decoupled," using the operator-splitting method (Section 7.2). To study this approach, it is perhaps best to consider the system in the Lagrangian framework (9.14). In this case the operators a, b of Equation (7.2) are given, respectively, by $-\frac{\partial}{\partial \xi} \Phi$ and Ψ. Unlike the two-dimensional case (7.16)–(7.18), where the splitting is "spatial," the splitting here is "functional." Its first part is identical to the reduced system (9.8), written in Lagrangian coordinates. The second part is just the reaction equation (9.2); it leaves the flow variables ρ, u, E unmodified while decreasing z along the particle paths. Formally, by using Strang's scheme (7.8), we retain the second-order accuracy for the combined algorithm. However, this algorithm may lead to nonphysical solutions. Such solutions exhibit the nonuniqueness built into the structure of weak solutions of the reacting flow system (9.5). They can be described as follows: Consider the Riemann problem for (9.5), where the initial data consist of an unburnt state (τ_+, p_+) on the right and a burnt state (τ_A, p_A) on the left (see Figure 9.2). As explained in Section 9.3, it is expected that the evolving solution approaches the Z–N–D profile, which is displayed in Figure 9.3. However, there exists another possibility, which incorporates the C–J model. Instead of starting with a leading shock in the unburnt gas (raising the pressure to p_A), this solution starts with a weak detonation, where the gas is fully burnt upon crossing the shock front, which brings it to the state B (Figure 9.2). This is then followed by a second shock, in the burnt species, raising the pressure to p_A. Such a trailing shock is possible since the weak detonation is supersonic with respect to the state behind it [Corollary 9.3(b)]. The solution thus obtained satisfies all the necessary jump conditions (see Figure 9.7). However, the weak detonation violates the entropy condition, as already noted. It is therefore labeled as a "nonphysical" solution, as opposed to the "physical" one, which approaches the Z–N–D profile.

The computational algorithm is found to be very sensitive to the details of the finite-differencing, as demonstrated in the papers by Colella, Majda, and Roytburd [29] and Ben-Artzi [3]. Numerical tests conducted there show that even under small variations of the scheme the solution switches from the physical to the nonphysical one. In particular, the paper by Colella, Majda, and Roytburd [29] shows that an application of the split scheme, as described here, generally produces the nonphysical solution. We remark that the coupling of the reaction equation to the conservation laws is clearly visible in Equation (9.19). The function Λ, which has a large value for a high reaction rate k, measures the deviation from the adiabatic law (4.68). When resolving the system (9.5) by a split scheme, the first part (conservation laws) ignores the term $-\Lambda$, while the second part (updating z) does not modify the pressure along a particle path. Although the overall algorithm is certainly consistent, this decoupling may be (at least partially) responsible for the appearance of the nonphysical solution (where the pressure jump across the leading weak detonation is lower than that of the physical solution). Observe, in addition, that the coupling of the reaction equation to the conservation laws is evident also in the solution to the linear GRP, even in the acoustic approximation (see Corollary 9.10).

Example 9.15 (Reacting-flow Riemann problem: Ozone decomposition)
As a numerical example, we take a Riemann problem for the reacting-flow system (9.5). The chemical reaction is that of ozone decomposition (an irreversible exothermic reaction by which ozone is converted into oxygen). The mass fraction z represents in this case the amount of "unburnt" ozone. The initial data correspond to a "C–J detonation." In other words, given a constant state of the ozone as a preshock state (in $x > 0$), the state for $x < 0$ is the one marked as "CJ" in Figure 9.1. It is fully determined by the preshock state and the equation of state, which we take as a γ-law equation [Equation (9.43)] with $\gamma = 1.4$ and $q_0 = 0.5196 \times 10^{10}$. We are using here cgs units (and K for temperature) for a realistic experiment. This example was used in Colella, Majda, and Roytburd [29], where the reader can find more information concerning the physical background. Our GRP computations follow Ben-Artzi [3].

The "reaction rate" function k in this example is taken to be a simplified Arrhenius equation,

$$k = KzH(T - T_c), \qquad K > 0, T = \frac{p}{\rho}, \tag{9.73}$$

where H is the Heaviside function,

$$H(\mu) = \begin{cases} 1, & \mu > 0, \\ 0, & \mu < 0. \end{cases}$$

The constant temperature T_c is a "threshold" (or "ignition") temperature. The reaction process takes place only when the gas temperature exceeds this value. Once started, the reaction rate is proportional to the mass fraction of the unburnt species z. In the present example we take

$$K = 0.5825 \times 10^{10}, \qquad T_c = 0.1155 \times 10^{10}.$$

It follows from Remark 9.7 that in this case the entropy $S(e, \rho, z)$ is given by (9.71); hence the functions f [see Equation (9.12)] and Λ [see Equation (9.17)] are given by

$$f(e, \rho, z) = -K q_0 \rho^{1-\gamma} z H(T - T_c), \qquad T = (\gamma - 1)(e - q_0 z),$$
$$\Lambda(e, \rho, z) = -(\gamma - 1) K q_0 \rho z H(T - T_c). \tag{9.74}$$

Recall (Remark 9.7) that q_0 is the specific chemical energy "stored" in the unburnt gas. We conclude from (9.74) that, for fixed values of p, ρ and z, the function Λ increases with the product $K q_0$. By (9.19) and the considerations above, this function measures the deviation from the adiabatic law. Thus, we may expect that a split scheme becomes more "problematic" as the product $K q_0$ is increased. This observation is in agreement with the numerical results in Colella, Majda, and Roytburd [29].

The initial values for the unburnt state are

$$p_0 = 8.321 \times 10^5 = 0.821 \, p_{atm}, \qquad \rho_0 = 1.201 \times 10^{-3} = 0.931 \rho_{atm}, \qquad z_0 = 1,$$

where $p_{atm} = 1.0135 \times 10^6$ and $\rho_{atm} = 1.29 \times 10^{-3}$ are the atmospheric pressure and density, respectively.

The corresponding C–J data are

$$p_{CJ} = 6.270 \times 10^6, \qquad \rho_{CJ} = 1.945 \times 10^{-3}, \qquad z_{CJ} = 0.$$

We can now compute the evolving solution, using the GRP scheme. As explained at the beginning of this section the solution is expected to approach the Z–N–D profile. Using the notation of Figure 9.3 (with "A" being the C–J point) the pressure and density at the "spike" A' are given by

$$p_{ZND} = 9.74987 \times 10^6, \qquad \rho_{ZND} = 3.26370 \times 10^{-3}.$$

The speed of the shock (which is equal to the speed of the CJ front in the C–J model, as they lie on the same Rayleigh line) is $\sigma = 1.088 \times 10^5$ cm/s and the width of the Z–N–D reaction zone [say, where $p > 0.5(p_{ZND} + p_{CJ})$] is approximately 5×10^{-5} cm.

As already mentioned, the GRP algorithm is the one presented in Section 5.2. However, a word is in order concerning the differencing of the fourth (chemistry) equation in (9.5). Following the general rule, its discretized form is

$$(\rho z)_j^{n+1} - (\rho z)_j^n = -\frac{\Delta t}{\Delta x} \left[(\rho z u)_{j+1/2}^{n+1/2} - (\rho z u)_{j-1/2}^{n+1/2} \right] - \Delta t \, [k\rho]_j^{n+1/2},$$

where $[k\rho]_j^{n+1/2}$ represents an average value of the source term in cell j (over the time interval $[t_n, t_{n+1}]$). Noting Equation (9.73) we see that the time step Δt should be restricted not only by the CFL condition imposed on the system of conservation laws (see the discussion following Definition 5.32) but also by the requirement that the product $K\Delta t$ is sufficiently small. As K increases, the "stiffness" of the chemistry equation becomes more pronounced, forcing a considerable reduction of Δt. (By stiffness we simply mean that explicit integration of the reaction equation requires a smaller Δt for higher K.) We shall not address this stiffness issue here; it has been extensively studied in a variety of numerical investigations, which are not directly related to the present monograph. Note that if Δx is roughly equal to the size of the reaction zone (or larger) the computed "spike" of the Z–N–D profile is "absorbed" into one cell, thus approaching the C–J model. In other words, the numerical computation displays an "effective reaction rate" (see Footnote 2 in Section 9.2) measured by $K\Delta x$.

The GRP calculations have been performed in the spatial domain $[0, 5 \times 10^{-4}]$, which was divided into 100 equal cells of $\Delta x = 5 \times 10^{-6}$. The discontinuity is initially located at $x = 50\Delta x$. As the leading shock advances, we "remove" one cell from the left end of the domain and add one cell at the right end (imposing on it the state $[\rho_0, p_0, u_0 = 0, z_0]$), so as to maintain the front roughly at its initial location. The time step was set to $\Delta t = 5 \times 10^{-12}$, and the computation was conducted to time $t = 5 \times 10^{-8}$, with the results shown in Figure 9.4. These results clearly display a Z–N–D spike, and the peak values of ρ and p agree with the previously quoted exact values of ρ_{ZND} and p_{ZND}. We now repeat the computation, multiplying the cell size and the time step by 4, that is, $\Delta x = 2 \times 10^{-5}$ and $\Delta t = 2 \times 10^{-11}$, with all other parameters remaining the same. The results shown in Figure 9.5 clearly display a Z–N–D spike of a reduced level. This trend continues when we repeat the computation with a tenfold higher step (relative to the first case); that is, $\Delta x = 5 \times 10^{-5}$ and $\Delta t = 5 \times 10^{-11}$. The results shown in Figure 9.6 resemble closely a C–J

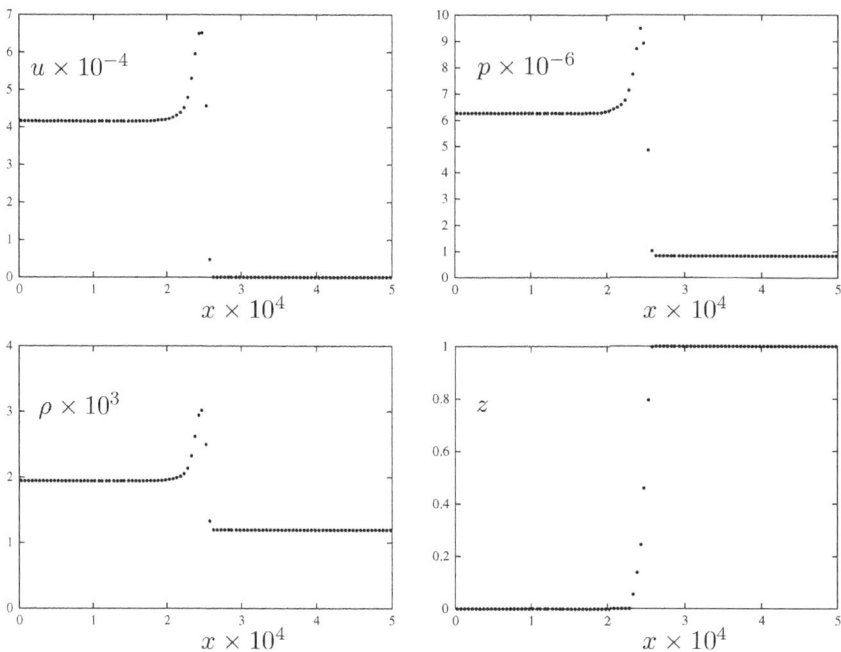

Figure 9.4. GRP computation of ozone decomposition wave (100 cells). $\Delta x = 5 \times 10^{-6} \approx 0.1 \times (\text{reaction zone})$.

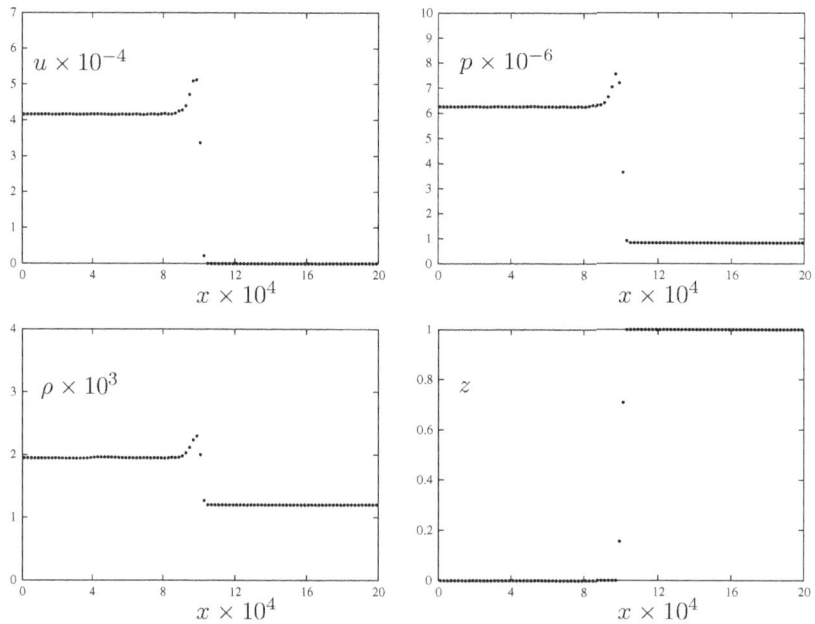

Figure 9.5. GRP computation of ozone decomposition wave (100 cells). $\Delta x = 2 \times 10^{-5} \approx 0.4 \times (\text{reaction zone})$.

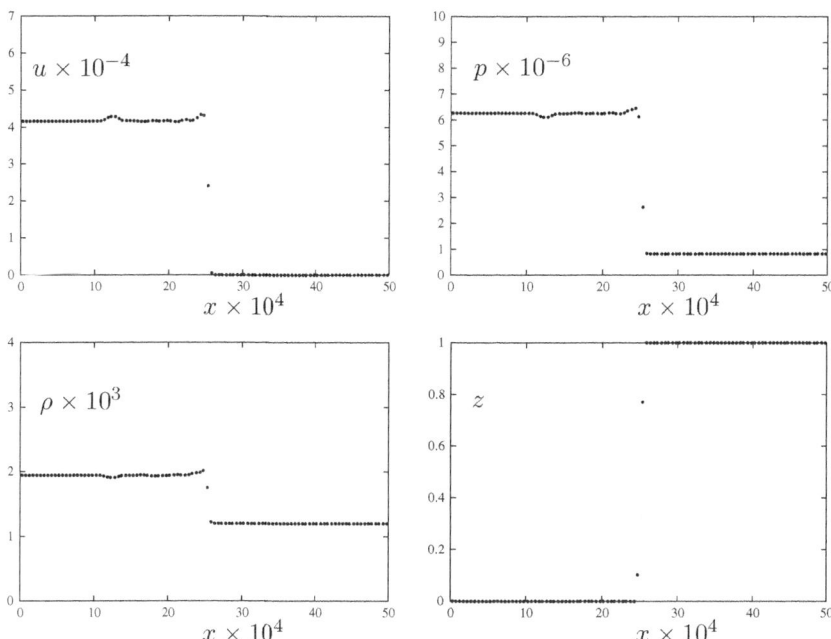

Figure 9.6. GRP computation of ozone decomposition wave (100 cells). $\Delta x = 5 \times 10^{-5} \approx 1.0 \times$ (reaction zone).

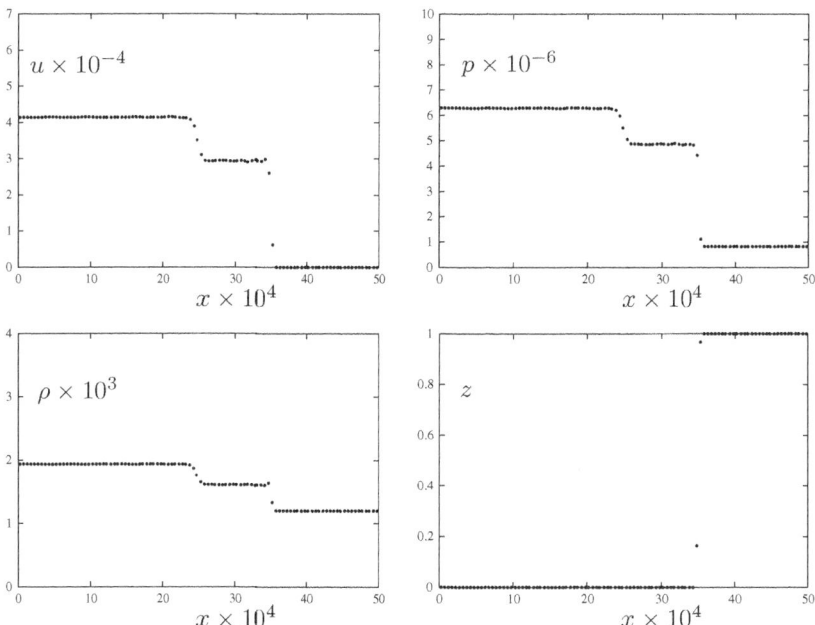

Figure 9.7. Split-scheme computation of ozone decomposition wave resulting in a weak detonation (100 cells). $\Delta x = 5 \times 10^{-5} \approx 1.0 \times$ (reaction zone).

detonation wave. The values at the front are slightly higher than the corresponding C–J values, and the wave is captured over 1–2 cells (i.e., the shock and the reaction zone are "smeared" over a comparable interval of 1–2 cells).

We now turn to the case of weak detonation, using a scheme that is closely related to the above-mentioned functionally split scheme. The data are the same as in the former case, except that the computation was continued to a tenfold longer time ($t = 10^{-7}$). This time was required to produce the "separated waves" pattern shown in Figure 9.7, where the burn reaction is completed behind the leading wave, and a second shock trails behind (at lower speed) in the burnt gas. We refer to Ben-Artzi [3] for more details. This result conforms to the split-scheme results obtained by Colella, Majda, and Roytburd [29].

Remark 9.16 We refer to Bourgeade [19] for GRP computations of detonations using a more complicated equation of state (and 2-D configurations). Also, we refer to Ben-Artzi and Birman [6] for GRP applications in quasi-1-D settings [as in (4.45)] and where external potentials are added to the chemical reactions. Such models are extensively studied in astrophysics.

10

Wave Interaction in a Duct – A Comparative Study

The GRP method was developed (Chapter 5) for compressible, unsteady flow in a duct of varying cross section. In the case of a planar two-dimensional duct, the quasi-1-D formulation is taken to be a reasonable approximation of the actual (2-D) flow. In this chapter we study a duct flow where an incident wave interacts with a short converging segment, producing interesting wave structures. An illustrative case is that of a rarefaction wave propagating through a "converging corridor," producing (at later times) a complex "reflected" wave pattern. Such a case is studied (numerically) in this chapter, using (a) the quasi-1-D approach of Chapter 5 and (b) the full two-dimensional computation as described in Section 8.3 (with the duct contour taken as a stationary boundary). The comparison between the two computations reveals some bounds of validity of the quasi-1-D approximation. We conclude the chapter by listing (Remark 10.1) several articles describing the application of the GRP to diverse fluid-dynamical problems, including well-known test cases, shock wave reflection phenomena compared to experimental observation, and even a case where a "moving boundary" experiment is favorably compared to the corresponding GRP solution.

Consider a centered rarefaction wave that propagates in a planar duct comprising two long segments of uniform cross-sectional area joined by a smooth converging nozzle. Such processes take place in numerous systems of industrial and scientific interest, for example in the air intake or the exhaust pipe of an internal combustion engine, and likewise in turbofan or turbojet engines. In addition to shedding light on the nature of the interaction between a rarefaction wave and a converging nozzle, the significance of this case lies in the comparison we make between a full multidimensional solution and the corresponding quasi-1-D approximation (Section 4.2). This is because the latter (often referred to as the "duct flow" approximation) is commonly employed as an engineering design tool. Hence, studying the bounds of its validity as a simplified approximation to the full multidimensional solution is highly significant to engineering design and analysis.

For the multidimensional computation of the wave interaction with a converging nozzle we employ the operator-split 2-D GRP scheme (Section 7.3), where the duct wall intersects an underlying two-dimensional Cartesian grid, using the conservation law scheme outlined in Section 8.3.

Inspection of the major physical characteristics of a variety of wave interaction flows (Igra, Wang, and Falcovitz [68]) revealed that at large times the solution produced by the quasi-1-D approximation, especially in the case of shock waves, was generally close to the full 2-D solution. However, this is not universally so, and as a specific example of wave interaction where the quasi-1-D solution and the 2-D solution differ, we consider the case of a 1:10 pressure ratio rarefaction wave in a fluid assumed to be a perfect gas with $\gamma = 1.4$. The CRW is initially located in the wider part of the duct, and it propagates toward a (short) converging nozzle of 2:1 cross-sectional area ratio. The initial data are that of a Riemann problem [Equation (4.100)] designed to produce a 3-CRW. It consists of two uniform states,

$$\mathbf{U}(r, 0) = \begin{cases} \mathbf{U}_L = [\rho_L, p_L, u_L] = [0.27030, 0.1, -1.4016], & r < 1.3, \\ \mathbf{U}_R = [\rho_R, p_R, u_R] = [1.4, 1, 0], & r > 1.3. \end{cases}$$

(10.1)

Here we use r as the spatial coordinate along the duct axis as in Section 4.2. The location of the initial discontinuity ($r = 1.3$) is just ahead of the converging segment that occupies the interval $1.6 \leq r \leq 2.6$. The cross-sectional area function of the duct $A(r)$ is given by

$$A(r) = \begin{cases} 2, & r < 1.6, \\ 2 \exp\left[-\frac{1-\cos(\pi(r-1.6))}{2} \ln 2\right], & 1.6 \leq r \leq 2.6, \\ 1, & r > 2.6. \end{cases}$$

(10.2)

We assume that the two-dimensional duct is symmetric and embedded in the (x, y) plane, so that the x axis, its centerline, coincides with the r axis. The upper contour of the duct [see Figure 10.2(a)] is thus $y(x) = \frac{1}{2}A(x)$. Owing to the duct symmetry, the 2-D computation is conducted in the upper half of the duct, embedding it in the (finite) rectangular domain $(x, y) \in [-1.6, 9.4] \times [0, 1]$, which is divided into a grid of 550×50 square cells ($\Delta x = \Delta y = 0.02$).

The computation is performed by the MBT method (Section 8.3), where the duct contour serves as a stationary rigid-wall boundary. A rigid-wall boundary condition is also imposed at the centerline ($y = 0$). On the left and right sides of the computational rectangle we impose "nonreflecting" boundary conditions (see Subsection 8.3.3), designed to allow waves to pass through these endplanes (almost) undisturbed. The computation was performed in the time interval $[0, 9]$, with time steps adjusted (at each integration cycle) to have a nearly constant CFL coefficient $\mu_{CFL} = 0.7$ (see Remark 3.17).

The quasi-1-D computation was conducted in the spatial interval $[-1.6, 9.4]$, which was divided into a grid of 550 cells of equal length $\Delta r = 0.02$. The cross-sectional area function is $A(r)$ [given in (10.2)]. The boundary conditions at either endpoints were of the same nonreflecting type as in the 2-D case. The computation was performed with time steps adjusted to have the same $\mu_{CFL} = 0.7$ and in the same time interval $[0, 9]$.

We now turn to the results of the 2-D computation, shown as time-sequence maps of isobars ($p = constant$) in Figure 10.1. When the right-propagating rarefaction wave enters the converging nozzle (Figure 10.1 at $t = 1.5$), the fluid in the nozzle is set in motion. It gradually evolves into a supersonic expansion flow in a diverging nozzle, as the fluid is moving from right to left, that is, in the direction of increasing duct cross-sectional area.[1]

Inspecting the time sequence of isobars plots (Figure 10.1), we observe that the entire flow field is progressively adjusting to the presence of a diverging nozzle in its midst. At time $t = 3$ the flow overexpands in the nozzle, so that an upstream-facing oblique shock wave is formed, raising the pressure of the expanding gas to match the downstream conditions (i.e., the conditions corresponding to U_L). This shock is marked by $*$ in Figure 10.1. It becomes more pronounced at $t = 4.5$, and at the next plot in the time sequence ($t = 6$) it stabilizes near the duct "corner point," which marks the transition from the diverging nozzle to the wider duct segment. In addition to pressure matching, the oblique shock serves to align the velocity of the flow along the diverging nozzle contour with the downstream duct wall. At this time, we also observe that another complex shock interaction forms at the centerline ($y = 0$), aligning the flow velocity with that boundary line. This latter shock system appears first as a short normal shock segment at $t = 7.5$. At the final time $t = 9$, that shock has evolved into a full Mach reflection. It exhibits the typical features of that pattern, namely, an oblique incident shock wave, a Mach stem (normal to the

---

[1] Note that in the initial state U_L the flow is already supersonic since $u_L = -1.947c_L$; that is, $|u_L| > c_L$.

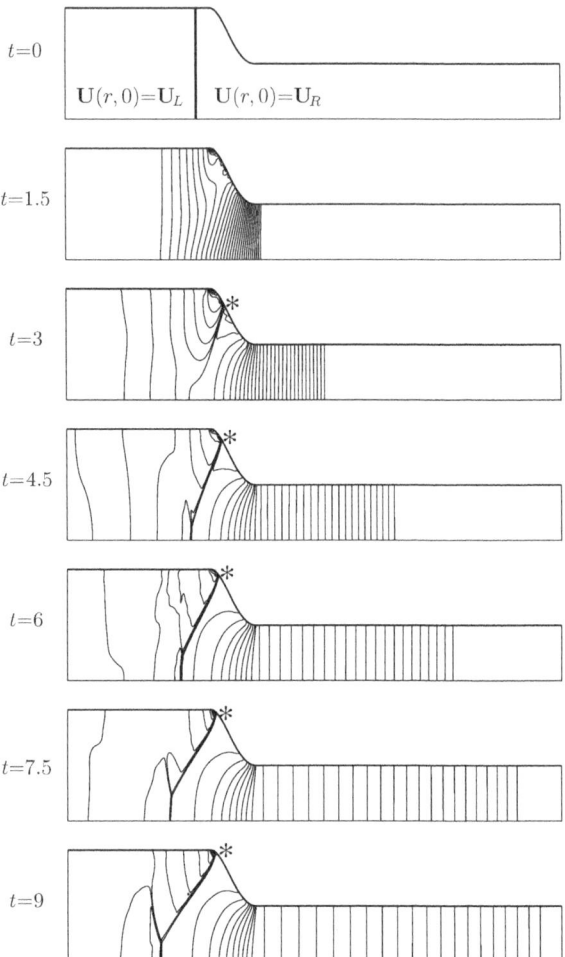

Figure 10.1. Isobar map time sequence of CRW interaction with a converging segment. The ∗ marks shock formation at duct wall. (Duct width here is twice the true size, for better visibility.)

centerline), a reflected shock wave, and a triple point where all three shock fronts meet (the downstream slip line starting at the triple point is not observed on the isobar plot since pressure is continuous across a slip line).

In Figure 10.2 we compare the results of the 2-D and quasi-1-D computations at the final time $t = 9$. First we show the 2-D isobars [Figure 10.2(a)] as in the last frame of Figure 10.1. This is followed by profiles of density, pressure, and flow Mach number, as functions of the centerline coordinate [Figures 10.2(b)–(d)]. They are extracted from the 2-D and the quasi-1-D

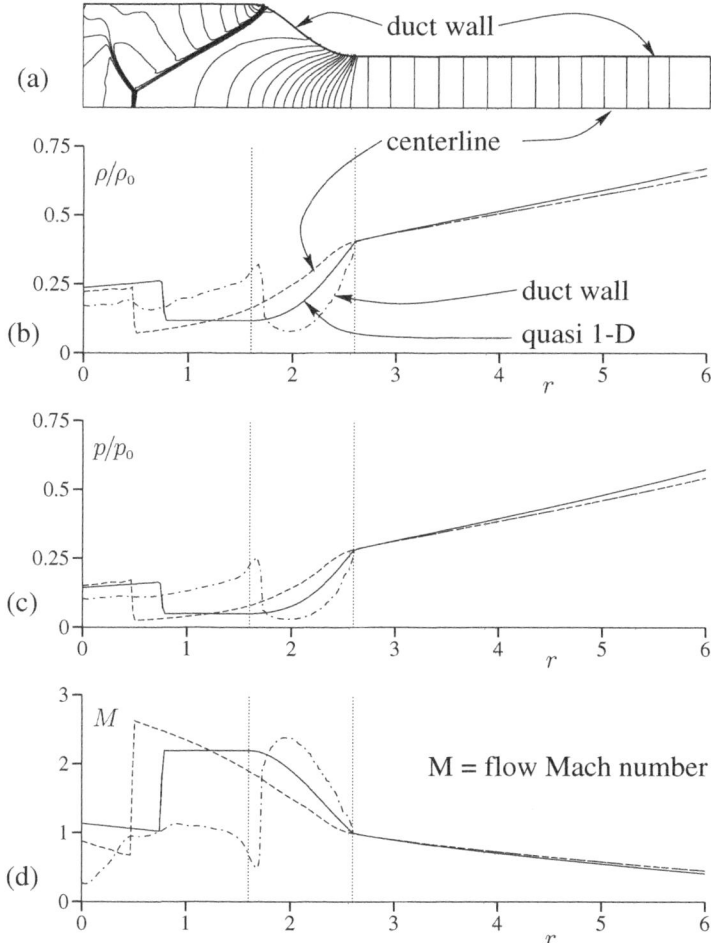

Figure 10.2. CRW interaction with a converging segment; comparison of 2-D and quasi-1-D results at time $t = 9$. (a) Isobars of 2-D calculation. (b)–(d) Quasi-1-D solution. Distribution of density, pressure, and flow Mach number (taken as positive).

solutions. The 2-D computation results are shown as dashed lines for the flow at the centerline and as dash-dot lines for the flow at the duct wall; the quasi-1-D profiles are shown as solid lines.

It is evident from the comparison (Figure 10.2) that the two solutions are in close agreement throughout the narrower duct segment ($r > 2.6$), but they disagree elsewhere. Also, it is observed that the "transmitted" part of the rarefaction wave propagates almost "one dimensionally" into the narrow part of the duct (where it naturally agrees well with the corresponding quasi-1-D solution).

Moreover, it is interesting to observe that at the entrance to the nozzle ($r = 2.6$), the flow speed is sonic [$M = 1$ in Figure 10.2(d)]. The nozzle flow at large times thus approaches a *steady* supersonic expansion flow, commencing at a virtual sonic plane ("nozzle entrance"), which serves to "match" the *unsteady* rarefaction wave on its right to the steady flow on its left.[2]

In other words, upon passing through the nozzle, the CRW is "truncated" into a "transmitted" part and a "reflected" part; the two are separated by a (nearly) steady flow through the diverging nozzle. With respect to the incident CRW, the first part corresponds to the sector between the leading characteristic $\frac{dr}{dt} = c_R$ and the "sonic characteristic" $\frac{dr}{dt} = u + c = 0$ (which is positioned at the nozzle entrance $r = 2.6$). The second part corresponds to the "tail" sector of the CRW, between the tail characteristic $\frac{dr}{dt} = u_L + c_L < 0$ and an unspecified inner characteristic (it is "unspecified" since here we focus our attention on the other parts of the flow field). The region of "nearly steady" flow comprises the entire diverging nozzle and extends to the left of the shock structure.

Now the source of disagreement between the quasi-1-D and the 2-D solutions is evident. The fluid expands as it (reversely) flows through the nozzle at supersonic speed, and a full 2-D description of this flow involves an oblique shock system at the nozzle exit [Figure 10.2(a)], which is poorly approximated by the cross-section-averaged quasi-1-D solution that relies on a normal shock for matching the overexpanded supersonic nozzle flow to the pressure ahead. Moreover, the flow passing through the Mach reflection shock structure is separated by the slip line into two streams having distinctly different thermodynamic properties. Naturally, the 1-D "averaging" of these two streams involves significant deviations from the 2-D flow field.

It is concluded that although quasi-1-D calculations may generally be adequate as an engineering approximation, a verification by comparison to the appropriate multidimensional solution is required to make sure that the disagreement between the two remains within acceptable bounds. Our test case analysis thus brings out the significance of full multidimensional numerical solutions; the (simpler) quasi-1-D solutions may not always serve as adequate approximations.

Remark 10.1 (Additional GRP fluid-dynamical studies) We briefly mention a series of fluid-dynamical studies using various GRP schemes.

---

[2] Recall that under the quasi-1-D approximation for steady compressible flow in a Laval nozzle ([84], [102]), a supersonic flow in the diverging part of the nozzle is possible only when the fluid enters it (through a minimum-area "throat") at sonic speed.

In an early application of GRP (see [42]), a quasi-1-D, spherically symmetric computation was used to obtain some flow features in the dilute air surrounding an exploding charge.

A comparative study on various fluid-dynamical schemes, including the 2-D GRP method, was conducted by Takayama and Jiang [107]; the test cases were shock reflection from a wedge, at angles just above and just below the transition from regular to Mach reflection. Another test case (involving, as the former, a self-similar solution) was the diffraction of a shock wave over an expansive 90° corner, studied by Hillier [61] using a 2-D GRP scheme.

Various shock wave studies combined experimental observations of complex shock interactions with the corresponding computational simulations by the 2-D GRP scheme. In [67] a time sequence of shock wave interaction with a square cavity was recorded and compared to GRP solutions. Similar studies are reported in [69] for a shock wave propagating in a branched duct and in [66] for the passage of a shock wave through a double-bend duct. In all these studies a remarkably good agreement was obtained between the observed (complex) shock structures and the corresponding GRP solutions.

Three additional studies involved comparison of experimental observations and solutions by the MBT method (Section 8.3). A shock wave regularly reflected from a wedge was subsequently reflected head-on from the duct endwall (see [38]), and the ensuing complex wave pattern was computed by representing the wedge as a rigid (stationary) wall. A nearly one-to-one agreement between experimental and computed isopycnic lines ($\rho = constant$) was obtained. The second study is based on a reduced version of MBT for the flow of gas–granular aggregates, accounting for both gas–grain and (elastic) grain–grain interactions [126]. Agreement with a "shock liftoff" experiment of a granular bed mounted in a vertical shock tube was obtained.

Finally, an experiment involving a truly "moving boundary" setup was conducted by Sasoh et al. [98] in a ram-accelerator facility, where a high-speed (1 km/s) conical projectile pierced through a thin diaphragm. The experimental results compared favorably to the corresponding MBT simulation.

Appendix A

Entropy Conditions for Scalar Conservation Laws

In Definition 2.15 we gave the most practical version of the entropy condition. It limits admissible shocks to those obtained by the intersection of "forward-moving" characteristics. These are therefore discontinuities that "cannot be avoided" or replaced by a rarefaction wave. In this Appendix we give some further insight into this concept of an "entropy satisfying" weak solution to (2.1), (2.2).

Our starting point is the physical notion of a "vanishing viscosity solution." In general terms, an equation leading to discontinuous solutions [such as (2.1)] is supplemented by "dissipative terms" (also referred to as "viscous terms"). In analogy to the physical situation, such terms have a "smoothing effect" on solutions with large gradients, thus replacing discontinuities by "transition zones" where the solution varies smoothly, albeit rapidly. As the viscous effects are diminished, those transition zones shrink to surfaces of zero width, across which the solution has a sharp jump. Mathematically speaking, the additional viscous terms are often represented by second-order derivatives with a small ("vanishing") coefficient.

To illustrate the situation, we consider the "moving step" problem for Burgers' equation (Example 2.12).

Example A.1 (Burgers' equation – a "viscous shock profile")   Instead of Equation (2.22), we now consider the "viscous Burgers' equation"

$$u_t^{(\epsilon)} + \left(\frac{1}{2}(u^{(\epsilon)})^2 \right)_x = \epsilon u_{xx}^{(\epsilon)}, \qquad \epsilon > 0, \qquad (A.1)$$

$$u_0^{(\epsilon)}(x) = u_0(x) = \begin{cases} 1, & x < 0, \\ 0, & x > 0. \end{cases} \qquad (A.2)$$

313

Equation (A.1) can be linearized by the Cole–Hopf transformation as follows: Define a new unknown function $v^{(\epsilon)}(x,t)$ by

$$v^{(\epsilon)}(x,t) = \exp\left(-\frac{1}{2\epsilon}\int_0^x u^{(\epsilon)}(y,t)\,dy\right) \tag{A.3}$$

(so that $u^{(\epsilon)}(x,t) = -2\epsilon\frac{v_x^{(\epsilon)}(x,t)}{v^{(\epsilon)}(x,t)}$). The reader can verify that $v^{(\epsilon)}$ satisfies the linear heat equation,

$$v_t^{(\epsilon)} = \epsilon v_{xx}^{(\epsilon)}, \tag{A.4}$$

$$v^{(\epsilon)}(x,0) = v_0^{(\epsilon)}(x) = \exp\left(-\frac{1}{2\epsilon}\int_0^x u_0(y)\,dy\right). \tag{A.5}$$

When $u_0(x)$ is given by (A.2), we get from (A.5)

$$v_0^{(\epsilon)}(x) = \begin{cases} \exp\left(-\frac{x}{2\epsilon}\right), & x < 0, \\ 1, & x \geq 0, \end{cases} \tag{A.6}$$

and the solution to (A.4) is given by the classical formula for the heat kernel,

$$v^{(\epsilon)}(x,t) = \int_{-\infty}^{\infty} G_\epsilon(x-y,t)v_0^{(\epsilon)}(y)\,dy, \tag{A.7}$$

where $G_\epsilon(x,t) = (4\pi\epsilon t)^{-\frac{1}{2}}\exp\left(-\frac{x^2}{4\epsilon t}\right)$. Equation (A.7) can be further written explicitly [using (A.6)] as

$$\begin{aligned}
v^{(\epsilon)}(x,t) &= (4\pi\epsilon t)^{-\frac{1}{2}}\left\{\int_{-\infty}^{0}\exp\left(-\frac{(x-y)^2}{4\epsilon t}-\frac{y}{2\epsilon}\right)dy\right. \\
&\quad \left. +\int_0^{\infty}\exp\left(-\frac{(x-y)^2}{4\epsilon t}\right)dy\right\} \\
&= (4\pi\epsilon t)^{-\frac{1}{2}}\int_{-\infty}^{0}\exp\left(-\frac{(x-y)^2}{4\epsilon t}\right)\left(\exp\left(-\frac{y}{2\epsilon}\right)+\exp\left(-\frac{xy}{\epsilon t}\right)\right)dy,
\end{aligned} \tag{A.8}$$

so that, in view of (A.3),

$$u^{(\epsilon)}(x,t)$$
$$= \frac{\displaystyle\int_{-\infty}^{0}\left[\frac{x-y}{t}\left(\exp\left(-\frac{y}{2\epsilon}\right)+\exp\left(-\frac{xy}{\epsilon t}\right)\right)+\frac{2y}{t}\exp\left(-\frac{xy}{\epsilon t}\right)\right]\exp\left(-\frac{(x-y)^2}{4\epsilon t}\right)dy}{\displaystyle\int_{-\infty}^{0}\exp\left(-\frac{(x-y)^2}{4\epsilon t}\right)\left(\exp\left(-\frac{y}{2\epsilon}\right)+\exp\left(-\frac{xy}{\epsilon t}\right)\right)dy}.$$
$$\tag{A.9}$$

For fixed $x \in \mathbb{R}, t > 0$, we are interested in the limit of $u^{(\epsilon)}(x, t)$ as $\epsilon \to 0$. Observe that if $x < \frac{1}{2}t$ then

$$\frac{\exp\left(-\frac{xy}{\epsilon t}\right)}{\exp\left(-\frac{y}{2\epsilon}\right)} \to 0 \quad \text{as } \epsilon \to 0 \quad \text{(for every } y < 0\text{)},$$

so that in this case

$$\lim_{\epsilon \to 0} u^{(\epsilon)}(x, t) = \lim_{\epsilon \to 0} \frac{\int\limits_{-\infty}^{0} \frac{x-y}{t} \exp\left(-\frac{(x-y)^2}{4\epsilon t}\right) \exp\left(-\frac{y}{2\epsilon}\right) dy}{\int\limits_{-\infty}^{0} \exp\left(-\frac{(x-y)^2}{4\epsilon t}\right) \exp\left(-\frac{y}{2\epsilon}\right) dy}. \tag{A.10}$$

Using the classical "steepest descent method" (Evans [36, Section 4.5.2]) we get

$$\lim_{\epsilon \to 0} u^{(\epsilon)}(x, t) = \frac{x - y_0}{t}, \qquad x < \frac{1}{2}t, \tag{A.11}$$

where y_0 is the maximum point of $\exp\left(-\frac{(x-y)^2}{4\epsilon t}\right) \exp\left(-\frac{y}{2\epsilon}\right)$, namely, $y_0 = x - t$. We conclude that

$$\lim_{\epsilon \to 0} u^{(\epsilon)}(x, t) = 1, \qquad x < \frac{1}{2}t. \tag{A.12}$$

In exactly the same way we get

$$\lim_{\epsilon \to 0} u^{(\epsilon)}(x, t) = 0, \qquad x > \frac{1}{2}t. \tag{A.13}$$

Thus, the "vanishing viscosity limit" (i.e., as $\epsilon \to 0$) of the solution to (A.1) yields the moving step (shock) solution of the Burgers' equation (2.22).

Example A.2 (Burgers' equation – a "viscous rarefaction wave") Turning to the case of a rarefaction wave (Example 2.13), we now need to study the solution of (A.1) subject to the initial condition

$$u_0^{(\epsilon)}(x) = \begin{cases} 0, & x < 0, \\ 1, & x > 0, \end{cases} \tag{A.14}$$

which yields

$$v_0^{(\epsilon)}(x) = \begin{cases} 1, & x < 0, \\ \exp\left(-\frac{x}{2\epsilon}\right), & x > 0. \end{cases} \tag{A.15}$$

Instead of (A.8) we get

$$v^{(\epsilon)}(x,t) = (4\pi\epsilon t)^{-\frac{1}{2}} \int_0^\infty \exp\left(-\frac{(x-y)^2}{4\epsilon t}\right)\left(\exp\left(-\frac{xy}{\epsilon t}\right)+\exp\left(-\frac{y}{2\epsilon}\right)\right) dy,$$

(A.16)

so that

$$u^{(\epsilon)}(x,t)$$
$$= \frac{\int\limits_0^\infty \left[\frac{x-y}{t}\left(\exp\left(-\frac{xy}{\epsilon t}\right)+\exp\left(-\frac{y}{2\epsilon}\right)\right)+\frac{2y}{t}\exp\left(-\frac{xy}{\epsilon t}\right)\right]\exp\left(-\frac{(x-y)^2}{4\epsilon t}\right) dy}{\int\limits_0^\infty \exp\left(-\frac{(x-y)^2}{4\epsilon t}\right)\left(\exp\left(-\frac{xy}{\epsilon t}\right)+\exp\left(-\frac{y}{2\epsilon}\right)\right) dy}.$$

(A.17)

For $x < 0$ we have (if $y > 0$)

$$\lim_{\epsilon\to 0} \frac{\exp\left(-\frac{y}{2\epsilon}\right)}{\exp\left(-\frac{xy}{\epsilon t}\right)} = 0,$$

so that, as before,

$$\lim_{\epsilon\to 0} u^{(\epsilon)}(x,t) = \lim_{\epsilon\to 0} \frac{\int_0^\infty \frac{x+y}{t}\exp\left[-\frac{(x-y)^2}{4\epsilon t}-\frac{xy}{\epsilon t}\right] dy}{\int_0^\infty \exp\left[-\frac{(x-y)^2}{4\epsilon t}-\frac{xy}{\epsilon t}\right] dy} = 0 \qquad (x < 0),$$

(A.18)

since the maximum of the function $-\frac{(x-y)^2}{4\epsilon t}-\frac{xy}{\epsilon t}$ is at $y_0 = -x$.

If $0 < x < t$ we note that both functions $-\frac{(x-y)^2}{4\epsilon t}-\frac{xy}{\epsilon t}$ and $-\frac{(x-y)^2}{4\epsilon t}-\frac{y}{2\epsilon}$ (as functions of $y \geq 0$) take their maximal values at $y_0 = 0$, so that we infer from (A.17)

$$\lim_{\epsilon\to 0} u^{(\epsilon)}(x,t) = \frac{x}{t}, \qquad 0 < x < t. \qquad (A.19)$$

Finally, when $x > t$, the function $-\frac{(x-y)^2}{4\epsilon t}-\frac{xy}{\epsilon t}$ has no stationary point for $y \geq 0$ whereas $-\frac{(x-y)^2}{4\epsilon t}-\frac{y}{2\epsilon}$ takes its maximum at $y = x - t$. In this case we obtain therefore

$$\lim_{\epsilon\to 0} u^{(\epsilon)}(x,t) = \frac{x-(x-t)}{t} = 1, \qquad x > t. \qquad (A.20)$$

Summary A.3 *Inspecting all limits in the preceding examples [(A.12), (A.13), and (A.18)–(A.20)] we conclude that for both initial conditions (A.2) and (A.14), the solution to the viscous model (A.1) converges, as $\epsilon \to 0$, to the correct entropy solution of the Burgers' equation (2.22), both in the case of a shock (Example 2.12) and in the case of a centered rarefaction wave (Example 2.13).*

We shall not go further into the study of "viscous conservation laws" [when $\frac{1}{2}u^2$ as in (A.1) is replaced by a general flux function $f(u)$] and their limiting behavior as $\epsilon \to 0$. We refer the reader to Godlewski and Raviart [55], Hörmander [64], and Evans [36] for more details. However, our derivation above of the moving step case for the Burgers' equation serves to validate the statement made at the beginning to this Appendix concerning the use of entropy conditions as a selection rule (among many possible weak solutions) for the unique physically relevant solution.

We shall now see yet another application of the zero-viscosity methodology for the derivation of an entropy condition equivalent to the Lax condition (Definition 2.15). We shall present it formally in the context of the general conservation law (2.1).

Take $\epsilon > 0$ and let [in analogy with (A.1)] $u^{(\epsilon)}$ be the solution to the "viscous conservation law"

$$u_t^{(\epsilon)} + f(u^{(\epsilon)})_x = \epsilon\, u_{xx}^{(\epsilon)},$$
$$u^{(\epsilon)}(x, 0) = u_0(x). \tag{A.21}$$

Under very general conditions on f and u_0 [certainly if $f \in C^1(\mathbb{R})$ and $u_0 \in L^\infty(\mathbb{R})$] the solution $u^{(\epsilon)}$ is classical (in particular $u_t^{(\epsilon)}, u_x^{(\epsilon)}, u_{xx}^{(\epsilon)}$ are all continuous for $t > 0$). We refer the reader to Godlewski and Raviart [55] and to Hörmander [64] for details and proofs.

Let $U(u)$ be a strictly convex smooth function of u and let $F(u)$ satisfy

$$F'(u) = f'(u)\, U'(u). \tag{A.22}$$

Multiplying Equation (A.21) by $U'u^{(\epsilon)}$ we obtain

$$U\left(u^{(\epsilon)}\right)_t + F\left(u^{(\epsilon)}\right)_x = \epsilon\, U'\left(u^{(\epsilon)}\right) u_{xx}^{(\epsilon)}$$
$$= \epsilon \left(U'\left(u^{(\epsilon)}\right) u_x^{(\epsilon)}\right)_x - \epsilon\, U''\left(u^{(\epsilon)}\right) (u_x^{(\epsilon)})^2. \tag{A.23}$$

Let $\phi(x, t) \geq 0$ be a test function supported in $\mathbb{R} \times (0, \infty)$ (in particular, $\phi(x, 0) \equiv 0$). Multiplying Equation (A.23) by ϕ and integrating by parts [compare (2.12)], we obtain

$$
-\int_{\mathbb{R}}\int_0^\infty \left[U\left(u^{(\epsilon)}\right)\phi_t + F\left(u^{(\epsilon)}\right)\phi_x - \epsilon U'\left(u^{(\epsilon)}\right)u_x^{(\epsilon)}\phi_x \right] dx\, dt
$$

$$
= -\epsilon \int_{\mathbb{R}}\int_0^\infty U''\left(u^{(\epsilon)}\right)\left(u_x^{(\epsilon)}\right)^2 \phi\, dx\, dt. \tag{A.24}
$$

Assume first that $u^{(\epsilon)}$, $u_x^{(\epsilon)}$ are bounded uniformly in $\epsilon > 0$, over the support of ϕ, and that $u^{(\epsilon)} \to u$ as $\epsilon \to 0$. Then, taking the limits in (A.23), as $\epsilon \to 0$, we get

$$
-\int_{\mathbb{R}}\int_0^\infty [U(u)\phi_t + F(u)\phi_x]\, dx\, dt = 0. \tag{A.25}
$$

However, the assumptions leading to (A.25), and in particular that of the uniform boundedness of $u_x^{(\epsilon)}$, are too restrictive in typical cases. For example, recall that we have already seen [Equations (A.12) and (A.13)] that in the case of a moving step [with u_0 as in (A.2)] the sequence $u^{(\epsilon)}$ converges to the discontinuous step, so that $u_x^{(\epsilon)}$ cannot remain bounded. We note, however, that assuming the total variation (in x) of $u^{(\epsilon)}$ to be bounded (uniformly in $\epsilon > 0$) we can still show that the third term in the left-hand side of (A.24) tends to 0 as $\epsilon \to 0$. Indeed, in this case $\int_{\mathbb{R}} |u_x^{(\epsilon)}|\, dx = $ total variation of $u^{(\epsilon)}$ (for fixed t). Thus, retaining the assumption $u^{(\epsilon)} \to u$ (boundedly) and the uniform boundedness of the total variation (in x) of $u^{(\epsilon)}$, and noting that $U'' > 0$ (convexity), we get from (A.24) the following inequality, which replaces (A.25):

$$
-\int_{\mathbb{R}}\int_0^\infty [U(u)\phi_t + F(u)\phi_x]\, dx\, dt \leq 0 \tag{A.26}
$$

[for all nonnegative test functions $\phi(x, t)$].

Note also that multiplying (A.21) by ϕ and using the same assumptions we obtain (2.12), so that $u(x, t)$ is a weak solution to (2.1), (2.2), in the sense of Definition 2.3.

The convex function $U(u)$ is called an "entropy function" and the associated [via (A.22)] function $F(u)$ is its "entropy flux" (see Lax [76]). It turns out that the inequality (A.26), when applied to all possible pairs (U, F), characterizes the unique weak solution that is obtained by the vanishing viscosity method, namely, as a limit of the solutions $u^{(\epsilon)}$. The reader is referred to Hörmander [64] and to Godlewski and Raviart [55] for full discussions of these matters. We also refer to Godlewski and Raviart [55] for a proof of the fact that, if the flux function $f(u)$ is convex, it suffices to take only *one* pair of (entropy, entropy flux) functions in (A.26). In the case of a Riemann problem, when $f(u)$ is not convex, the structure of the entropy solution is more complicated than that outlined in

Section 2.1 (see the paragraph following Definition 2.18) and consists of several waves. We refer the reader to Godlewski and Raviart [55] and to Chorin and Marsden [28] for a description of the "Oleinik constructions" of these solutions.

Finally, we show how the Lax entropy condition (Definition 2.15) is obtained from (A.26), in the case of the Burgers' equation (see LeVeque [82, Example 3.4]).

In this case $f(u) = \frac{1}{2}u^2$. Taking $U = u^2$ we have, from (A.22), $F(u) = \frac{2}{3}u^3$. As in the derivation of the balance law [Claim 2.5(b)] we obtain from (A.26), for every rectangle $Q^T_{x_1, x_2}$,

$$\int_{x_1}^{x_2} u^2(x, T)\, dx - \int_{x_1}^{x_2} u^2(x, 0)\, dx$$

$$+ \frac{2}{3}\left[\int_0^T u^3(x_2, t)\, dt - \int_0^T u^3(x_1, t)\, dt\right] \le 0. \qquad (A.27)$$

If

$$u(x, t) = \begin{cases} u_1, & x \le x(t), \\ u_r, & x > x(t), \end{cases}$$

is a moving step solution, where $x(t) = St$ [$S = \frac{1}{2}(u_1 + u_r)$ by the Rankine–Hugoniot condition], then (A.27) yields, when $x_1 = 0$, $x_2 = ST$,

$$ST(u_1^2 - u_r^2) + \frac{2}{3}T(u_r^3 - u_1^3) \le 0,$$

or, by $S = \frac{1}{2}(u_1 + u_r)$,

$$T(u_1 - u_r)\left[2S^2 - \frac{2}{3}(u_r^2 + u_r u_1 + u_1^2)\right] = -\frac{T}{6}(u_1 - u_r)^3 \le 0.$$

Thus a moving step satisfying the entropy condition (A.27) is possible only with $u_1 > u_r$.

Appendix B

Convergence of the Godunov Scheme

In this appendix we outline the proof of Theorem 3.6, concerning the convergence of the Godunov scheme (3.11) to the entropy solution of the (nonlinear) scalar conservation law. As in the case of Theorem 2.28, where convergence was proved in the *linear* case, we try to avoid as much as possible the use of general (more advanced) mathematical facts. Instead, we rely on specific details pertaining to the Godunov scheme. We refer the reader to Smoller [104] for Oleinik's proof of the convergence of the Lax–Friedrichs scheme and to Godlewski and Raviart [55] for a broader discussion of convergence of discrete schemes.

Recall that $\tilde{u}(x, t)$ is the exact solution of the scalar conservation law subject to the initial condition $\tilde{u}(x, t_n) = U^n(x)$, where $U^n(x)$ is the piecewise constant function given by (3.2). At the next time level $t_{n+1} = t_n + k$ the function $U^{n+1}(x)$ is computed as in Definition 3.5. For the sequence of averages $\left\{U_j^n\right\}_{j=-\infty}^{\infty}$ we define the total variation by

$$TV\left(U^n\right) = \sum_{j=-\infty}^{\infty} \left|U_{j+1}^n - U_j^n\right|. \tag{B.1}$$

Clearly, $TV\left(U^n\right) = TV\left(U^n(x)\right)$, where, for any function $g(x)$ vanishing at infinity, the total variation is defined by,

$$TV\left(g\right) = \sup\left\{\sum_{j=-\infty}^{\infty} \left|g(y_{j+1}) - g(y_j)\right|\right\}, \tag{B.2}$$

and the supremum is taken over all monotonically increasing sequences

$\{y_j\}_{j=-\infty}^{\infty} \subseteq \mathbb{R}$. If $g(x) \in L^1(\mathbb{R}) \cap L^\infty(\mathbb{R})$ it can be shown that

$$TV(g) = \sup_{h>0} \frac{1}{h} \int_{-\infty}^{\infty} |g(y+h) - g(y)| \, dy.$$

In particular, our assumption that $u_0(x)$ is of finite total variation implies that

$$TV(U^0) = \sum_{j=-\infty}^{\infty} |U_{j+1}^0 - U_j^0| = \sum_{j=-\infty}^{\infty} \frac{1}{\Delta x} \left| \int_{x_{j-1/2}}^{x_{j+1/2}} [u_0(x + \Delta x) - u_0(x)] \, dx \right|$$

$$\leq \frac{1}{\Delta x} \sum_{j=-\infty}^{\infty} \int_{x_{j-1/2}}^{x_{j+1/2}} |u_0(x + \Delta x) - u_0(x)| \, dx \qquad (B.3)$$

$$= \frac{1}{\Delta x} \int_{-\infty}^{\infty} |u_0(x + \Delta x) - u_0(x)| \, dx \leq TV(u_0) < \infty.$$

Our first claim is that the sequence $\{TV(U^n)\}_{n=-\infty}^{\infty}$ is nonincreasing.

Claim B.1 *For the Godunov scheme,*

$$TV(U^{n+1}) \leq TV(U^n), \qquad n = 0, 1, 2, \ldots.$$

Proof The solution $\tilde{u}(x, t)$ is constant along characteristic (straight) lines [see (2.10)]. The lines starting at time $t = t_n$ from points in cell j have uniform slopes $f'(U_j^n)$. It can be easily seen, by the CFL condition, that all points in cell j are reached, at time $t = t_{n+1}$, by characteristic lines emanating at $t = t_n$ from points in cells $j - 1$, j, $j + 1$.[1] In particular, let $(\tilde{x}_{j+1/2}, t_n)$ be the point lying on the characteristic line through $(x_{j+1/2}, t_{n+1})$, namely, $\tilde{x}_{j+1/2} + f'(U^n(\tilde{x}_{j+1/2})) \cdot k = x_{j+1/2}$ [if the point is sonic we get $\tilde{x}_{j+1/2} = x_{j+1/2}$, $U^n(\tilde{x}_{j+1/2}) = u_{\min}$ as in (2.28)]. All points in cell j are thus reached, at time $t = t_{n+1}$, by characteristic lines emanating at $t = t_n$ from points in $[\tilde{x}_{j-1/2}, \tilde{x}_{j+1/2}]$ at time $t = t_n$. The range of values of $\tilde{u}(x, t_{n+1})$ is therefore contained in the interval $[\min U^n(x), \max U^n(x)]$, $\tilde{x}_{j-1/2} < x < \tilde{x}_{j+1/2}$. Observe that a section of cell j may be covered by (part of) a rarefaction fan centered,

---

[1] We include points (at most two) that lie on a shock trajectory [starting at $(x_{j\pm1/2}, t_n)$] and are reached by two characteristic lines. This case is entirely incorporated in the ensuing argument and will not be mentioned explicitly.

say, at $x_{j+1/2}$ (see Figure 3.1). The values of $\tilde{u}(x, t_{n+1})$ there lie between U_j^n and U_{j+1}^n and $\tilde{x}_{j+1/2} \geq x_{j+1/2}$.

Note also that by the CFL condition $x_{j-3/2} < \tilde{x}_{j-1/2} \leq \tilde{x}_{j+1/2} < x_{j+3/2}$, so that, for $\tilde{x}_{j-1/2} \leq x \leq \tilde{x}_{j+1/2}$,

$$\min \left(U_j^n, \, U_{j\pm1}^n \right) \leq \min U^n(x) \leq \max U^n(x) \leq \max \left(U_j^n, \, U_{j\pm1}^n \right).$$

We conclude that the average U_j^{n+1} satisfies, for some point $\tilde{x}_j$,

$$U_j^{n+1} = U^n \left(\tilde{x}_j \right), \qquad \tilde{x}_{j-1/2} \leq \tilde{x}_j \leq \tilde{x}_{j+1/2}. \tag{B.4}$$

In Equation (B.4) we use the observation that if there is no point x in cell j such that $U_j^{n+1} = \tilde{u}(x, t_{n+1})$, then U_j^{n+1} is an "intermediate value in a jump"; namely, there is a shock trajectory passing through $(\bar{x}, t_{n+1})$, $x_{j-1/2} < \bar{x} < x_{j+1/2}$, such that $\tilde{u}(\bar{x}+, t_{n+1}) < U_j^{n+1} < \tilde{u}(\bar{x}-, t_{n+1})$. This shock originates necessarily at one of the cell endpoints, say, $x_{j-1/2}$, and satisfies

$$U_{j-1}^n = \tilde{u}(\bar{x}-, t_{n+1}) > U_j^{n+1} > \tilde{u}(\bar{x}+, t_{n+1}) = U_j^n. \tag{B.5}$$

We then set $\tilde{x}_j = x_{j-1/2}$ and define $U^n(\tilde{x}_j) = U_j^{n+1}$. This construction yields a monotonically increasing sequence $\tilde{x}_j < \tilde{x}_{j+1}$, $-\infty < j < \infty$. We have

$$TV \left(U^{n+1} \right) = \sum_{j=-\infty}^{\infty} \left| U_{j+1}^{n+1} - U_j^{n+1} \right|$$

$$= \sum_{j=-\infty}^{\infty} \left| U^n(\tilde{x}_{j+1}) - U^n(\tilde{x}_j) \right| \leq TV \left(U^n(x) \right) = TV \left(U^n \right),$$

which concludes the proof of the claim.   □

Corollary B.2

$$TV \left(U^n \right) \leq TV \left(U^0 \right) < \infty.$$

Next we inspect the variation in time of the approximating sequences $\{U_j^n\}_{j=-\infty}^{\infty}$.

Claim B.3 *For any $n \geq p \geq 0$,*

$$\sum_{j=-\infty}^{\infty} \left| U_j^n - U_j^p \right| \leq 2\lambda S_0(n - p) \, TV \left(U^0 \right). \tag{B.6}$$

Proof In view of the defining equation (3.11), we have

$$\left| U_j^{n+1} - U_j^n \right| \leq \lambda \left| f\left(R\left(0;\, U_j^n, U_{j+1}^n\right)\right) - f\left(R\left(0;\, U_{j-1}^n, U_j^n\right)\right) \right|$$

$$\leq \lambda S_0 \left| R\left(0;\, U_j^n, U_{j+1}^n\right) - R\left(0;\, U_{j-1}^n, U_j^n\right) \right|, \tag{B.7}$$

where in the last step we have used the mean value theorem and (3.7) (to estimate $\left| f'(v) \right| \leq S_0$ for $|v| \leq M_n \leq M_0$). An easy inspection of the Riemann solution (2.28) yields

$$\left| R\left(0;\, U_j^n, U_{j+1}^n\right) - R\left(0;\, U_{j-1}^n, U_j^n\right) \right| \leq \left| U_{j+1}^n - U_j^n \right| + \left| U_j^n - U_{j-1}^n \right|, \tag{B.8}$$

so that summation over j in (B.7) yields (B.6), when going $(n - p)$ time steps backward and noting Claim B.1. $\qquad\square$

We can now conclude the proof of Theorem 3.6 as follows. Fix $T > 0$. For a given time step $k > 0$ we extend the sequence of approximating functions $U^0(x), U^1(x), \dots, U^n(x), \dots, nk \leq T$ (these functions depend, of course, on k), so that we get a function defined on the entire strip $\mathbb{R} \times [0, T]$ by

$$U_k(x, t) = U^n(x) \qquad \text{for} \quad t \in [nk, (n + 1)k). \tag{B.9}$$

Thus, $U_k(x, t)$ is piecewise constant in space (with jumps at $x_{j+1/2} = (j + 1/2)\Delta x$, $-\infty < j < \infty$) and in time (with jumps at $t_n = nk \leq T$, $n = 1, 2, \dots$). It takes the constant value U_j^n in the rectangle $(x_{j-1/2}, x_{j+1/2}) \times [nk, (n + 1)k)$ (see Figure B.1).

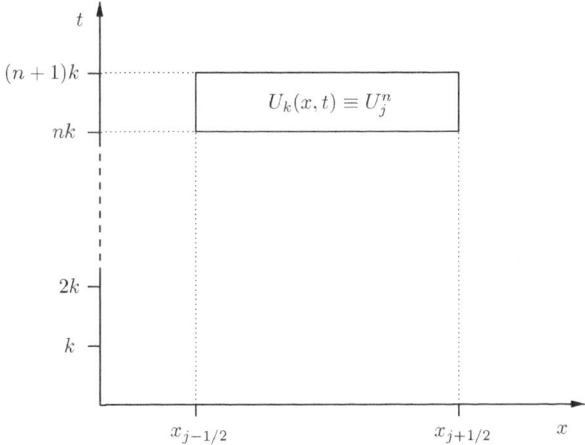

Figure B.1. The piecewise-constant function $U_k(x, t)$.

The reader should note that our notation is somewhat ambiguous. Indeed, for different k's the functions $U^n(x)$ in (B.9), for the same integer n, represent two different functions (at two different time levels) constructed by the Godunov scheme using the two different time steps.

Pick any time level $\tau \in [0, T]$. By Claim 3.7 the family $\{U_k(x, \tau)\}_{k>0}$ (viewed as a family of functions of the space variable x) is uniformly bounded, and by Corollary B.2 all total variations of these functions are also uniformly bounded [by $TV(u_0)$, see (B.3)]. Thus, applying Helly's Selection Theorem (see Nathanson [92]) we can extract a sequence of time steps, $k_i^{(1)} \to 0$, such that the corresponding sequence of functions $\left\{U_{k_i^{(1)}}(x, \tau)\right\}_{i=1}^{\infty}$ converges at every $x \in \mathbb{R}$, and in $L_{\text{loc}}^1(\mathbb{R})$ (see Definition 2.25), to a limit function, which we denote by $v(x, \tau)$. Picking another value $\bar{\tau} \in [0, T]$, we obtain, by the same argument, a further subsequence $\left\{k_i^{(2)}\right\}_{i=1}^{\infty} \subseteq \left\{k_i^{(1)}\right\}_{i=1}^{\infty}$ such that $\left\{U_{k_i^{(2)}}(x, \bar{\tau})\right\}_{i=1}^{\infty}$ converges [for all x and in $L_{\text{loc}}^1(\mathbb{R})$] to a limit function $v(x, \bar{\tau})$. Because it is a subsequence of $\left\{U_{k_i^{(1)}}\right\}_{i=1}^{\infty}$, $\left\{U_{k_i^{(2)}}(x, \bar{\tau})\right\}_{i=1}^{\infty}$ clearly converges to $v(x, \tau)$. We can now apply the familiar "diagonal process" to obtain a sequence $k_i \to 0$ such that the corresponding functions $\left\{U_{k_i}(x, \tau_l)\right\}$ converge [for every $x \in \mathbb{R}$ and in $L_{\text{loc}}^1(\mathbb{R})$] to $v(x, \tau_l)$, where the sequence $\{\tau_l\}_{l=1}^{\infty}$ is dense in $(0, T)$. But now we claim that the sequence $\left\{U_{k_i}(x, t)\right\}_{i=1}^{\infty}$ converges [pointwise and in $L_{\text{loc}}^1(\mathbb{R})$] for *every* $t \in [0, T]$. To this end we first note that, multiplying (B.6) by Δx and noting that $\lambda(n - p)\Delta x = k(n - p) = nk - np$, and using the definition of the function $U_k(x, t)$, we have, with $nk \le t_1 < (n + 1)k$, $pk \le t_2 \le (p + 1)k$,

$$\int_{-\infty}^{\infty} |U_k(x, t_1) - U_k(x, t_2)| \, dx \le 2S_0 \, (|t_1 - t_2| + k) \cdot TV(u_0). \tag{B.10}$$

Taking $t_1 = t$ and $t_2 = \tau_l$, for some index l, and letting $k = k_i$ in (B.10), we get

$$\int_{-\infty}^{\infty} |U_{k_i}(x, t) - U_{k_i}(x, \tau_l)| \, dx \le 2S_0 \, (k_i + |t - \tau_l|) \cdot TV(u_0). \tag{B.11}$$

Let $\epsilon > 0$ be given, and take a finite interval $[-X, X] \subseteq \mathbb{R}$. Let l be such that $|t - \tau_l| < \epsilon$ ($\{\tau_l\}_{l=1}^{\infty}$ is dense in $(0, T)$). Since $\left\{U_{k_i}(x, \tau_l)\right\}_{i=1}^{\infty}$ converges to $v(x, \tau_l)$ in $L^1(-X, X)$ we can find an index $J = J(\epsilon, l)$, such that

$$\int_{-X}^{X} |U_{k_i}(x, \tau_l) - v(x, \tau_l)| \, dx \le \epsilon, \qquad i > J. \tag{B.12}$$

We may further assume that $k_i < \epsilon$ for $i > J$. Inserting all this in (B.11) we obtain, for $i > J$,

$$\int_{-X}^{X} |U_{k_i}(x, t) - v(x, \tau_l)| \leq \int_{-X}^{X} \{|U_{k_i}(x, t) - U_{k_i}(x, \tau_l)|$$

$$+ |U_{k_i}(x, \tau_l) - v(x, \tau_l)|\} \, dx \leq (4 \cdot TV(u_0) + 1) \epsilon; \quad \text{(B.13)}$$

hence also, for $i > j > J$,

$$\int_{-X}^{X} |U_{k_i}(x, t) - U_{k_j}(x, t)| \, dx \leq \int_{-X}^{X} \{|U_{k_i}(x, t) - v(x, \tau_l)|$$

$$+ |U_{k_j}(x, t) - v(x, \tau_l)|\} \, dx \leq 2(4 \cdot TV(u_0) + 1) \epsilon. \quad \text{(B.14)}$$

We conclude that the sequence $\{U_{k_i}(x, t)\}_{i=1}^{\infty}$ is a Cauchy sequence in $L^1(-X, X)$ and thus converges to some function $v(x, t)$. Since X was arbitrary, the function $v(x, t)$ is defined for all $x \in \mathbb{R}$ and the convergence $U_{k_i}(x, t) \to v(x, t)$ is in $L^1_{\text{loc}}(\mathbb{R})$. Passing to the limit $i \to \infty$ in (B.13) we now have

$$\int_{-X}^{X} |v(x, t) - v(x, \tau_l)| \, dx \leq 4(TV(u_0) + 1)\epsilon, \quad \text{(B.15)}$$

under the assumption $|t - \tau_l| < \epsilon$. In other words, taking any subsequence $\tau_{l_j} \to t$ and noting that X is arbitrary, we get, from (B.15), [2]

$$v(x, t) = \lim_{\tau_{l_j} \to t} v(x, \tau_l) \quad \text{in} \quad L^1_{\text{loc}}(\mathbb{R}). \quad \text{(B.16)}$$

In other words, for every $t \in (0, T)$ the function $v(x, t)$ is uniquely determined as a limit of a subsequence of $\{v(x, \tau_l)\}_{l=1}^{\infty}$. However, by Helly's theorem, the sequence $\{U_{k_i}(x, t)\}_{i=1}^{\infty}$, being uniformly bounded and with uniformly bounded total variations, possesses a subsequence that converges pointwise and hence necessarily to $v(x, t)$. Thus, since all converging subsequences have the same limit function $v(x, t)$, the whole sequence $\{U_{k_i}(x, t)\}_{i=1}^{\infty}$ converges pointwise to $v(x, t)$.

We have therefore obtained a limit function $v(x, t)$ of the sequence $\{U_{k_i}(x, t)\}_{i=1}^{\infty}$ in the strip $\mathbb{R} \times [0, T]$. It remains to show that $v(x, t)$ is a weak solution (see Definition 2.3) and that it satisfies the entropy condition (see Appendix A), thus ensuring its uniqueness.

---

[2] The reader may notice that in these arguments we have just reproduced the classical proof of the Arzela–Ascoli theorem [for functions of t valued in $L^1_{\text{loc}}(\mathbb{R})$].

Recall (Definition 2.3) that to prove that $v(x, t)$ is a weak solution we need to show that, for every test function $\phi(x, t) \in C_0^1$,

$$\int\limits_{\mathbb{R}} \int\limits_0^\infty (v\phi_t + f(v)\phi_x)\, dx\, dt + \int\limits_{\mathbb{R}} \phi(x, 0)u_0(x)\, dx = 0. \tag{B.17}$$

Fix $k = \Delta t > 0$ and let $\{U_j^n\}_{j,n}$ be the values determined by the Godunov scheme (3.10), (3.11) [these are the values taken on by the function $U_k(x, t)$, by (B.9)]. Let $\phi_{j,n} = \phi(x_j, t_n)$ and multiply (3.10) by $\phi_{j,n}$ to obtain

$$\phi_{j,n}\left(U_j^{n+1} - U_j^n\right) = -\lambda\phi_{j,n}\left(f_{j+1/2}^{G,n+1/2} - f_{j-1/2}^{G,n+1/2}\right). \tag{B.18}$$

Summing over $-\infty < j < \infty$ we readily obtain

$$\sum_{j=-\infty}^\infty \phi_{j,n}\left(U_j^{n+1} - U_j^n\right) = \lambda \sum_{j=-\infty}^\infty f_{j+1/2}^{G,n+1/2}\left(\phi_{j+1,n} - \phi_{j,n}\right). \tag{B.19}$$

Recall [see (3.9)] that $f_{j+1/2}^{G,n+1/2} = f\left(R(0;\, U_j^n,\, U_{j+1}^n)\right)$ and that [see (2.28)] in all cases $V_j^n := R(0;\, U_j^n,\, U_{j+1}^n)$ is a value that lies between U_j^n and U_{j+1}^n. We can therefore rewrite (B.19) as

$$\sum_{j=-\infty}^\infty \phi_{j,n}\left(U_j^{n+1} - U_j^n\right) = \lambda \sum_{j=-\infty}^\infty f(V_j^n)\left(\phi_{j+1,n} - \phi_{j,n}\right),$$

and summing over $0 \le n < \infty$ and multiplying both sides by Δx we get

$$k\Delta x \left\{ \sum_{n=1}^\infty \sum_{j=-\infty}^\infty \frac{\phi_{j,n} - \phi_{j,n-1}}{k} U_j^n + \sum_{n=1}^\infty \sum_{j=-\infty}^\infty f(V_j^n)\frac{\phi_{j+1,n} - \phi_{j,n}}{\Delta x} \right\}$$

$$+ \Delta x \sum_{j=-\infty}^\infty \phi_{j,0}\, U_j^0 = 0. \tag{B.20}$$

Observe that these sums are all finite, since $\phi \in C_0^1$ means that $\phi_{j,n}$ vanishes for large $|j|, n$.

Returning to the notation U_k [in (B.9)], let us introduce a similar piecewise-constant function $\Phi_{k,t}$ by

$$\Phi_{k,t}(x, t) = \frac{\phi_{j,n} - \phi_{j,n-1}}{k} \quad \text{if} \quad x \in (x_{j-1/2}, x_{j+1/2}),\, t \in [nk, (n+1)k).$$

$$\tag{B.21}$$

Clearly, since $\phi \in C_0^1$, $\Phi_{k,t}(x, t) \to \phi_t(x, t)$ uniformly in $\mathbb{R} \times [0, T]$ (simply write the difference in terms of Taylor's expansion up to second order in t). Specializing to the sequence $k = k_i$,

$$k_i \Delta x \cdot \sum_{n=1}^{\infty} \sum_{j=-\infty}^{\infty} \frac{\phi_{j,n} - \phi_{j,n-1}}{k} U_j^n = \int_{\mathbb{R}} \int_0^T \Phi_{k_i,t}(x, t) U_{k_i}(x, t) \, dx \, dt$$

$$\xrightarrow[k_i \to 0]{} \int_{\mathbb{R}} \int_0^T \phi_t(x, t) v(x, t) \, dx \, dt, \tag{B.22}$$

where in the last step we have used $U_{k_i} \to v$ and Lebesgue's dominated convergence theorem. We can treat similarly the second sum in (B.20). Indeed, define piecewise-constant functions

$$\left.\begin{array}{l} G_k(x, t) = f(V_j) \\[2mm] \Phi_{k,x}(x, t) = \dfrac{\phi_{j+1,n} - \phi_{j,n}}{\Delta x} \end{array}\right\} (x, t) \in (x_{j-1/2}, x_{j+1/2}) \times [nk, (n+1)k),$$

$$\tag{B.23}$$

so that

$$k \Delta x \cdot \sum_{n=1}^{\infty} \sum_{j=-\infty}^{\infty} f(V_j) \frac{\phi_{j+1,n} - \phi_{j,n}}{\Delta x} = \int_{\mathbb{R}} \int_0^T G_k(x, t) \Phi_{k,x} \, dx \, dt.$$

Clearly, $\Phi_{k,x} \to \phi_x$ as $k \to 0$ uniformly in $\mathbb{R} \times [0, T]$. However, the value V_j^n is always between U_j^n and U_{j+1}^n, so the continuity of f and the convergence pointwise of U_{k_i} to v imply that $G_{k_i}(x, t) \to f(v)$, pointwise. All functions $G_{k_i}(x, t)$ are uniformly bounded [by $f(M_0)$; see Claim 3.7], so, once again by the dominated convergence theorem, we have

$$k_i \Delta x \cdot \sum_{n=1}^{\infty} \sum_{j=-\infty}^{\infty} f(V_j^n) \frac{\phi_{j+1,n} - \phi_{j,n}}{\Delta x} \xrightarrow[k_i \to 0]{} \int_{\mathbb{R}} \int_0^T f(v) \phi_x \, dx \, dt.$$

$$\tag{B.24}$$

The fact that the last sum in (B.20) converges to $\int_{\mathbb{R}} \phi(x, 0) u_0(x) \, dx$ as $\Delta x_i \to 0$ (note that $\lambda = \frac{k_i}{\Delta x_i} = constant$ and $k_i \to 0$ implies $\Delta x_i \to 0$) is left to the reader as an easy exercise. Inserting the limits (B.22), (B.24), along with the last observation, in (B.20), yields (B.17); hence $v(x, t)$ is indeed a weak solution.

Remark B.4 The preceding argument, using the convergence $U_{k_i} \to v$ in the proof that v is a weak solution, can be generalized to other schemes. It is then known as the "Lax–Wendroff theorem." We refer the reader to the books by LeVeque [82] and by Godlewski and Raviart [55] for a general statement of this theorem.

Remark B.5 Strictly speaking, we have proven that $v(x, t)$ is a weak solution in the strip $\mathbb{R} \times [0, T]$. However, an inspection of our arguments shows that we could replace $[0, T]$ by $[0, \infty)$, by taking a sequence $\{T_l\}_{l=1}^\infty$ that is dense in $[0, \infty)$. The estimates needed in the proof [most notably (B.11)] hold uniformly in $t \in [0, \infty)$. Thus, $v(x, t)$ is actually a weak solution in $\mathbb{R} \times [0, \infty)$.

Finally, it remains to prove that $v(x, t)$ is the (unique) entropy solution to (2.1), (2.2). In dealing with the Godunov scheme, it is very easy to do the proof in the context of the "entropy-function entropy-flux" formalism, as given in (A.26).

So, let $\Psi(u)$ be a strictly convex smooth function, and let $F(u)$ be the associated flux, $F'(u) = \Psi'(u) f'(u)$. We need to show that, for all nonnegative test functions $\phi \in C_0^1$,

$$-\int_{\mathbb{R}} \int_0^\infty [\Psi(v)\phi_t + F(v)\phi_x] \, dx \, dt \le 0. \tag{B.25}$$

This inequality is satisfied by the solution $\tilde{u}(x, t)$, $t_n \le t \le t_{n+1}$. Let $U_k(x, t)$ be the piecewise-constant function constructed by the Godunov scheme, as in (B.9). Repeating the argument in the proof of Claim 2.5(b) (by using the same sequence of nonnegative test functions) in the rectangle $(x_{j-1/2}, x_{j+1/2}) \times (t_n, t_{n+1})$ and passing to the limit in (B.25) we get [instead of the balance equation (2.3)] the inequality

$$\int_{x_{j-1/2}}^{x_{j+1/2}} \left[\Psi(\tilde{u}(x, t_{n+1})) - \Psi(U_j^n) \right] dx \le -\int_{t_n}^{t_{n+1}} \left[F(R(0; U_j^n, U_{j+1}^n)) \right.$$

$$\left. - F(R(0; U_{j-1}^n, U_j^n)) \right] dt. \tag{B.26}$$

In this inequality we have already used that $\tilde{u}(x, t_n) \equiv U_j^n$ in cell j and that $\tilde{u}(x_{j+1/2}, t) \equiv R(0; U_j^n, U_{j+1}^n)$, $t_n \le t \le t_{n+1}$ [see (3.8)]. Since Ψ is strictly convex, we get by Jensen's inequality

$$
\frac{1}{\Delta x} \int_{x_{j-1/2}}^{x_{j+1/2}} \Psi \left(\tilde{u}(x, t_{n+1}) \right) dx \leq \Psi \left(\frac{1}{\Delta x} \int_{x_{j-1/2}}^{x_{j+1/2}} \tilde{u}(x, t_{n+1}) \, dx \right)
$$

$$
= \Psi \left(U_j^{n+1} \right), \tag{B.27}
$$

where (3.10) was used in the last step. Thus, the inequality (B.26) can be rewritten as

$$
\Psi \left(U_j^{n+1} \right) - \Psi \left(U_j^n \right) \leq -\lambda \left[F \left(R(0; U_j^n, U_{j+1}^n) \right) - F \left(R(0; U_{j-1}^n, U_j^n) \right) \right].
$$

$$
\tag{B.28}
$$

We can now repeat exactly the proof of (B.17), replacing U_k by $\Psi \left(U_k \right)$ and f by F [i.e., $f_{j+1/2}^{G,n+1/2}$ by $F \left(R(0; U_j^n, U_{j+1}^n) \right)$] to obtain from the *discrete inequality* (B.28) the *integrated inequality (B.25)*. This concludes the proof that v is indeed the (unique) entropy solution to (2.1), (2.2).

Appendix C

Riemann Solver for a γ-Law Gas

Numerous schemes for fluid dynamics (including the GRP scheme) are based on solving a Riemann problem at cell interfaces, requiring an efficient and robust algorithm for solving this problem. Here we briefly present a Newton–Raphson (iterative) algorithm for solving the RP for perfect gases, which typically converges in about three iterations, producing an accurate solution of the RP in the (u, p) plane.

The Riemann problem is the IVP given in terms of Equation (4.47) subject to the initial data (4.100). The procedure for solving a RP, as outlined in Construction 4.46, consists in finding the intersection point (u^*, p^*) of the interaction curves I_1^l, I_3^r, as depicted in Figure 4.18. The equations we use here for the (u, p) interaction curves are those given in Summary 4.49, with the transformation $\zeta = p^{\frac{\gamma-1}{2\gamma}}$, so that in the (u, ζ) plane the rarefaction branch of either interaction curve is a straight line, as observed in Remark 4.51. For the Newton–Raphson iterations we also require the derivative functions (slopes of the respective interaction curves), which are given by [see Equations (4.115) and (4.116)]

$$
\left(\frac{du}{d\zeta}\right) =
\begin{cases}
-\left(\frac{2}{\gamma-1}\right)\frac{c_L}{\zeta_L}, & \zeta \leq \zeta_L, \\[2ex]
-\left(\frac{2}{\gamma-1}\right)\frac{c_L}{\zeta}\left(\frac{\zeta}{\zeta_L}\right)^{\frac{2\gamma}{\gamma-1}} \times \left\{1 + \left(\frac{\gamma+1}{4\gamma}\right)\left[\left(\frac{\zeta}{\zeta_L}\right)^{\frac{2\gamma}{\gamma-1}} - 1\right]\right\} \\[2ex]
\quad \times \left\{1 + \left(\frac{\gamma+1}{2\gamma}\right)\left[\left(\frac{\zeta}{\zeta_L}\right)^{\frac{2\gamma}{\gamma-1}} - 1\right]\right\}^{-3/2}, & \zeta > \zeta_L,
\end{cases}
$$

$$\tag{C.1}$$

for the I_1^l interaction curve and by

$$
\left(\frac{du}{d\zeta}\right) =
\begin{cases}
\left(\frac{2}{\gamma-1}\right)\frac{\alpha_R}{\zeta_R}, & \zeta \le \zeta_R, \\[2ex]
\left(\frac{2}{\gamma-1}\right)\frac{\alpha_R}{\zeta}\left(\frac{\zeta}{\zeta_R}\right)^{\frac{2\gamma}{\gamma-1}} \times \left\{1 + \left(\frac{\gamma+1}{4\gamma}\right)\left[\left(\frac{\zeta}{\zeta_R}\right)^{\frac{2\gamma}{\gamma-1}} - 1\right]\right\} \\[2ex]
\qquad \times \left\{1 + \left(\frac{\gamma+1}{2\gamma}\right)\left[\left(\frac{\zeta}{\zeta_R}\right)^{\frac{2\gamma}{\gamma-1}} - 1\right]\right\}^{-3/2}, & \zeta > \zeta_R,
\end{cases}
$$

$$\text{(C.2)}$$

for the I_3^r interaction curve.

The solution algorithm starts by "sorting out" the solution branch on each interaction curve. The idea is to determine *in advance* the shock/rarefaction branch on which the intersection point (u^*, p^*) is located. This can be done by exploiting the monotonicity property of the interaction curves, which is unaltered by the transformation from (u, p) to (u, ζ). The idea is depicted graphically in Figure C.1 for the right interaction curve I_3^r. Let u_{LR} be the velocity value on I_1^l corresponding to the pressure p_R (see Figure C.1). If $u_{LR} > u_R$ (as actually seen in Figure C.1) then the intersection value u^* satisfies $u_R < u^* < u_{LR}$. Hence the intersection point (u^*, p^*) lies on the shock branch of I_3^r. In other words, the wave Γ_3 (see Construction 4.46) is a 3-shock. Conversely, if $u_{LR} < u_R$, the intersection value u^* satisfies $u_{LR} < u^* < u_R$ and Γ_3 is a 3-CRW. An analogous procedure is repeated to determine the solution branch on I_1^l. Once we know which branch the solution is located on, the iterations for

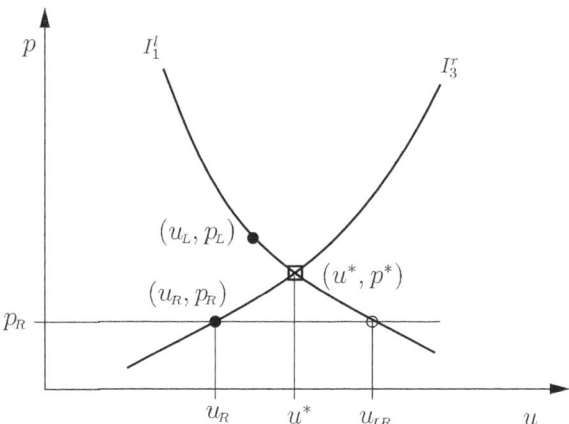

Figure C.1. Determination of the shock/rarefaction branch on the right interaction curve I_3^r.

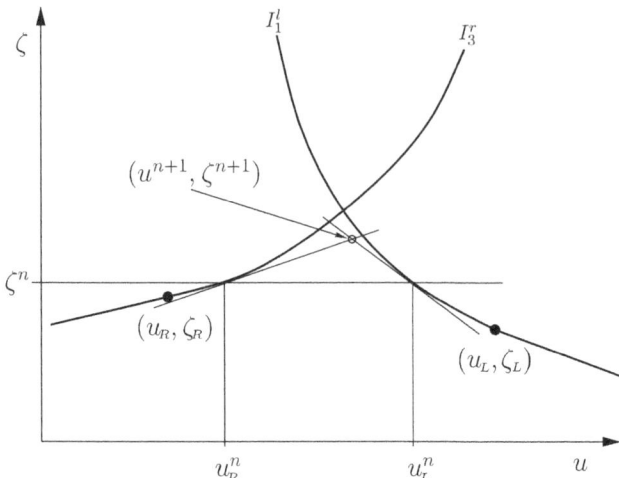

Figure C.2. Iterative procedure for determining the intersection point between I_1^l and I_3^r.

finding (u^*, p^*) are made with the correct shock/rarefaction branch on either side.

Turning to the actual solution, we start by the "zero iteration," which is obtained by calculating the intersection of the two rarefaction branches in (u, ζ). This is trivial since it consists in finding the intersection point of two straight lines. We note that this first estimate of (u^*, p^*) is akin to an acoustic approximation [which is obtained as a linear intersection in the (u, p) plane]. When this approximation is not sufficiently accurate, we proceed as follows.

Denote the solution obtained at the n th iteration by (u^n, ζ^n), with $n = 0, 1, 2, \ldots$ ($n = 0$ being the "zero iteration"). Let u_L^n, u_R^n be the velocities at the points of intersection of $\zeta = \zeta^n$ with the curves I_1^l, I_3^r, respectively (see Figure C.2). Then (u^{n+1}, ζ^{n+1}) is obtained as the intersection point of the tangent lines to I_1^l, I_3^r at (u_L^n, ζ^n), (u_R^n, ζ^n), respectively. This iterative procedure is terminated when $|u_L^n - u_R^n|$ is sufficiently small. The last intersection point is then taken to designate the state (u^*, p^*) at the contact discontinuity resolving the given Riemann problem.

Appendix D

The MUSCL Scheme

In this appendix we prove the claim made in Section 5.2 that van Leer's MUSCL scheme is an L_2 scheme (see the discussion following Definition 5.41). As explained there it means that, given piecewise-linear distributions as in (5.93), the numerical fluxes $\mathbf{F}_{j+1/2}^{n+1/2}$, $\mathbf{G}_{j+1/2}^{n+1/2}$ are computed within $O(k^2)$ error of the exact ones (for the linear GRP). This, in turn, will follow if we show that the time derivatives of flow variables along the contact discontinuity are evaluated within $O(k^2)$ error. This observation brings us back to the main theorem of the Lagrangian treatment, Theorem 5.7 (Section 5.1). In this context, let us consider the left coefficients a_L, b_L, d_L, which, in the setup of Section 5.1, were obtained by the resolution of the CRW (Claim 5.17). For simplicity, we assume planar symmetry, so that $A(0) = 1$ and $\lambda = 0$ in Equation (5.48). In regions of smooth flow, the small parameter k of the numerical setting is replaced here by the small parameter $1 - \beta^*$, which measures the jump in (Lagrangian) characteristic slopes across the CRW [and hence is $O(\Delta r)$]. Note that, setting $\beta^* = 1$ in Equation (5.48), we obtain the coefficients for the acoustic (L_1) approximation, as in Remark 5.21 and Construction 5.39. We conclude, therefore, that for an L_2 scheme we need to find an approximate coefficient $\widetilde{d_L}$ such that

$$\left| \widetilde{d_L} - d_L \right| = O\left((1 - \beta^*)^2\right), \qquad \beta^* = \frac{g_L^*}{g_L}. \tag{D.1}$$

By (5.24), we have

$$a(\beta^*) = a(1) + H(1) \cdot (1 - \beta^*) + O\left((1 - \beta^*)^2\right), \tag{D.2}$$

333

where, by (5.30) and (5.36),

$$H(1) = \frac{1}{2} g_L^{-1} I(1) \frac{\partial u}{\partial \beta}(0, 1),$$

$$I(1) = G_{\rho_l} (p_L; \rho_L, p_L) \rho_L' + G_{p_l} (p_L; \rho_L, p_L) p_L' \qquad (D.3)$$

(where ρ_L' and p_L' are the initial Lagrangian slopes).

The states $(\rho(0, \beta), p(0, \beta))$ are all isentropic with respect to (ρ_L, p_L), owing to the isentropic character of the CRW in the associated Riemann solution. In particular, we get the identity

$$g_L = G (p_L; \rho(0, \beta), p(0, \beta)), \qquad (D.4)$$

for G as in (5.28). When (D.4) is differentiated with respect to β, we have at $\beta = 1$

$$0 = G_{\rho_l} (p_L; \rho_L, p_L) \frac{\partial \rho}{\partial \beta}(0, 1) + G_{p_l} (p_L; \rho_L, p_L) \frac{\partial p}{\partial \beta}(0, 1)$$

$$= \left[G_{\rho_l} (p_L; \rho_L, p_L) + c_L^2 G_{p_l} (p_L; \rho_L, p_L) \right] \frac{\partial \rho}{\partial \beta}(0, 1), \qquad (D.5)$$

which implies the vanishing of the expression in square brackets. It follows that

$$I(1) = (\rho_L' - c_L^{-2} p_L') G_{\rho_l}(p_L; \rho_L, p_L). \qquad (D.6)$$

Invoking the characteristic relation (5.26)(ii), we get by (D.3)

$$H(1) = -\frac{1}{2} g_L^{-2} (\rho_L' - c_L^{-2} p_L') G_{\rho_l} (p_L; \rho_L, p_L) \frac{\partial p}{\partial \beta}(0, 1)$$

$$= -\frac{1}{2} \rho_L^{-2} (\rho_L' - c_L^{-2} p_L') G_{\rho_l} (p_L; \rho_L, p_L) \frac{\partial \rho}{\partial \beta}(0, 1)$$

$$= -\frac{1}{2} \rho_L^{-2} (\rho_L' - c_L^{-2} p_L') \frac{d}{d\beta} G (p_L; \rho(0, \beta), p_L)_{\beta=1}.$$

We therefore get, within $O\left((1 - \beta^*)^2\right)$ error,

$$H(1)(1 - \beta^*) = \frac{1}{2} \rho_L^{-2} (\rho_L' - c_L^{-2} p_L') \cdot \left[G \left(p_L; \rho_L^*, p_L \right) - g_L \right]. \qquad (D.7)$$

We now make the following structural assumption on the function $G (p; \rho_l, p_l)$:

$$G_{\rho_l} (p_L; \rho_l, p_L)_{\rho_l=\rho_L} = \frac{1}{2} \frac{g_L}{\rho_L} = \frac{1}{2} c_L. \qquad (D.8)$$

For a perfect (γ-law) gas, this equation is satisfied [see Equation (5.39)]. In the general case, it means that the isentropic curve $p = p(\rho; \rho_i, p_i)$ is locally a power function. We now evaluate $-(g_L g_L^*)^{1/2} H(1)(1 - \beta^*)$ from (D.7), noting (D.8) and using, with $\tau = 1/\rho$,

$$\rho_L' = -\rho_L^2 \tau_L', \qquad \rho_L^* - \rho_L = -\rho_L^2 (\tau_L^* - \tau_L) + O\big((1 - \beta^*)^2\big).$$

The following equalities hold within an $O\big((1 - \beta^*)^2\big)$ error:

$$-(g_L g_L^*)^{1/2} H(1)(1 - \beta^*) = \frac{1}{4} g_L \cdot g_L^{-2} (g_L^2 \tau_L' + p_L') \cdot \frac{g_L}{\rho_L} (\rho_L^* - \rho_L)$$

$$= -\frac{1}{4} \rho_L (g_L^2 \tau_L' + p_L')(\tau_L^* - \tau_L). \tag{D.9}$$

However, by (5.25) and the characteristic relation (5.26)(ii) we have, again within an $O\big((1 - \beta^*)^2\big)$ error,

$$-(g_L g_L^*)^{1/2} a(1) = \frac{p_L^* - p_L}{u_L^* - u_L} (u_L' + g_L^{-1} p_L'). \tag{D.10}$$

Using (D.9) and (D.10) in (D.2) and noting (5.48) we have

$$\widetilde{d}_L = -(g_L g_L^*)^{1/2} \big[a(1) + H(1) \cdot (1 - \beta^*)\big] = \frac{p_L^* - p_L}{u_L^* - u_L} (u_L' + g_L^{-1} p_L')$$

$$- \frac{1}{4} \rho_L (g_L^2 \tau_L' + p_L')(\tau_L^* - \tau_L) = d_L + O\big((1 - \beta^*)^2\big), \tag{D.11}$$

so that $\widetilde{d}_L$ satisfies (D.2).

The coefficient $\widetilde{d}_L$ is exactly the coefficient obtained by van Leer [113, Equation (65)]. As in the acoustic case (Proposition 5.9) we can use the "reflected" expression $\widetilde{d}_R$ as an approximation to d_R, within the same error bound.

Remark D.1 Van Leer obtains $\widetilde{d}_L$ by treating the CRW as a "weak shock" and using the Rankine–Hugoniot jump conditions between the states V_L and V_L^*. This is the reason for the appearance of the term $W = \frac{p_L^* - p_L}{u_L^* - u_L}$ in (D.11); it is the "shock" speed by (4.89)(ii). From the derivation made here it is clear that $\widetilde{d}_L$ of (D.11) is by no means the unique possible approximation to d_L, within $O\big((p^* - p_L)^2\big)$. Following van Leer's idea, we could use instead the coefficients in (5.58) and approximate them to within an $O\big((p^* - p_R)^2\big)$ error. In so doing, we shall need again some information about the equation of state Φ, which is equivalent to assumption (D.8).

Bibliography

[1] F. Alcrudo and P. Garcia-Navarro. A high-resolution Godunov-type scheme in finite volumes for the 2-D shallow water equations. *Int. J. Numerical Methods Fluids*, 16:489–505, 1993.

[2] M. Ben-Artzi. The generalized Riemann problem in compressible duct flow. *Contemporary Math.*, 60:11–18, 1987.

[3] M. Ben-Artzi. The generalized Riemann problem for reactive flows. *J. Comput. Phys.*, 81:70–101, 1989.

[4] M. Ben-Artzi and A. Birman. Application of the "generalized Riemann problem" method to 1-D compressible flows with material interfaces. *J. Comput. Phys.*, 65:170–178, 1986.

[5] M. Ben-Artzi and A. Birman. A GRP scheme for reactive duct flows in external fields. In R. L. Dwoyer, M. Y. Hussaini, and R. G. Voigt, editors, *Proceedings of the 11th International Conference on Numerical Methods in Fluid Dynamics*, volume 323 of *Springer Lecture Notes in Physics*, pages 147–150. Springer-Verlag, New York, 1989.

[6] M. Ben-Artzi and A. Birman. Computation of reactive duct flows in external fields. *J. Comput. Phys.*, 86:225–255, 1990.

[7] M. Ben-Artzi and J. Falcovitz. A second-order Godunov-type scheme for compressible fluid dynamics. *J. Comput. Phys.*, 55:1–32, 1984.

[8] M. Ben-Artzi and J. Falcovitz. GRP – An analytic approach to high resolution upwind schemes for compressible fluid flow. In Soubaramayer and J. P. Boujot, editors, *Proceedings of the 9th International Conference on Numerical Methods in Fluid Dynamics*, volume 218 of *Springer Lecture Notes in Physics*, pages 87–91. Springer-Verlag, New York, 1985.

[9] M. Ben-Artzi and J. Falcovitz. A high-resolution upwind scheme for quasi-1-D flows. In R. Glowinski, F. Angrand, A. Dervieux, and J. A. Desideri, editors, *Numerical Methods for the Euler Equations of Fluid Dynamics*, pages 66–83. SIAM, Philadelphia, 1985.

[10] M. Ben-Artzi and J. Falcovitz. An upwind second-order scheme for compressible duct flows. *SIAM J. Sci. Stat. Comput.*, 7:744–768, 1986.

[11] M. Ben-Artzi and J. Falcovitz. GRP schemes for time-dependent compressible flows. In *Papers Presented at the Beijing Workshop on Computational Fluid Dynamics*, pages 67–80. Academia Sinica, Institute of Mechanics, October 1995.

337

[12] M. Ben-Artzi, J. Falcovitz, and A. Birman. The GRP treatment of flow singularities. In D. Leutloff and R. C. Srivastava, editors, *Computational Fluid Dynamics – Selected Topics*, pages 233–244. Springer-Verlag, New York, 1995.

[13] M. Ben-Artzi, J. Falcovitz, and U. Feldman. Remarks on high-resolution split schemes computation. *SIAM J. Sci. Comput.*, 22:1008–1015, 2000.

[14] M. Ben-Artzi, J. Falcovitz, and J. Li. Wave interactions and numerical approximations for two-dimensional scalar conservation laws. Preprint, Institute of Mathematics, Hebrew University of Jerusalem, 2001.

[15] G. Ben-Dor. *Shock Wave Reflection Phenomena*. Springer-Verlag, New York, 1992.

[16] G. Ben-Dor, O. Igra, and T. Elperin, editors. *Handbook of Shock Waves*. Academic Press, New York, 2001.

[17] A. Birman, N. Y. Har'el, J. Falcovitz, M. Ben-Artzi, and U. Feldman. Operator-split computation of 3-D symmetric flow. *Comput. Fluid Dynamics*, 10:37–43, 2001.

[18] F. Bouchut, C. Bourdarias, and B. Perthame. A MUSCL method satisfying all the numerical entropy inequalities. *Math. Comput.*, 65:1439–1461, 1996.

[19] A. Bourgeade. Second-order scheme for reacting flows – Application to detonation. Technical Report CEA-N-2570, Centre d'Etude de Limeil-Valenton, 1988.

[20] A. Bourgeade, P. LeFloch, and P.-A. Raviart. Approximate solution of the generalized Riemann problem and applications. In C. Carasso, P.-A. Raviart, and D. Serre, editors, *Nonlinear HyperBolic problems*, volume 1270 of *Springer Lecture Notes in Mathematics*, pages 1–9. Springer-Verlag, New York, 1987.

[21] A. Bourgeade, P. LeFloch, and P.-A. Raviart. An asymptotic expansion for the solution of the generalized Riemann problem. Part 2: Application to the equations of gas dynamics. *Ann. Inst. H. Poincaré, Nonlinear Anal.*, 6:437–480, 1989.

[22] J. M. Burgers. A mathematical model illustrating the theory of turbulence. *Adv. Appl. Mech.*, 1:171–199, 1948.

[23] G.-Q. Chen, D. Li, and D. Tan. Structure of Riemann solutions for 2-dimensional scalar conservation laws. *J. Differential Equations*, 127:124–147, 1996.

[24] I. L. Chern, J. Glimm, O. McBryan, B. Plohr, and S. Yaniv. Front tracking for gas dynamics. *J. Comput. Phys.*, 62:83–110, 1986.

[25] A. J. Chorin. Random choice solution of hyperbolic systems. *J. Comput. Phys.*, 22:517–533, 1976.

[26] A. J. Chorin. Random choice methods with applications to reacting gas flow. *J. Comput. Phys.*, 25:253–272, 1977.

[27] A. J. Chorin, T. J. R. Hughes, M. F. McCracken, and J. E. Marsden. Product formulas and numerical algorithms. *Comm. Pure Appl. Math.*, 31:205–256, 1978.

[28] A. J. Chorin and J. E. Marsden. *A Mathematical Introduction to Fluid Mechanics*, 2nd ed., volume 4 of *Texts in Applied Mathematics*. Springer-Verlag, New York, 1990.

[29] P. Colella, A. Majda, and V. Roytburd. Theoretical and numerical structure for reacting shock waves. *SIAM J. Sci. Stat. Comput.*, 7:1059–1080, 1986.

[30] R. Courant and K. O. Friedrichs. *Supersonic Flow and Shock Waves*. Springer-Verlag, New York, 1976.

[31] R. Courant and D. Hilbert. *Methods of Mathematical Physics*, volume II. Interscience, New York, 1962.

[32] M. G. Crandall and A. Majda. The method of fractional steps for conservation laws. *Numer. Math.*, 34:285–314, 1980.

[33] M. G. Crandall and A. Majda. Monotone difference approximations for scalar conservation laws. *Math. Comput.*, 34:1–21, 1980.

[34] C. Dafermos. Polygonal approximations of solutions of the initial value problem for a conservation law. *J. Math. Analysis Appl.*, 38:33–41, 1972.

[35] W. Dai and P. R. Woodward. A simple Riemann solver and high-order Godunov schemes for hyperbolic systems of conservation laws. *J. Comput. Phys.*, 121:51–65, 1995.

[36] L. C. Evans. *Partial Differential Equations.* American Mathematical Society, Providence, RI, 1997.

[37] J. Falcovitz. Startup flow in a shock tube operated by a sliding partition. In *Proceedings of International Conference on Fluid Engineering, Tokyo International Forum, Japan Society of Mechanical Engineers, Tokyo, Japan*, I:403–406, 1997.

[38] J. Falcovitz, G. Alfandary, and G. Ben-Dor. Numerical simulation of the head-on reflection of a regular reflection. *Int. J. Numerical Methods Fluids*, 17:1055–1077, 1993.

[39] J. Falcovitz, G. Alfandary, and G. Hanoch. A two-dimensional conservation laws scheme for compressible flows with moving boundaries. *J. Comput. Phys.*, 138:83–102, 1997.

[40] J. Falcovitz and M. Ben-Artzi. Recent developments of the GRP method. *JSME Int. J., Ser. B*, 38(4):497–517, 1995.

[41] J. Falcovitz and A. Birman. A singularities tracking conservation laws scheme for compressible duct flows. *J. Comput. Phys.*, 115(2):431–439, 1994.

[42] J. Falcovitz and A. E. Fuhs. Impulsive loading from a bare explosive charge in space. *AIAA J. Spacecraft Rockets*, 25:400–404, 1988.

[43] J. Falcovitz and K. Takayama. Vortex generation in a partitioned shock tube. In K. Takayama, editor, *Proceedings of the 2nd International Workshop on Shock Wave Vortex Interaction, Tohoku University, Tohoku print, Sendai, Japan*, 173–182, 1997.

[44] J. Falcovitz, K. Yamada, H. Nagoya, and K. Takayama. Computation and validation of shock wave interaction with fluid inhomogeneities. In B. Sturtevant, J. E. Shepherd, and H. G. Hornung, editors, *Proceedings of the 20th International Symposium on Shock Waves, Pasadena, USA*, pages 489–494. World Scientific, Singapore, 1995.

[45] W. Feller. *An Introduction to Probability Theory and Its Applications*, volume I. Wiley, New York, 1957.

[46] E. Fermi. *Thermodynamics.* Dover, New York, 1956.

[47] P. Garcia-Navarro, M. E. Hubbard, and A. Priestley. Genuinely multidimensional upwinding for the 2-D shallow water equations. *J. Comput. Phys.*, 121:79–93, 1995.

[48] M. Gehmeyr, B. Cheng, and D. Mihalas. Noh's constant velocity shock problem revisited. *Shock Waves*, 7:255–274, 1997.

[49] J. Glimm. Solutions in the large for nonlinear hyperbolic systems of equations. *Comm. Pure Appl. Math.*, 18:697–715, 1965.

[50] J. Glimm, M. J. Graham, J. W. Grove, X. L. Li, T. M. Smith, D. Tan, F. Tangerman, and Q. Zhang. Front tracking in two and three dimensions. *Comput. Math. Appl.*, 35:1–11, 1998.

[51] J. Glimm, J. W. Grove, X. L. Li, and N. Zhao. Simple front tracking. In G.-Q. Chen and E. DiBenedetto, editors, *Contemporary Mathematics*, volume 238.

340 *Bibliography*

American Mathematical Society, Providence, RI, 1999.

[52] J. Glimm, C. Klingenberg, O. McBryan, B. Plohr, D. Sharp, and S. Yaniv. Front tracking and two-dimensional Riemann problems. *Adv. Appl. Math.*, 6:259–290, 1985.

[53] J. Glimm, G. Marshall, and B. Plohr. A generalized Riemann problem for quasi one-dimensional gas flows. *Adv. Appl. Math.*, 5:1–30, 1984.

[54] E. Godlewski and P.-A. Raviart. *Hyperbolic Systems of Conservation Laws.* Ellipses, Paris, 1991.

[55] E. Godlewski and P.-A. Raviart. *Numerical Approximation of Hyperbolic Systems of Conservation Laws.* Springer-Verleg, New York, 1996.

[56] S. K. Godunov. A difference scheme for the numerical computation of discontinuous solutions of the equations of fluid dynamics. *Mat. Sbornik*, 47:271–306, 1959.

[57] J. Goodman and R. LeVeque. A geometric approach to high resolution TVD schemes. *SIAM J. Numerical Anal.*, 25:268–284, 1988.

[58] D. Gottlieb. Strang-type difference schemes for multidimensional problems. *SIAM J. Numerical Anal.*, 9:650–661, 1972.

[59] J. Guckenheimer. Shocks and rarefactions in two space dimensions. *Arch. Rational Mech. Anal.*, 59:281–291, 1975.

[60] E. Harabetian. A convergent series expansion for hyperbolic systems of conservation laws. *Trans. AMS*, 294:383–424, 1986.

[61] R. Hillier. Computation of shock wave diffraction at a ninety degrees convex edge. *Shock Waves*, 1:89–98, 1991.

[62] C. W. Hirt and A. A. Amsden. An arbitrary Lagrangian Eulerian computing method for all flow speeds. *J. Comput. Phys.*, 14:227–253, 1974.

[63] L. Hörmander. *Lectures on Nonlinear Hyperbolic Differential Equations.* Springer-Verlag, New York, 1997.

[64] O. Igra, T. Elperin, J. Falcovitz, and B. Zmiri. Shock wave interaction with area change in ducts. *Shock Waves*, 3(3):233–238, 1994.

[65] O. Igra and J. Falcovitz. Numerical solution to rarefaction or shock wave/duct area-change interaction. *AIAA J.*, 24:1390–1394, 1986.

[66] O. Igra, J. Falcovitz, T. Meguro, K. Takayama, and W. Heilig. Experimental and theoretical studies of shock wave propagation through double-bend ducts. *J. Fluid Mech.*, 437:255–282, 2001.

[67] O. Igra, J. Falcovitz, H. Reichenbach, and W. Heilig. Experimental and numerical study of the interaction between a planar shock wave and a square cavity. *J. Fluid Mech.*, 313:105–130, 1996.

[68] O. Igra, L. Wang, and J. Falcovitz. Non-stationary compressible flow in ducts with varying cross-section. *Proc. Inst. Mech. Eng., Part G*, 212:225–243, 1998.

[69] O. Igra, L. Wang, J. Falcovitz, and W. Heilig. Shock wave propagation in a branched duct. *Shock Waves*, 8:375–381, 1998.

[70] Z. Jiang, J. Falcovitz, and K. Takayama. Numerical simulation of detonation in converging chambers. *JSME Int. J., Ser. B*, 40(3):422–431, 1997.

[71] A. Kalma. Computation of reacting flows with high activation energy. In H. Dwyer, editor, *Proceedings of the 4th International Symposium on Computational Fluid Dynamics*, volume I, pages 563–569. University of California at Davis, 1991.

[72] D. Kröner. *Numerical Schemes for Conservation Laws.* Wiley–Teubner, Hoboken, NJ, 1997.

[73] S. Kruzkov. First order quasilinear equations with several space variables. *Mat. Sbornik, Engl. Transl. Math USSR—Sbornik*, 10:217–243, 1970.

[74] C. D. Landau and E. M. Lifshitz. *Fluid Mechanics,* second edition. Pergamon, New York, 1987.

[75] P. D. Lax. *Hyperbolic Systems of Conservation Laws and the Mathematical Theory of Shock Waves.* SIAM, Philadelphia, 1973.

[76] P. D. Lax and B. Wendroff. Systems of conservation laws. *Comm. Pure Appl. Math.,* 13:217–237, 1960.

[77] P. D. Lax and B. Wendroff. Difference schemes for hyperbolic equations with high order of accuracy. *Comm. Pure Appl. Math.,* 17:381–398, 1964.

[78] P. LeFloch and Ta. T. Li. A global asymptotic expansion for the solution to the generalized Riemann problem. *Asymptotic Anal.,* 3:321–340, 1991.

[79] P. LeFloch and J. G. Liu. Generalized monotone schemes, discrete paths of extrema, and discrete entropy conditions. *Math. Comput.,* 68:1025–1055, 1999.

[80] P. LeFloch and P.-A. Raviart. An asymptotic expansion for the solution of the generalized Riemann problem. Part I: General theory. *Ann. Inst. H. Poincaré, Nonlinear Anal.,* 5:179–207, 1988.

[81] R. J. LeVeque. *Numerical Methods for Conservation Laws,* second edition. Lectures in Mathematics ETH Zürich. Birkhäuser, Basel, 1992.

[82] A. Levy. On Majda's model for dynamic combustion. *Commun. PDE,* 17:657–698, 1992.

[83] J. Li, S. Yang, and T. Zhang. *The Two-Dimensional Riemann Problem in Gasdynamics.* Pitman, London, 1998.

[84] H. W. Liepmann and A. Roshko. *Elements of Gasdynamics.* Wiley, New York, 1957.

[85] P. L. Lions and P. E. Souganidis. Convergence of MUSCL and filtered schemes for scalar laws and Hamilton–Jacobi equations. *Numer. Math.,* 69:441–470, 1995.

[86] T. P. Liu. Asymptotic behavior of solutions of general systems of nonlinear hyperbolic conservation laws. *Indiana Univ. Math. J.,* 27:211–253, 1978.

[87] X. D. Liu and E. Tadmor. Third order nonoscillatory central scheme for hyperbolic conservation laws. *Numer. Math.,* 79:397–425, 1998.

[88] A. Majda. A qualitative model for dynamic combustion. *SIAM J. Appl. Math.,* 41:70–93, 1981.

[89] I. S. Men'shov. Increasing the order of approximation of Godunov's scheme using solutions of the generalized Riemann problem. *USSR Comput. Math. Math. Phys.,* 30(5):54–65, 1990.

[90] G. Miron, O. Igra, S. Rosenwacks, and J. Falcovitz. Parametric study of a supersonic unsteady flow in a nozzle for a potential lead azide laser. *Int. J. Fluid Mech. Res.,* 26(2):146–168, 1999.

[91] I. R. Nathanson. *Theory of Functions of a Real Variable,* volume 1. Ungar, New York, 1955.

[92] H. Nessyahu and E. Tadmor. Non-oscillatory central differencing for hyperbolic conservation laws. *J. Comput. Phys.,* 87:408–463, 1990.

[93] W. F. Noh. CEL: A time dependent two-space-dimensional, coupled Eulerian–Lagrangian code. In B. Alder, S. Fernbach, and M. Rotenberg, editors, *Methods in Computational Physics,* volume 3, pages 117–179. Academic Press, New York, 1964.

[94] W. F. Noh. Errors for calculations of strong shocks using artificial viscosity and an artificial heat flux. *J. Comput. Phys.,* 72:78–120, 1987.

[95] S. Osher and E. Tadmor. On the convergence of difference approximations to scalar conservation laws. *Math. Comput.,* 50:19–51, 1988.

[96]   R. D. Richtmyer and K. W. Morton. *Difference Methods for Initial-Value Problems*. Interscience, New York, 1967.

[97]   T. Saito and I. I. Glass. Application of random-choice method to problems in gas dynamics. *Progr. Aerospace Sci.*, 21:201–247, 1984.

[98]   A. Sasoh, J. Maemura, S. Hirakata, J. Falcovitz, and K. Takayama. Gasdynamics of diaphragm rupture. Impingement by a conically-nosed, ram-accelerator projectile. *Shock Waves*, 9:19–30, 1999.

[99]   R. Saurel, M. Larini, and J.-C. Loraud. Exact and approximate Riemann solvers for real gases. *J. Comput. Phys.*, 112:126–137, 1994.

[100] C. Schulz-Rinne. Classification of the Riemann problem for two-dimensional gas dynamics. *SIAM J. Math. Anal.*, 24:76–88, 1993.

[101] D. Serre. *Systems of Conservation Laws*, volume I. Cambridge University Press, London, 1999.

[102] A. H. Shapiro. *The Dynamics and Thermodynamics of Compressible Fluid Flow*. Ronald, New York, 1953.

[103] J. Smoller. *Shock Waves and Reaction–Diffusion Equations*. Springer-Verlag, New York, 1983.

[104] G. A. Sod. A survey of several finite difference methods for systems of non-linear hyperbolic conservation laws. *J. Comput. Phys.*, 27:1–31, 1978.

[105] G. Strang. On the construction and comparison of difference schemes. *SIAM J. Numerical Anal.*, 5:506–517, 1968.

[106] E. G. Tabak. A second-order Godunov method on arbitrary grids. *J. Comput. Phys.*, 124:383–395, 1996.

[107] K. Takayama and Z. Jiang. Shock wave reflection over wedges: A benchmark test for CFD and experiments. *Shock Waves*, 7:191–203, 1997.

[108] G. I. Taylor. *The Air Wave Surrounding an Expanding Sphere*, volume III of *The Scientific Papers of G. I. Taylor*. Cambridge University Press, London, 1963.

[109] Z. H. Teng. On the accuracy of fractional step methods for conservation laws in two dimensions. *SIAM J. Numerical Anal.*, 31:43–63, 1994.

[110] Z. H. Teng, A. J. Chorin, and T. P. Liu. Riemann problems for reacting gas, with applications to transition. *SIAM J. Appl. Math.*, 42:964–981, 1982.

[111] E. F. Toro. *Riemann Solvers and Numerical Methods for Fluid Dynamics. A Practical Introduction*, second edition. Springer-Verlag, New York, 1999.

[112] B. van Leer. Towards the ultimate conservative difference scheme V. *J. Comput. Phys.*, 32:101–136, 1979.

[113] J. P. Vila. An analysis of a class of second-order accurate Godunov-type schemes. *SIAM J. Numerical Anal.*, 26:830–853, 1989.

[114] A. I. Vol'pert. The space BV and quasilinear equations. *Mat. Sbornik, Engl. Transl. Math USSR—Sbornik*, 2:225–267, 1967.

[115] D. H. Wagner. The Riemann problem in two space dimensions for a single conservation law. *SIAM J. Math. Anal.*, 14:534–559, 1983.

[116] D. H. Wagner. Equivalence of the Euler and Lagrangian equations of gas dynamics for weak solutions. *J. Differential Equations*, 68:118–136, 1987.

[117] B. Y. Wang, Q. S. Wu, C. Wang, O. Igra, and J. Falcovitz. Shock wave diffraction by a square cavity filled with dusty gas. In R. Hillier, editor, *Proceedings of the 22nd International Symposium on Shock Waves, Imperial College, London, University of Southampton, Southampton, UK*, 1405–1410, 1999.

[118] B. Y. Wang and R. S. Wu. Numerical investigation of dusty gas shock wave propagation along a variable cross-section channel. In K. Takayama, editor,

Shock Waves: Proceedings of the 18th International Symposium on Shock Waves, Sendai, Japan, pages 521–526. Springer-Verlag, New York, 1991.

[119] M. Wierse. Solving the compressible Euler equations in time dependent geometries. In *Proceedings of the 5th International Conference on Hyperbolic Problems, University at Stony Brook, NY, USA, 13–17 June, 1994*, pages 485–492. World Scientific, Singapore, 1994.

[120] H. Yang. Convergence of Godunov type schemes. *Appl. Math. Lett.*, 9:63–67, 1996.

[121] H. Yang. One wavewise entropy inequality for high-resolution schemes II: Fully-discrete MUSCL schemes with exact evolution in small time. *SIAM J. Numerical Anal.*, 36:1–31, 1999.

[122] H. Q. Yang and A. J. Przekwas. A comparative study of advanced shock-capturing schemes applied to Burgers' equation. *J. Comput. Phys.*, 102:139–159, 1992.

[123] L.-A. Ying and Z. H. Teng. Riemann problem for a reacting and convection hyperbolic system. *Approximation Theory Appl.*, 1:95–122, 1984.

[124] L.-A. Ying and C.-H. Wang. The discontinuous initial value problem of a reacting gas flow system. *Trans. AMS*, 266:361–387, 1981.

[125] P. Zhang and T. Zhang. Generalized characteristic analysis and Guckenheimer structure. *J. Differential Equations*, 152:409–430, 1999.

[126] B. Zingerman, R. Alimi, S. Wald, and J. Falcovitz. Lagrange–Euler scheme for heterogeneous gas–grain flow. In R. Hillier, editor, *Proceedings of the 22nd International Symposium on Shock Waves, Imperial College, London, University of Southampton, Southampton, UK*, pages 1423–1428, 1999.

Glossary

CRW centered rarefaction wave
IVP initial value problem
MBT moving boundary tracking

Flow Variables and Thermodynamic Quantities (Section 4.2)

S entropy
p pressure
$\rho, \tau = 1/\rho$ density and specific volume
c speed of sound
$g = \rho c$ "Lagrangian" speed of sound

Coordinates

r spatial (Section 4.2)
x planar (Sections 2.1, 4.1, 4.2)
ξ Lagrangian (Section 4.2)
t time

Riemann and Generalized Riemann Problem

$\mathbf{R}\left(\frac{x}{t}; \mathbf{U}_L, \mathbf{U}_R\right)$ Eulerian solution, planar conservation form (Sections 4.1, 4.2)
$\mathbf{R}^A\left(\frac{x}{t}; \mathbf{U}_L, \mathbf{U}_R\right)$ Eulerian solution, associated Riemann problem (Section 4.2)
$\mathbf{R}\left(\frac{\xi}{t}; \mathbf{V}_L, \mathbf{V}_R\right)$ Lagrangian solution, planar conservation form (Section 4.2)
$\mathbf{R}^A\left(\frac{\xi}{t}; \mathbf{V}_L, \mathbf{V}_R\right)$ Lagrangian solution, associated Riemann problem (Section 4.2)
$\mathbf{V}^*, \mathbf{V}_L^*, \mathbf{V}_R^*$ contact solution in Lagrangian coordinates (Figure 5.2)

$\left(\frac{\partial \mathbf{V}}{\partial t}\right)^*$ time derivative along contact at $t = 0$ (Table 5.5)

$\mathbf{U}_0$ $\mathbf{R}^A\left(0; \mathbf{U}_L, \mathbf{U}_R\right)$ in Eulerian coordinates (Section 5.1)

$\left(\frac{\partial \mathbf{U}}{\partial t}\right)_0$ solution to linear GRP, Eulerian coordinates (Section 5.1)

Γ_1, Γ_2, Γ_3 waves in solution to the GRP (Section 5.1)

RP   Riemann problem

GRP   generalized Riemann problem

I_1^l interaction curve: all right states connected to a given left state by
1-shock or 1-CRW (Section 4.2)

I_3^r interaction curve: all left states connected to a given right state by
3-shock or 3-CRW (Section 4.2)

1-wave   a shock or a CRW associated with the C_1 characteristic family (i.e.,
propagating in the $-x$ direction (left-facing wave)) (Section 4.2)

3-wave   a shock or a CRW associated with the C_3 characteristic family (i.e.,
propagating in the $+x$ direction (right-facing wave)) (Section 4.2)

Functional Spaces

Compactly supported function in $\mathbb{R} \times (0, \infty)$ a function that vanishes
outside some rectangle $[-A, A] \times [0, T]$.

C_0^1 continuously differentiable and compactly supported functions

C^1 continuously differentiable functions

$L_{\mathrm{loc}}^1(\mathbb{R})$ functions integrable on every finite interval in $\mathbb{R}$

Index